NUCLEI AND MESOSCOPIC PHYSICS

NUCLEI AND MESOSCOPIC PHYSICS

Workshop on Nuclei and Mesoscopic Physics

WNMP 2007

East Lansing, Michigan *20 – 22 October 2007*

EDITORS
Pawel Danielewicz
Piotr Piecuch
Vladimir Zelevinsky
Michigan State University
East Lansing, Michigan

SPONSORING ORGANIZATIONS
National Superconducting Cyclotron Laboratory, Michigan State University
Office of Vice Provost for Libraries, Computing, and Technology, Michigan State University
AT&T Lectureships in Information Technology Endowment

AMERICAN INSTITUTE OF PHYSICS

Melville, New York, 2008
AIP CONFERENCE PROCEEDINGS ■ VOLUME 995

Editors:

Pawel Danielewicz
National Superconducting Cyclotron Laboratory
Michigan State University
East Lansing, MI 48824-1321
USA

E-mail: danielewicz@nscl.msu.edu

Piotr Piecuch
Department of Chemistry
Michigan State University
East Lansing, MI 48824-1322
USA

E-mail: piecuch@chemistry.msu.edu

Vladimir Zelevinsky
National Superconducting Cyclotron Laboratory
Michigan State University
East Lansing, MI 48824-1321
USA

E-mail: zelevinsky@nscl.msu.edu

Photo Layout: Photographs courtesy of Prof. F. M. Izrailev

L.C. Catalog Card No. 2008923560
ISBN 978-0-7354-0514-1
ISSN 0094-243X

Printed in the United States of America

CONTENTS

Preface ... vii
Conference Photographs ... ix

How Changing Physical Constants and Violation of Local Position Invariance May Occur? ... 1
V. V. Flambaum and E. V. Shuryak
A Cluster Model of ^{6}He and ^{6}Li ... 12
J. R. Armstrong, A. A. Sakharuk, and V. G. Zelevinsky
Equation of State and Pairing Gaps in Cold Atoms and Low-Density Neutron Matter ... 17
J. Carlson, A. Gezerlis, and S. Reddy
Two-Component Bose Gases under Rotation ... 25
S. Bargi, K. Kärkkäinen, J. Christensson, G. M. Kavoulakis, M. Manninen, and S. M. Reimann
Cold Atom Ballistics by Coherent Control ... 34
L. E. Reichl and B. P. Holder
Nondynamic Correlation and Coupled-Cluster Methods ... 43
J. Paldus and X. Li
A Compact Spin-Free Coupled-Cluster Theory for Open-Shell Systems ... 53
D. Datta and D. Mukherjee
Extrapolating Potential Energy Surfaces by Scaling Electron Correlation: Isomerization of Bicyclobutane to Butadiene ... 62
J. J. Lutz and P. Piecuch
Some Recent Developments of Quantum Monte Carlo to Molecular Systems ... 72
W. A. Lester, Jr. and D. Domin
Continuum Shell Model: From Ericson to Conductance Fluctuations ... 75
G. L. Celardo, F. M. Izrailev, S. Sorathia, V. G. Zelevinsky, and G. P. Berman
Decay Rates Statistics of Unstable Classically Chaotic Systems ... 85
V. V. Sokolov
Doorway States and the Super-Radiant Mechanism ... 92
N. Auerbach
Time-Dependent Green's Functions Approach to Nuclear Reactions ... 98
A. Rios and P. Danielewicz
Quantum Transport in Mesoscopic Semiconductor Devices: Vortices, Flows and Atomistic Effects ... 104
J. R. Barker
Self-Consistent Hartree-Fock Approximation for Non-Equilibrium Electron Transport through Nanostructures ... 115
M. F. Gelin and D. S. Kosov
Electron Correlations and the Conductance of a Quantum Point Contact at Finite Temperature ... 122
C. Sloggett, A. I. Milstein, and O. P. Sushkov
Scanning Charge Accumulation Probe of Semiconductor Donor Molecules ... 130
S. H. Tessmer, I. Kuljanishvili, C. Piermarocchi, T. A. Kaplan, and J. F. Harrison
Electron Transport through Protein Fragments ... 135
D. M. Cardamone and G. Kirczenow
Photochemistry of Biological Chemosensors, Organic Light-Emitting Diodes, and Inner-Shell Electronic Processes ... 145
M. Ehara and H. Nakatsuji
Interplay of Orbital Paramagnetism and Diamagnetism in Nanometer Scale 2DEG Strips ... 152
M. J. Harrison
Instabilities of Doped Mott Insulators and the Properties of High-Temperature Superconductors ... 160
M. Guidry, Y. Sun, and C.-L. Wu
Mesoscopic Fluctuations of the Pairing Gap ... 173
S. Åberg, H. Olofsson, and P. Leboeuf
Thermodynamic Analysis of Spectra ... 185
G. E. Mitchell and J. F. Shriner, Jr.

Interaction Matrix Element Fluctuations in Quantum Dots 192
L. Kaplan and Y. Alhassid
Superscars, Doorway States and Nodal Domains in a Barrier Billiard 202
A. Richter
Symmetry Breaking Study with Random Matrix Ensembles 212
M. S. Hussein, J. X. de Carvalho, M. P. Pato, and A. J. Sargeant

Conference Program 221
List of Participants 225
Author Index 231

Preface

The volume contains the Proceedings of the *Second WORKSHOP on NUCLEI and MESOSCOPIC PHYSICS*, WNMP07, held on October 20-22, 2007, at the National Superconducting Cyclotron Laboratory (NSCL) of Michigan State University, in East Lansing, Michigan.

The first workshop of this type, WNMP04, was organized at the NSCL in October 2004 and attracted many physicists, theoreticians and experimentalists, working in various subfields of physics and chemistry, including nuclear physics, quantum chromodynamics, condensed matter, atomic and molecular physics and quantum information. Proceedings of that workshop have been published by the American Institute of Physics (AIP Conference Proceedings, vol. 777, ed. V. Zelevinsky, Melville, New York, 2005).

The main goal of this workshop series is to bring together scientists studying systems that are mesoscopic in nature, which reveal generic features and allow for similar approaches, and, thus, open the way for exchange of ideas, practical tools and possible collaborations. Organizing the workshops, as well as other inter-collegial contacts, such as periodic lectures of well known scientists, the Mesoscopic Theory Center (MTC) has started its work under the umbrella of the NSCL and MSU Department of Physics and Astronomy. The International Advisory Committee for the MTC includes Y. Imry (Weizmann Institute, Israel), D. Mattis (University of Utah), D. Mukherjee (Indian Association for the Cultivation of Science), L. Reichl (University of Texas at Austin), S. Reimann (Lund University, Sweden), T. Seligman (UNAM, Mexico) and H.A. Weidenmüller (Heidelberg, Germany).

As has been stressed in the Preface to the Proceedings WNMP04, "Historically, many ideas currently forming the core of the mesoscopic area have been for the first time addressed in nuclear science. For example, the concept of spectral manifestations of quantum chaos and random matrix theory as its extreme limit were originally developed on the foundation of nuclear spectroscopy. Nowadays, one of the main driving forces in development of the mesoscopic physics is certainly the tempestuous growth of nanoscience and related technology. The computational methods of many-body nuclear theory were successfully accepted and significantly advanced in modern quantum chemistry". Consequentially, the NSCL, with its unique *Coupled Cyclotron Facility* used for engineering and studying exotic nuclear species, serves as a natural venue for such meetings and for the activities of the MTC.

This volume contains the majority of presentations made at the Workshop; we also include few works presented by postdoctoral associates and students at the poster session. The topics covered include variation of fundamental constants, the many-body problem of nuclei and nuclear matter, physics of cold gases, computational methods for quantum many-body systems in physics, chemistry and biology, physics of open mesoscopic systems, quantum transport, phase transitions in magnetism and superconductivity, and quantum chaos. The diverse topics nevertheless turn out to be internally related and many common themes can be traced across the articles presented in the volume. The resulted multicolored *quilt* (old Michiganian speciality of patterned stitching) well represents richness and harmony of mesoscopic science.

The Organizing Committee included Pawel Danielewicz, Piotr Piecuch, Alexander Volya, and Vladimir Zelevinsky. With much gratitude we can repeat what has been said in the Proceedings for WNMP04: "The lion's share of work was done by Shari Conroy in her usual quiet, pleasant, and extremely efficient manner". We are very thankful to Jeremy Armstrong, Chasity Fudella, Laura Gonzales, Arnau Rios, and Roman Sen'kov for their help. The support by the AT&T Lectureships in Information Technology Endowment, Office of Vice Provost for Libraries, Computing, and Technology at Michigan State University, and National Superconducting Cyclotron Laboratory at Michigan State University is gratefully acknowledged.

Pawel Danielewicz
Piotr Piecuch
Vladimir Zelevinsky

NSCL/MSU

VICTOR
FLAMBAUM
UNIVERSITY OF
NEW SOUTH WALES

Fermi gas
172Yb
171Yb
Moment of

Funding

STEVE & BARRY'S
UNIVERSITY SPORTSWEAR
ALL CLOTHING
IN THE STORE
$19 98
OR
LESS

GSS
N=200

How changing physical constants and violation of local position invariance may occur?

V.V. Flambaum* and E.V. Shuryak†

*School of Physics, The University of New South Wales, Sydney NSW 2052, Australia and Institute for Advanced Study, Massey University (Albany campus), Privite Bag 102904, North Shore MSC, Auckland, New Zealand
†Department of Physics and Astronomy, State University of New York, Stony Brook NY 11794-3800, USA

Abstract. Light scalar fields very naturally appear in modern cosmological models, affecting such parameters of Standard Model as electromagnetic fine structure constant α, dimensionless ratios of electron or quark mass to the QCD scale , $m_{e,q}/\Lambda_{QCD}$. Cosmological variations of these scalar fields should occur because of drastic changes of matter composition in Universe: the latest such event is rather recent (redshift $z \sim 0.5$), from matter to dark energy domination. In a two-brane model (we use as a pedagogical example) these modifications are due to changing distance to "the second brane", a massive companion of "our brane". Back from extra dimensions, massive bodies (stars or galaxies) can also affect physical constants. They have large scalar charge Q_d proportional to number of particles which produces a Coulomb-like scalar field $\phi = Q_d/r$. This leads to a variation of the fundamental constants proportional to the gravitational potential, e.g. $\delta\alpha/\alpha = k_\alpha \delta(GM/rc^2)$. We compare different manifestations of this effect, which is usually called violation of local position invariance. The strongest limits $k_\alpha + 0.17k_e = (-3.5 \pm 6) * 10^{-7}$ are obtained from the measurements of dependence of atomic frequencies on the distance from Sun (the distance varies due to the ellipticity of the Earth's orbit).

Keywords: Variation Fundamental constants
PACS: 06.20.Jr , 06.30.Ft, 11.10.Kk, 11.15.-q

INTRODUCTION

Changing parameters of the standard model are usually associated with the effect of massless (strictly speaking, very light) scalar fields. One candidate, much discussed in literature, is the so called *dilaton*: a very special scalar which appears in string models together with a graviton, in a massless multiplet of closed string excitations. Other scalars naturally appear in string-theory-inspired cosmological models, in which our Universe is a "brane" floating in a space of larger dimensions. The scalars are simply brane coordinates in extra dimensions.

At the other hand, available observational limits on physical constant variations at present time are quite strict, allowing only scalar coupling tiny in comparison with gravity. The only relevant scalar field recently discovered – the cosmological dark energy – also so far does not show visible variations. So one may wander why should one discuss time or space-depending physics at all?

One motivation was provided by Damour et al [1, 2] who pointed out that cosmological evolution of scalars naturally leads to their self-decoupling. Damour and Polyakov have further suggested that variations should happen when the scalars get excited by some physical change in the Universe, such as the phase transitions or other drastic change in the Equation of State (EoS) of the Universe. They considered few of them, but since the time of their paper a new fascinating EoS transition has been discovered: from matter dominated (decelerating) era to dark energy dominated (accelerating) era. It is relatively recent event, corresponding to cosmological redshift $z \approx 0.5$.

The time dependence of the disturbance related to it can be calculated, and it turned out [4, 5] that the self-decoupling process is effective enough to explain why after this transition the variation of constants is as small as observed in laboratory experiments at the present time, as well as at Oklo ($\sim$ 2 billion years ago or $z = 0.14$) and isotopes ratios in meteorites (4.6 billion years to now, $z = 0.45 - 0$), while being at the same time consistent with possible observations [7, 8] of the variations of the electromagnetic fine structure constant at $z \sim 1$.

In this paper we discuss why and how such excitation works in some modern cosmological models at the same cosmological time of this transition. Another vast area we will address here is similar variations of constants in space, near massive bodies such as stars (Sun), pulsars, Galaxy. We will compare possible sensitivities related with different possible objects, point out limitations following from some recent experiments with atomic clocks and suggest new measurements.

CP995, *Nuclei and Mesoscopic Physics - WNMP 2007,* edited by P. Danielewicz, P. Piecuch, and V. Zelevinsky

Although both authors are neither cosmologists nor string theorists, we selected two-brane model as our main example, for pedagogical reasons. Variations of physical constants with distance to Sun and in cosmological time have a similar nature: in both cases it is the interaction with a massive body (Sun or heavy "second brane", respectively) via scalar fields. We hope it will be helpful to get the message across for a non-specialized reader.

WHICH SCALARS?

The dilaton and its coupling functions

Large distance interactions in Universe are dominated by gravity, which cannot be screened and get cumulative effects from large bodies. Einstein's description – general relativity – uses spin-2 metric tensor $g_{\mu\nu}$ coupled to energy-momentum tensor $T^{\mu\nu} \sim \delta S_m / \delta g_{\mu\nu}$ of all matter. Thinking about large distance modification of gravity, people for a long time asked why not also add a massless (or very light) spin-0 scalar field in a similar way.

The question is what is their nature and how are they are coupled to matter. One choice (originating from literature on general relativity tests) is to work in the so called "Einstein frame" and keep gravitational part of the action in the usual form, while adding a scalar field in all other terms via a "coupling function" $A(\phi)$ in the metric

$$\tilde{g}_{\mu\nu} = A^2(\phi) g^E_{\mu\nu} \tag{1}$$

where g^E is the usual "Einsteinian" metric.

$$S = \frac{1}{16\pi G} \int d^4x (g^E)^{1/2} \left[R(g^E) - 2 g^E_{\mu\nu} \partial_\mu \phi \partial_\nu \phi \right] + S_{matter}(\tilde{g}_{\mu\nu}, \psi ...) \tag{2}$$

If so, there is a universal source of this field ϕ,

$$\phi = -4\pi G T \frac{\partial A(\phi)}{\partial \phi} \tag{3}$$

containing the trace of total matter energy-momentum tensor $T = T^\mu_\mu$. This operator T has a special meaning in physics, being a generator of the so called dilatational transformation – a change of units of all fields. That is why a field coupled to it is called a dilaton field.

Thus this field can be seen as a scale parameter: it plays the same role as e.g. a rotation angles $\vec{n}$ for the rotation operator $exp(i\vec{n}\vec{J})$ (where $\vec{J}$ is angular momentum operator), or coordinate $\vec{x}$ in displacement operator $exp(i\vec{x}\vec{P})$ (where $\vec{P}$ is momentum). So, if $\phi(x) \neq \phi(y)$ at two points x, y it means that clocks and meters *made of matter fields* (such as atoms or nuclei) are rescaled differently there. Note that this statement makes sense only because the purely geometrical Einsteinian metric g^E of space-time itself was fixed independently of $\phi(x)$, and gravitational meters - e.g. the Schwarzschield radius of a black hole – are *not* changed.

This field ϕ described so far is interacting with the total $T = \rho - 3P$ of matter (here P is the pressure). Note that it is enough to violate the equivalence principle: it is say coupled to "dust" $P = 0$ in the same way as gravity, to the mass, but not to a bag filled by black-body radiation $\rho = 3P$. Furthermore, the energy-momentum tensor trace can be expressed in terms of fundamental gauge and fermionic fields

$$T = \sum_{i=1,2,3} \frac{\beta_i(g)}{2g_i} (G^i_{\mu\nu})^2 + \sum_f m_f \bar{\psi}_f \psi_f \tag{4}$$

Recall that Maxwellian energy-momentum tensor has zero trace and thus the gauge fields of electromagnetic, weak and strong interactions ($i = 1,2,3$ in the first sum) appear in it only at the quantum level, due to the so called scale anomaly induced by running coupling constant, with $\beta_i \approx -b_i g_i^2 / 32\pi^2$ (in the one loop order) being the corresponding beta functions. The weight of say proton or nuclei include say the electromagnetic part proportional to $\vec{E}^2 + \vec{B}^2$ while the operator above contains scalar combination $\vec{E}^2 - \vec{B}^2$ with a different coefficient, they are *not* the same.

However, even the string-theory dilaton is only simple at high string scale, where supersymmetries hold together massless supermultiplet which includes both the spin-2 graviton and spin-0 dilaton, together with a bunch of other

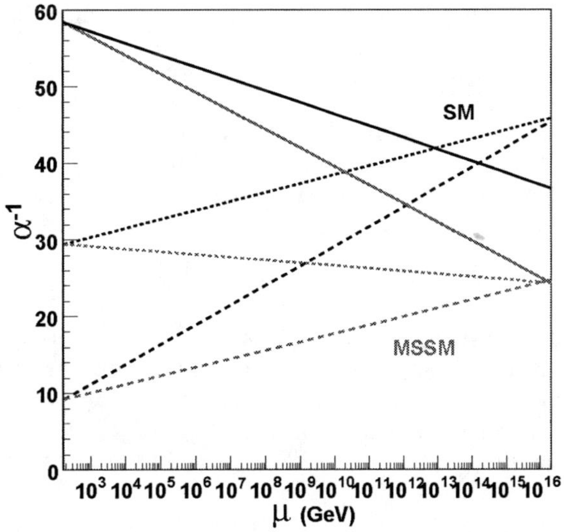

FIGURE 1. From [9]. Evolution of the three gauge couplings α_1 ($SU(5)$ normalized, as the dashed line), α_2 (dotted line) and α_3 (solid line) in the SM (upper set of curves) and MSSM (lower set of curves), assuming superpartner masses are all close to the top quark mass.

spin 3/2,1,1/2,0 fields. By the time one goes down to the scale at which we discuss Standard Model of particle physics these symmetries should be broken by complicated dynamical phenomena nobody understands. This may lead to a dilaton mass, and renormalize its interaction with different kind of matter differently.

In desperation, one may thus start from the opposite extreme: assume *different* coupling functions $A_i(\phi)$. in *all* term of effective low energy Lagrangian. Fortunately one can still get rid of some of them by rescaling all the matter field (we already discussed removing scalar from gravity part above). Let us think about QED or QCD action, as examples. Let us start with the fermionic kinetic/interaction term $\bar{\psi}(i\partial_\mu - A_\mu)\gamma_\nu \psi \tilde{g}_{\mu\nu}$ be modified by modifying ψ, eA_μ to absorb $A^2(\phi)$ in it: but now we have no freedom over the gauge field term

$$\frac{A_e(\phi)}{4e^2}\int d^4x\sqrt{\tilde{g}}F_{\mu\nu}F_{\mu'\nu'}\tilde{g}^{\mu\mu'}\tilde{g}^{\nu\nu'} \tag{5}$$

because we have already fixed A_μ at the previous step. The lesson is this: ϕ dependent "coupling functions" can be delegated into the gauge field coupling constants, as Bekenstein [3] did for QED (and extensively used in discussions of the variations such as [4, 5]).

A number of authors [15] (see also review[16]) suggested variations of em,weak and strong couplings should be related provided the variation comes from the high energy (right in Fig.1) end. The strong (i=3), and electroweak (i=1,2) inverse coupling constants have the following dependence on the scale μ and normalization point μ_0:

$$\alpha_i^{-1}(\mu) = \alpha_i^{-1}(\mu_0) + b_i ln(\mu/\mu_0) \tag{6}$$

In the Standard Model $2\pi b_i = 41/10, -19/6, -7$ and the couplings are related as $\alpha^{-1} = (5/3)\alpha_1^{-1} + \alpha_2^{-1}$.

Fig.1 shows two popular scenarios of Grand Unification: with the standard model as well as for its minimal supersymmetric extension (MSSM). In the latter case one can see that 3 curves cross at one point, believed to be a "root" of the three branches (electromagnetic, weak and strong). One may select the unification point for μ_0, and for example, $\mu = m_Z$ is the Z-boson mass.

(String theories lead to more complicated "trees", which however also have a singly "root", at a string scale Λ_s and bare string coupling g_s.)

So, what happens when one subject the theory to a nonzero value of the scalar field (dilaton)? A number of authors [15] (see also review[16]) suggested variations of em,weak and strong couplings should be related provided the variation comes from the high energy end. There is quite extended discussion of what may happen, but basically there are two possibilities.

If one assumes that only $\alpha_{GUT} \equiv \alpha_i(\mu_0)$ varies, the eqn (6) gives us the same shifts for all inverse couplings

$$\delta\alpha_1^{-1} = \delta\alpha_2^{-1} = \delta\alpha_3^{-1} = \delta\alpha_{GUT}^{-1} \tag{7}$$

as the whole MSSM "tree" (Fig.1) moves up or down. If so, the variation of the strong interaction constant $\alpha_3(m_z)$ is much larger than the variation of the em constant α, $\delta\alpha_3/\alpha_3 = (\alpha_3/\alpha_1)\delta\alpha_1/\alpha_1$.

Another option is the variation of the GUT scale (μ/μ_0 in eqn (6). If so the whole "tree" at Fig.1 moves left-right and quite different relations between variations of the three coupling follows

$$\delta\alpha_1^{-1}/b_1 = \delta\alpha_2^{-1}/b_2 = \delta\alpha_3^{-1}/b_3 \tag{8}$$

Note that now variations have different sign since the one loop coefficients b_i have different sign for 1 and 2,3.

However, it is hard to see how this relation appears dynamically, at least in cosmological models on the market. One would rather think the influence of the "coupling functions" depending on scalars in the matter terms of the Lagrangian would actually mean "curving" of the lines of the "tree" at Fig.1. Another unclear issue is the modification of lepton/quark masses, which are proportional to Higgs VEV and thus depend on the mechanism of electroweak symmetry breaking.

Introducing cosmological two-brane model

It would be premature at this moment to subscribe to any model, of course. However we still find some of them more dynamically natural and possibly useful, at least pedagogically, to explain our intuitive ideas in relating distinct physics together. Field and string theory provide examples of topological objects of different dimensions – branes – on which effective fermionic, gauge and scalar fields live. One of them may be our 4-d Universe with our Standard model fields, floating in a larger-dimensional space [1].

It inspired many specific cosmological models providing some dynamical realizations of Brane World Cosmology (see e.g. [13, 14] for reviews). More specifically, our pedagogical point is best served by the two-brane models, in which our 3-d brane (coordinates $\vec{x}$) is complemented by another 3-d brane, parallel to it but shifted in the 5-th dimension (we will call y). The Big Bang in this model is thus supplemented not only by the influence of the "bulk fields" living on and in between branes (gravitons and dilatons among them) but also by the second brane.

Without matter, symmetries of the model naturally balance all forces between the branes – gravity and scalar attraction is balanced by Coulomb-like vector repulsion [2]. It means that in this approximation the positions of two branes y_1, y_2 are irrelevant: branes can "levitate" in presence of each other, anywhere they chose. With extra matter added to branes this symmetry is broken and branes start their motion (making y_i a function of time t). If matter is not homogeneous, as shown in Fig.2 for the left brane, the brane bends, making it a function of 3-space point $y_i(\vec{x})$. In general, there appear 4-d scalar field $y_i(\vec{x},t)$, describing the shape of the moving brane.

Unlike models mentioned above, nothing unusual happens with physics at high energy (right) part of Fig.1: all changes are restricted to low energy (left) end. Since gauge bosons are generated dynamically – by open strings with both ends on our brane – one can calculate how much this affect their effective Lagrangian and in principle *predict* how they are affected by the relative dilaton field, or y_1, y_2 scalars.

These two branes were originally introduced by Randall and Sundrum [6] to explain why gravity can be strong in general but exponentially weak at our brane [3] due to warping of the space-time by gravity of the companion brane.

(Another fascinating aspect of the two-brane model which we will not discuss at all is a possibility to explain the dark matter phenomenon by ascribing it to the *other* brane.)

---

[1] Note the similarity with expanding the view from the 2-d Earth surface, our natural habitat, to 4d Einsteinian Universe. The difference is: we can see stars and galaxies, but we dont know anything definite about these extra dimensions yet.

[2] There are also some other stability conditions we dont mention, see [6, 14].

[3] our brane one may call "the brane Earth" and the other "the brane Sun", to enhance analogy to Copernicus/Kepler system of the world. It also explains which one has the dominant mass.

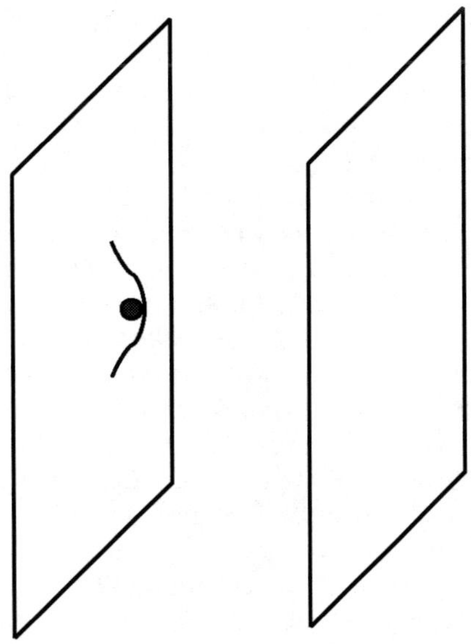

FIGURE 2. Schematic picture of 2-brane construction. The coordinates inside two parallel branes are called $\vec{x}$ and the one normal to them y. A massive body (shaded circle) on a brane can cause its deformation, as shown, due to extra attraction to the second brane.

THE COSMOLOGICAL CHANGE

Self-decoupling mechanism

The cosmological equations follow from Einstein's equation and are well known ($q = 8\pi G$)

$$-3\ddot{a}/a = 2\dot{\phi}^2 + 8\pi G * (\rho + 3P) \tag{9}$$

$$3(\frac{\dot{a}}{a})^2 + 3K/a^2 = 16\pi G\rho + \dot{\phi}^2 \tag{10}$$

where $a(t)$ is the scale factor, K is related to spatial curvature (below we will use K=0 flat space) and ρ, P are the total energy density and pressure. The evolution of the dilaton field is given by

$$\ddot{\phi} + 3(\frac{\dot{a}}{a})\dot{\phi} = 4\pi G \sum_A A'_A(\phi)(\rho_A - 3P_A) \tag{11}$$

where the sum in the r.h.s. is over various components of matter in Universe, with corresponding contributions to energy density and pressure[4].

In order to understand what this equation is describing, let us start following the original Damour-Polyakov paper [2] and consider a single-species Universe (no sum in the r.h.s.). It is also useful to introduce "logarithmic cosmic time" $\tau = ln(a(t)) + const$ we will denote derivative over it by a prime[5]. The following single eqn for $\phi(\tau)$ then follows

$$\frac{2}{3-\phi'^2}\phi'' + (1 - P/\rho)\phi' = A'(\phi)(1 - 3P/\rho) \tag{12}$$

which can be seen as an oscillator with exciting force (r.h.s) and the damping (the second term in the l.h.s.) containing the EoS ratio P/ρ. If its evolution can be put in as a function of time τ, this can be treated as an independent eqn.

---

[4] Note that simple additivity of contributions to energy-momentum tensor is ambiguous for non-gravitationally interacting species, but we dont expect it to be a problem for the case we aimed at, in which one component is baryonic/dark matter and the other the dark energy.

[5] A prime in $A'(\phi)$ is still a derivative over ϕ though: we hope it will not confuse the reader.

For radiation-dominated Universe the r.h.s. ("excitation force") is zero ($\rho = 3P$), while l.h.s. corresponds to damped oscillator which rapidly relaxes to rest. For other single-component eras the r.h.s. is not zero, but is dominated by a single value of P/ρ (e.g. zero for matter-dominated one) and a single function $A(\phi)$: the evolution is a relaxation toward the minimum of this function, which say happens at some $\phi = \phi_m$. At the minimum A' is zero and all interaction of the small-amplitude dilaton with matter vanishes.

The only time when this argument does not work is when EoS of the Universe is changing. In a simplest case, there are two comparable contributions of two types of matter. In particular, let us focus directly at the latest cosmological event of this change in which components of the energy-momentum tensor have comparable contributions of matter (luminous and dark) and the dark energy (cosmological constant). A single equation for ϕ (12) is no longer valid and one should solve all coupled equations (given above) together. One may find explicit solutions e.g. in [4, 5], but it is easy to guess what is happening without numerical solutions: a damped oscillator slides from one equilibrium position (the minimum of coupling function to matter at $\phi = \phi_m$) to a new minimum of another coupling function to the dark energy at $\phi = \phi_d$. This slide is dominated by strong damping of cosmic oscillator, which is universal. As a result, one is able to predict the *time dependence* of the cosmological scalar field (if not the magnitude itself.). In summary: the mechanism leads to a ladder-type evolution pattern of the scalar field.

As shown in [4, 5], the last slide is such that there is actually no contradiction between limits from Oklo and meteorite isotope compositions with nonzero claim in quasar spectra [7, 8].

There is another version of scalar motion, in which people want it to slide not toward a constant value now (usually taken to be zero) but to infinity, reproducing the so called "tracking" solutions. In other words, these authors want to hit two birds with one stone and to solve the "dark energy" puzzle by the same scalar. If so, the rolling field cannot do it sufficiently quickly, and in this case the meteorite-based limits and quasar data cannot be reconciled [20].

As we will discuss in the next section, however, there is no shortage of scalar fields in modern models, and it is quite natural that different fields would be responsible for these two phenomena.

Cosmic evolution of scalars in two-brane cosmology

The three scalars of the model are the bulk dilaton and 2 brane shapes $y_1(\vec{x},t), y_2(\vec{x},t)$ as shown in Fig. 2 : their combinations are two scalars to appear in the Lagrangian. The supersymmetric potential ensures "levitation" and defines the basic low energy Lagrangian. The parameters of the model include tensions (masses per area) of two branes and bulk cosmological constant, to be tuned as explained in [6] for static construction. To make it dynamically moving, there is additional model potential expressing the brane tension as a function of bulk dilaton scalar which is arbitrarily selected to be exponential $U \sim exp(\alpha_d \phi)$. As it turns out, the parameter α_d is restricted from above by cosmological limitations on the variations of the gravitational constant and electromagnetic fine structure constant [11]. In simple terms, it means that the two branes should be sufficiently far apart, not to cause too much disturbance in our world.

As it stands, the factors depending on scalars appear in the action in front of the Einstein-Hilbert R term. In the language of relativity practitioners it would be called "Jordan frame", which is deposited by a conformal transformation of the metric to the so called "Einstein frame" in which there are no scalars in front of the R term any more, with the cost of getting them in matter Lagrangian instead. The corresponding total low energy effective action of this model becomes of the form

$$
\begin{aligned}
S_{\mathrm{EFF}} = \frac{1}{16\pi G} \int d^4x \sqrt{-g} [\mathscr{R} - \frac{12\alpha_d^2}{1+2\alpha_d^2} (\partial\phi)^2 \\
- \frac{6}{2\alpha_d^2+1} (\partial R)^2] - \int d^4x \sqrt{-g} (V_{\mathrm{eff}}(\phi,R) + W_{\mathrm{eff}}(\phi,R)) \\
+ S_m^{(1)}(\Psi_1, A^2(\phi,R) g_{\mu\nu}) + S_m^{(2)}(\Psi_2, B^2(\phi,R) g_{\mu\nu}).
\end{aligned}
\tag{13}
$$

where one finds two tensions (or cosmological constants) of the two branes $V_{\mathrm{eff}}, W_{\mathrm{eff}}$ and two coupling functions A, B, all of the depending on the dynamical scalar fields ϕ, R (in turn, related to the original three scalars, the dilaton and 2 shapes). Writing down the evolution equations for gravity and both scalars, and going to the cosmological setting via standard steps one gets the following set of equations.

The so called Friedmann equation for Hubble parameter $H = \dot{a}(t)/a(t)$ where $a(t)$ is the Universe scale factor reads

$$H^2 = \frac{8\pi G}{3}(\rho_1+\rho_2+V_{\text{eff}}+W_{\text{eff}}) \tag{14}$$

$$+\frac{2\alpha_d^2}{1+2\alpha_d^2}\dot{\phi}^2+\frac{1}{1+2\alpha_d^2}\dot{R}^2. \tag{15}$$

The second Einstein equation is

$$\dot{H}+H^2 = -\frac{4\pi G}{3}(\rho_1+3p_1+\rho_2+3p_2-2V_{\text{eff}}-2W_{\text{eff}})$$

$$-\frac{4\alpha_d^2}{1+2\alpha_d^2}\dot{\phi}^2-\frac{2}{1+2\alpha_d^2}\dot{R}^2 \tag{16}$$

The field equations for R and ϕ read

$$\ddot{R}+3H\dot{R} = -8\pi G\frac{1+2\alpha_d^2}{6}[\frac{\partial V_{\text{eff}}}{\partial R}+\frac{\partial W_{\text{eff}}}{\partial R}$$

$$+\alpha_R^{(1)}(\rho_1-3P_1)+\alpha_R^{(2)}(\rho_2-3P_2)] \tag{17}$$

$$\ddot{\phi}+3H\dot{\phi} = -8\pi G\frac{1+2\alpha_d^2}{12\alpha_d^2}[\frac{\partial V_{\text{eff}}}{\partial \phi}+\frac{\partial W_{\text{eff}}}{\partial \phi}$$

$$+\alpha_\phi^{(1)}(\rho_1-3P_1)+\alpha_\phi^{(2)}(\rho_2-3P_2)]. \tag{18}$$

In these expressions the matter on both branes has energy densities and pressures ρ_i, P_i entering for scalars is the trace of the energy–momentum tensor for each brane's matter.

The coupling functions $\alpha_\phi^{(i)}$ and $\alpha_R^{(i)}$ are defined as

$$\alpha_\phi^{(1)} = \frac{\partial \ln A}{\partial \phi}, \qquad \alpha_\phi^{(2)} = \frac{\partial \ln B}{\partial \phi}; \tag{19}$$

$$\alpha_R^{(1)} = \frac{\partial \ln A}{\partial R}, \qquad \alpha_R^{(2)} = \frac{\partial \ln B}{\partial R}. \tag{20}$$

and in the model with exponential potential they have a particular form, depending on the parameter α_d of the model. These functions describe mutual attraction of the two branes, and thus the distance between them.

In Fig.3(a) of preprint [12] one finds cosmic evolution of matter composition of this model [10], which is tuned to reproduce durations of radiation, matter and scalar-dominated eras. Fig.3(b) shows the corresponding motion of the two branes. As one should expect from discussion in the preceding section, one finds some sliding behavior, from equilibrium at radiation dominated era to a gentle slide during matter era to another change recently (related to scalar dominance and accelerated Universe, as now observed).

The main lesson from this is that the two-brane model leads to nearly the same cosmological eqns as people were discussing on more general grounds for ad hoc scalars. Therefore, the ideas about cosmological attractor leading to self decoupling of the scalars work. In particular, in radiation-dominated era there is no “force" term in the r.h.s. as $\rho = 3P$ and rolling to the minimum kills evolution fast. Matter dominated period start changes in the scalars again, as well as recent transition from matter to cosmological constant dominance.

An ideal model would (i) produce the potential at the current minimum equal to the observed cosmological constant; (ii) make relaxation to it so good that all current and recent (Oklo, meteorites) strong limits on any modifications of the constants be naturally satisfied; and maybe (iii) produce interesting modifications at redshift $z \sim 1$ where the matter composition at our brane had changed the last time.

A particular potentials studied in [10] numerically have not achieved these goals (and in fact the authors found some unexpected theoretical problems on the way). But the game is too recent to give up on it, and the settings itself is clearly sufficiently flexible to try many variants.

CHANGING PHYSICS NEAR MASSIVE BODIES

The reason gravity is so important at large scales is that its effect is additive. The same should be true for massless (or very light) scalars: its effect near large body is proportional to the number of particles in it.

Unlike in the brane models, in which one can only hope to feel motion of the second brane via constant variations, we do see the Sun. But suppose we cannot (e.g. there are always clouds) and the temperature is also unchanged. Can one still feel that the Sun gets closer or further, in a periodic fashion? By an oscillating physical constants, perhaps.

Scalars coupled to gravity in static case

For static star (or Galaxy) the eqn for the dilaton is

$$\Delta\phi = 4\pi G A'(\phi) T \tag{21}$$

to be of course complimented by Einstein's eqn to which it is coupled

$$R_{\mu\nu} = 2\partial_\mu\phi\partial_\nu\phi + 8\pi G(T_{\mu\nu} - (T/2)g_{\mu\nu}) \tag{22}$$

It is a nonlinear eqn inside the body, which as usual for stars should be solved starting from r=0 outward. But what one should expect to find for dilaton outside the body?

If there is no matter there, $T = 0$ at $r > R$, then the only possibility is simply a combination of a Coulomb-like field and constant

$$\phi = Q_d/r + \phi_0 \tag{23}$$

Constant part is the cosmological dilaton expectation value we discussed above: the Coulomb-like field 1/r is the subject of this section.

But before we discuss it, let us think what would happen at constant density, say if one wish to include some matter such as interstellar gas or cosmological constant, which produces small but nonzero T everywhere? If one writes $A'(\phi) = \beta\phi$ the r.h.s. is then a mass term with the dilaton mass

$$m_d^2 = 4\pi G\beta T \tag{24}$$

If T is given by the cosmological constant this mass becomes $m_d^2 = 6\beta\Omega_c H^2$ where $\Omega_c = 0.7$ is its current balance in total density and H is current Hubble constant. Ignoring factors of the order one one finds that this mass is only important for sizes as large as the visible Universe itself. Thus, the first lesson thing we learned from this is that the sign of β should be such that there is no instability present in the Universe: we have to be near the minimum, not maximum of the potential.

The other lesson is that for any body the effective dilaton mass can only gets important if its size is close to its own relativistic radius. It is indeed precisely in this case, for which Damour and Esposito-Farese [23] argued that large dilaton-induced effect is possible inside neutron stars, provided the coupling β happen to be specially tuned to the star parameters making it into a kind or resonator amplifying small cosmological dilaton field outside.

In fact we know for sure that *outside* neutron stars the dilaton field is small. It was shown in multiple works (see refs e.g. in [23]) that the absence of scalar-induced corrections for binary pulsar motion can get significant limits on that, although they are still somewhat weaker than the limit $A' < 10^{-3}$ following from solar system experiments related to post-Newtonian gravity effects.

However impressive is the accuracy of the available data on motion of some binary pulsars, we think that potentially spectroscopic studies of scalar-induced modification of physics constants is a more promising way to go for future studies, as accuracy of spectral line frequencies is still much higher.

For not-too-relativistic objects, like the usual stars or planets, both their total mass M and the total dilaton charge Q are simply proportional to the number of nucleons in them, and thus the dilaton field is simply proportional to the gravitational potential

$$\phi - \phi_0 = \kappa(GM/rc^2) \tag{25}$$

and thus we expect that the fundamental constants would also depend on the position via the gravitational potential at the the measurement point.

Comparison between potential and real experiments

Naively, one may think that the larger is the dimensionless gravity potential (GM/rc^2) of the object considered, the better. However, different objects allow for quite different accuracy.

Let us mention few possibilities, using as a comparison parameter the product of gravity potential divided by the tentative relative accuracy

$$P = (GM/rc^2)/(accuracy) \tag{26}$$

(i) Gravity potential on Earth is changing due to ellipticity of its orbit: the corresponding variation $\delta(GM/rc^2) = 3.3 * 10^{-10}$. The accuracy of atomic clocks in laboratory conditions approaches 10^{-16}, and so $P \sim 3 * 10^6$. However, comparing a clock on Earth and distant satellite one may get $\delta(GM/rc^2) \sim 10^{-9}$ and $P \sim 10^7$. The space mission was recently discussed, e.g. in the proposal [19] and references therein.

(ii) Sun (or other ordinary stars) has $GM/rc^2 \sim 2 * 10^{-7}$. Assuming accuracy 10^{-7} in the measurements of atomic spectra near the surface we get $P \sim 1$. However, a mission with modern atomic clocks sent to the Sun would have $P \sim 10^8$ or so, see details in the proposal [18].

(iii) The stars at different positions inside our (or other) Galaxy have gravitational potential difference of the order of 10^{-7}, and (like for the Sun edge) one would expect $P \sim 1$. Clouds which give the observable absorption lines in quasar spectra have also different gravitational potentials (relative to Earth), of comparable magnitude.

(iv) White/brown dwarfs have $GM/rc^2 \sim 3 * 10^{-4}$, and in some cases rather low temperature. We thus get $P \sim 3 * 10^3$.

(v) Neutron stars have very large gravitational potential $GM/rc^2 \sim .1$, but high temperature and magnetic fields make accuracy of atomic spectroscopy rather problematic, we give tentative accuracy 1 percent. $P \sim 10$.

(vi) Black holes, in spite of its large gravitational potential, have no scalar field outside the Shwartzschield radius, and thus are not useful for our purpose.

Accuracy of the atomic clocks is so high because they use extremely narrow lines. At this stage, therefore, star spectroscopy seem not to be competitive: the situation may change if narrow lines be identified.

Now let us see what is the best limit available today. As an example we consider recent work [21] who obtained the following value for the half-year variation of the frequency ratio of two atomic clocks: (i) optical transitions in mercury ions $^{199}Hg^+$ and (ii) hyperfine splitting in ^{133}Cs (the frequency standard). The limit obtained is

$$\delta ln(\frac{\omega_{Hg}}{\omega_{Cs}}) = (0.7 \pm 1.2) * 10^{-15} \tag{27}$$

For Cs/Hg frequency ratio of these clocks the dependence on the fundamental constants was evaluated in [22] with the result

$$\delta ln(\frac{\omega_{Hg}}{\omega_{Cs}}) = -6\frac{\delta\alpha}{\alpha} + 0.04\frac{\delta(m_q/\Lambda_{QCD})}{(m_q/\Lambda_{QCD})} - \frac{\delta(m_e/\Lambda_{QCD})}{(m_e/\Lambda_{QCD})} \tag{28}$$

Another work [24] compare H and ^{133}Cs hyperfine transitions. The amplitude of the half-year variation found were

$$\delta ln(\omega_H/\omega_{Cs}) < 7 * 10^{-15} \tag{29}$$

The sensitivity [22]

$$\delta ln(\frac{\omega_H}{\omega_{Cs}}) = 0.83\frac{\delta\alpha}{\alpha} + 0.11\frac{\delta(m_q/\Lambda_{QCD})}{(m_q/\Lambda_{QCD})} \tag{30}$$

There is no sensitivity to m_e/Λ_{QCD} because they are both hyperfine transitions.

As motivated above, we assume that scalar and gravitational potentials are proportional to each other, and thus introduce parameters k_i as follows

$$\frac{\delta\alpha}{\alpha} = k_\alpha \delta(\frac{GM}{rc^2}) \tag{31}$$

$$\frac{\delta(m_q/\Lambda_{QCD})}{(m_q/\Lambda_{QCD})} = k_q \delta(\frac{GM}{rc^2}) \tag{32}$$

$$\frac{\delta(m_e/\Lambda_{QCD})}{(m_e/\Lambda_{QCD})} = \frac{\delta(m_e/m_p)}{(m_e/m_p)} = k_e \delta(\frac{GM}{rc^2}) \tag{33}$$

where in the r.h.s. stands half-year variation of Sun's gravitational potential on Earth.

In such terms, the results of both experiments can be rewritten as

$$k_\alpha + 0.17k_e = (-3.5 \pm 6) * 10^{-7} \tag{34}$$

$$|k_\alpha + 0.13k_q| < 2.5 * 10^{-5} \tag{35}$$

The sensitivity coefficients for other optical clocks can be found in Refs. [25, 26], for the hyperfine clocks in Ref. [22]. The sensitivity coefficients may be very large in transitions between very close levels in Dysprosium atom [25, 27] ($\delta ln(\omega_{Dy}/\omega_{Cs}) \sim 10^8 \delta\alpha/\alpha$, see similar cases in [28]), molecules ($\sim 10 - 1000$) [29], and optical UV transition in ^{229}Th nucleus ($\sim 10^5$). The sensitivity coefficients may also be very large in collision of cold molecules near Feshbach resonance ($\sim 10^2 - 10^{12}$) [31].

ACKNOWLEDGMENTS

This work is partially supported by the Australian Research Council.

REFERENCES

1. T. Damour and K. Nordtvedt, Phys.Rev.Lett. **70**, 2217 (1993) Phys.Rev.D **48**, 3436 (1993)
2. T. Damour and A. M. Polyakov, Nucl. Phys. B **423**, 532 (1994) [arXiv:hep-th/9401069].
3. J.D.Bekenstein, Phys.Rev.D25 (1982) 1527.
4. H.Sandwick, J.D.Barrow and J.Magueijo, Phys.Rev.Lett 88 , 03102 (2002)
5. K.Olive and M.Pospelov, Phys.Rev.D65 085044 (2002)
6. L. Randall and R. Sundrum, Phys.Rev.Lett. **83**, 4690 (1999)
7. J.K Webb, V.V. Flambaum, C.W. Churchill, M.J. Drinkwater and J.D. Barrow, Phys.Rev.Lett. **82**, 884 (1999)
8. J.K. Webb, M.T. Murphy, V.V. Flambaum, V.A. Dzuba, J.D. Barrow, C.W. Churchill, J.X. Prochaska and A.M. Wolfe, Phys.Rev.Lett. **87**, 091301 (2001)
9. D.Rainwater and T.M.P.Tait, hep-th/0701093.
10. P. Brax, C. van de Bruck, A. C. Davis and C. S. Rhodes, Phys. Rev. D **67**, 023512 (2003) [arXiv:hep-th/0209158].
11. P. Brax, C. van de Bruck, A. C. Davis and C. S. Rhodes, Astrophys. Space Sci. **283**, 627 (2003) [arXiv:hep-ph/0210057].
12. V.V. Flambaum, E.V. Shuryak, physics/0701220.
13. E. Kiritsis, Fortsch. Phys. **52**, 200 (2004) [Phys. Rept. **421**, 105 (2005 ERRAT,429,121-122.2006)] [arXiv:hep-th/0310001].
14. P. Brax, C. van de Bruck and A. C. Davis, Rept. Prog. Phys. **67**, 2183 (2004) [arXiv:hep-th/0404011].
15. W.J. Marciano, *Phys. Rev. Lett.*, **52**, 489 (1984); B.A. Campbell , K.A. Olive , *Phys. Lett. B* **345**, 429 (1995); X. Calmet and H. Fritzsch, *Eur. Phys. J* C24, 639 (2002); P. Langacker, G. Segre and M.J. Strassler, *Phys. Lett. B* **528**, 121 (2002); T.Dent, M. Fairbairn.*Nucl. Phys.B*, **653** , 256 (2003).
16. J-P. Uzan, *Rev. Mod. Phys.* **75**, **403** (2003).
17. H. Georgi, H. R. Quinn and S. Weinberg, Phys. Rev. Lett. **33**, 451 (1974).
18. L.Maleki and J.Prestage, Lecture Notes in Physics, 648, 341 (2004)
19. S.Schiller et al, gr-qc/0608081.
20. T. Damour, F. Piazza and G. Veneziano, Phys. Rev. Lett. **89**, 081601 (2002) [arXiv:gr-qc/0204094].
21. T.M.Fortier et al, submitted to Phys.Rev.Lett.
22. V.V.Flambaum and A.F.Tedesco, Phys.Rev.C73 , 055501 (2006)
23. T.Damour and G.Esposito-Farese, Phys.Rev.Lett. 70 2220 (1993)
24. A.Bauch and S.Weyers, Phys.Rev.D65, 081101R (2002)
25. V.A. Dzuba, V.V. Flambaum, J.K. Webb. *Phys. Rev. A* **59**, 230 (1999).V. A. Dzuba, V.V. Flambaum, J. K. Webb , *Phys. Rev. Lett.* **82**, 888-891, 1999.
26. V.A. Dzuba, V.V. Flambaum, *Phys. Rev. A* **71**, 052509 (2005); V.A. Dzuba, V.V. Flambaum, M.G. Kozlov, and M. Marchenko. *Phys. Rev. A* **66**, 022501-1-8 (2002); J.C. Berengut, V.A. Dzuba, V.V. Flambaum, and M.V. Marchenko. *Phys. Rev. A* **70**, 064101 (2004). E.J. Angstmann, V.A. Dzuba, V.V. Flambaum, physics/0407141.
27. V. A. Dzuba, V. V. Flambaum, M. V. Marchenko, *Phys. Rev. A* **68**, 022506 1-5 (2003). A T Nguyen, D Budker, S K Lamoreaux and J R Torgerson *Phys. Rev. A.* **69**, 022105 (2004).
28. V. A. Dzuba and V. V. Flambaum, *Phys. Rev. A* **72**, 052514 (2005); E. J. Angstmann, V. A. Dzuba, V. V. Flambaum, S. G. Karshenboim, A. Yu. Nevsky, *J. Phys. B* **39**, 1937-1944 (2006) ; physics/0511180,

29. V.V. Flambaum, *Phys. Rev. A* **73**, 034101 (2006).
30. V.V. Flambaum, Phys. Rev. Lett. 97, 092502 1-3 (2006). E. Peik, Chr. Tamm. *Europhys. Lett.* **61**, 181 (2003).
31. Cheng Chin, V.V. Flambaum, Phys. Rev. Lett. 96, 230801 1-4 (2006)

A Cluster Model of ^{6}He and ^{6}Li

Jeremy R. Armstrong*, Alexander A. Sakharuk† and Vladimir G. Zelevinsky*

*NSCL and Department of Physics and Astronomy, Michigan State University, East Lansing, MI, 48824, USA
†College of Arts and Sciences, Florida Gulf Coast University, Fort Myers, FL 33965, USA

Abstract. Small nuclei provide an ideal testing ground of few-body theories. ^{6}He is particularly interesting in that it shows an extended particle distribution similar to a halo nucleus, is loosely bound, and is a Borromean system. We apply a fully antisymmetric model with correct projection into eigenstates of angular momentum. Binding energies, charge radii, matter radii, and $B(E2;0^+ \to 2^+)$ were obtained for the $\alpha + 2n$ and cigar configurations. The same configurations were used to obtain the same observables for ^{6}Li. Were were then able to calculate the $\log ft$ value for the beta decay of ^{6}He.

INTRODUCTION

Helium-6 is has received much attention in recent years from experiment[1, 2, 3, 4, 5, 6], *ab initio* theory[7, 8, 9, 10, 11], 3-body model[12, 13, 14, 15, 16] , and molecular dynamics[17, 18, 19]. It has an extended spatial structure, where the external neutrons seem to exist primarily in the classically-forbidden region, outside the range of the nuclear attraction from the core. It is also a Borromean system, that is a three-body system where none of the two-body subsystems (in this case ^{5}He and the dineutron) are bound. ^{6}Li, as one of the two stable isotopes of lithium, remains an interesting nucleus for theories of light nuclei.

Cluster models of nuclei have a long history. They first appeared in 1937[20]. The early models pictured the nucleus as a molecule of alpha particles. The models met with some success, but struggled to describe nuclei that were not made of an integer number of alpha particles, or an integer number of alpha particles plus or minus one particle. More realistic cluster models appeared in the 1960s, one of which is described in great detail in [21]. These models treated the clusters as being composed of individual nucleons, and took into account the Pauli principle with either explicit anti-symmetrization, or some method of simulating the Pauli principle such as projection out Pauli-forbidden states. In the 1980s, three-body models appeared which treated a system of a large core plus two few valence nucleons. The structure of the core was simplified into something like a rigid rotor, with the Pauli principle treated approximately. The calculations were carried out in the hyperspherical expansion, in order to obey the proper asymptotics for Borromean systems and obtain the correct energy thresholds necessary for accurate reaction calculations[22]. Concurrently, the molecular dynamics models appeared [23]. These models, of which there are two main versions: Anti-symmetrized Molecular Dynamics (AMD), and Fermionic Molecular Dynamics (FMD), treat each nucleon as a Gaussian wave-packet (or the superposition of Gaussian wave-packets). The wave-packet contains both position and momentum information. In the FMD wave function, they then minimize the expectation value of their effective Hamiltonian with respect to the parameters of all their single particle states in a single Slater determinant. They then minimize the energy of a combination of Slater determinants with respect to collective variables such as the radius and multipole moments. In the AMD method, they solve the frictional cooling equation[24] in order to determine the optimum values of their single-particle parameters, then add constraints in order to superpose multiple Slater determinants. These models are computationally heavy, but produce results that agree well with experiment.

The goal of our work is to use a physically transparent model to describe the structure and some dynamics of light nuclei, specifically ^{6}He and ^{6}Li. The model, though simple and straightforward, is quantum mechanically rigorous and avoids approximations related to the neglect of the Pauli principle. The model should reproduce the properties of the studied nuclei reasonably well.

METHOD

The formalism employed in the present calculations is the Brink formalism[25] in secondary quantization with non-orthogonal orbitals. Figure 1 illustrates the two configurations that are used as a basis. The wave function for the

CP995, *Nuclei and Mesoscopic Physics - WNMP 2007*, edited by P. Danielewicz, P. Piecuch, and V. Zelevinsky

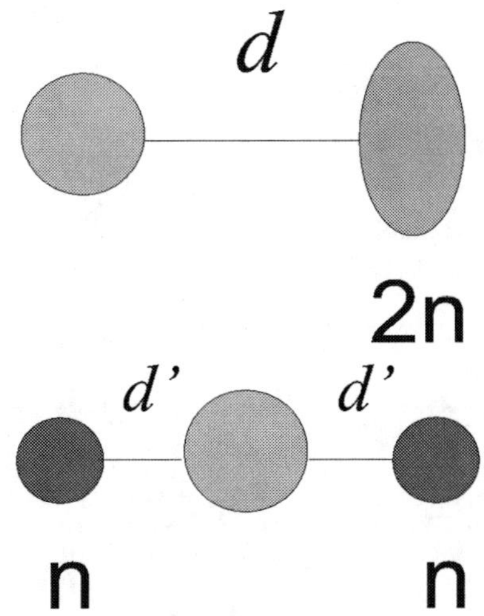

FIGURE 1. Diagram showing the two basis configurations used in the calculations after projecting out the eigenstate of angular momentum. The top picture is an alpha particle and a two-nucleon cluster, while the bottom picture is a nucleon-alpha-nucleon chain (cigar configuration). The parameters d and d' reflect the distance between the alpha particle and an external particle.

alpha-dineutron configuration of ^{6}He is:

$$\psi_1 = N p^\dagger_\alpha p^\dagger_{-\alpha} n^\dagger_\alpha n^\dagger_{-\alpha} n^\dagger_2 n^\dagger_{-2} |0\rangle, \tag{1}$$

where $p^\dagger_\alpha$ creates a spin-up proton in the alpha particle, $n^\dagger_\alpha$ creates a spin-up neutron in the alpha, and $n^\dagger_2$ creates a spin-up neutron in the dineutron. The minus signs in the subscripts refer to a spin-down particle. For ^{6}Li, with a ground state of 1^+, the wave function has an external proton and neutron, and for simplicity we work in the state where the external spins are parallel. For the cigar configuration, the wave function is:

$$\psi_2 = \frac{N}{2} p^\dagger_\alpha p^\dagger_{-\alpha} n^\dagger_\alpha n^\dagger_{-\alpha} \sum_m (-1)^{1/2-m} r^\dagger_m l^\dagger_{-m} |0\rangle, \tag{2}$$

where r refers to the particle on the right, and l is the particle on the left side of the alpha particle. In the case of ^{6}Li, the sum is over isospin projection, as the deuteron is an isosinglet. These operators are Fermi operators, and therefore have anti-commutation relations between them:

$$\{a_i, a_j\}_+ = \{a^\dagger_i, a^\dagger_j\}_+ = 0; \qquad \{a_i, a^\dagger_j\}_+ = \theta_{ij} = \langle \phi_i | \phi_j \rangle,$$

where θ_{ij} is the overlap of the single particle wave functions.

The trial single- particle wave functions in the body-fixed frame are of the form:

$$\phi_i(\mathbf{r}) = \left(\frac{v_i}{\pi}\right)^{3/4} \exp\left[-v_i(\mathbf{r} - \mathbf{z}_i)^2/2\right], \tag{3}$$

where $\mathbf{z}_i$ is the position vector (either the position of the alpha particle or the external particles) of the particle and v_i is the variational oscillator parameter, which is a variational parameter and is kept common within the alpha or among the external particles, but can be different between the two types of particles (alpha and external).

Projection into states where angular momentum is a good quantum number is done by following the prescription outlined in [26]. We take the overlap of a wave function in the body-fixed frame, with the symmetry axis along the z-axis, and find the overlap between it and another wave function that was been rotated about the y-axis:

$$\langle \psi^I_M | O | \psi^I_M \rangle = \frac{\int d^I_{KK} \langle \psi | O e^{\iota \vartheta J_y} | \psi \rangle d(\cos\vartheta)}{\int d^I_{KK} \langle \psi | e^{\iota \vartheta J_y} | \psi \rangle d(\cos\vartheta)}, \tag{4}$$

where the projection along the symmetry axis, K, is taken to be equal to the laboratory frame projection, M. In ^{6}He, this simplifies to an integral over a Legendre polynomial since the spin contribution is absent. Since the deuteron component of ^{6}Li is spin-one, we must use the completely general expression when doing calculations with ^{6}Li.

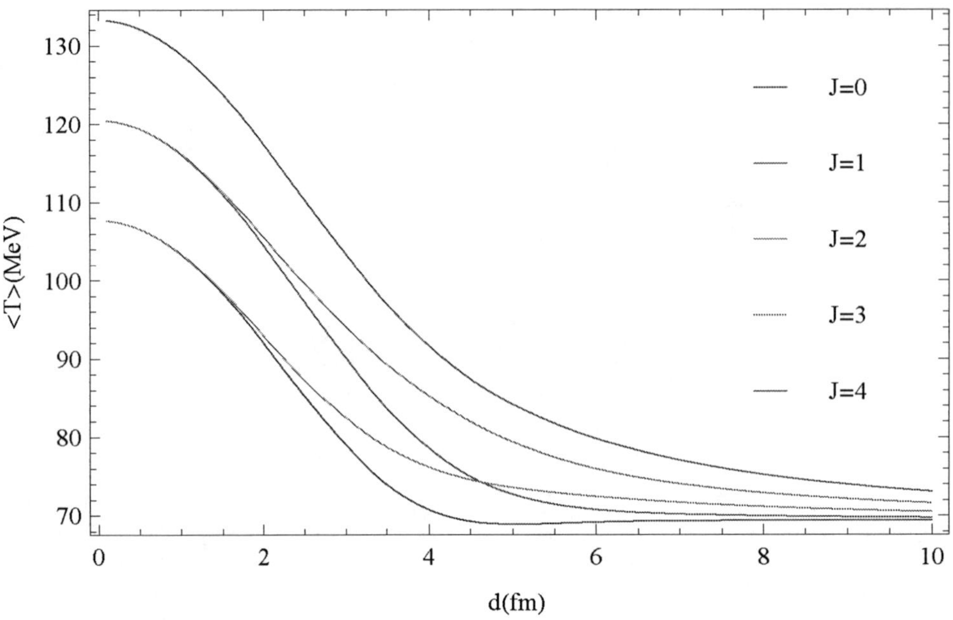

FIGURE 2. The expectation value of the kinetic energy of the alpha-dineutron configuration of ^{6}He as a function of d. In this plot, the oscillator parameters of all particles was set equal to 0.53 fm^{-2}.

The anti-commutation relations are used to compute the normalization and expectation value of any operator. For example, the normalization is:

$$\begin{aligned}\langle\psi_1|\psi_1\rangle &= N^2\langle 0|n_{-2}n_2n_{-\alpha}n_\alpha p_{-\alpha}p_\alpha p^\dagger_\alpha p^\dagger_{-alpha} n^\dagger_\alpha n^\dagger_{-\alpha} n^\dagger_2 n^\dagger_{-2}|0\rangle, \\ &= \theta_\alpha^2\left(\theta_\alpha\theta_d - \theta_{\alpha d}^2\right)^2,\end{aligned}$$

where θ_α and θ_d are the overlaps within the alpha particle and dineutron centers, and $\theta_{\alpha d}$ is the overlap between the two centers. The overlaps within the centers are not equal to one only because of the rotation necessary for angular momentum projection. If the oscillator parameters in eq.(3) are equal, then when the centers coincide, all overlaps are equal to one, which makes the normalization vanish. This is because if the centers coincide, there are four s-wave neutrons (in the case of ^{6}He) at the same point in space, which is forbidden by Fermi statistics. Expectation values are computed by sandwiching an operator between the wave functions, and then again making the expression normal ordered by way of the anti-commutation relations, then dividing by the normalization. A correct calculation of the matrix elements will take care of the 0/0 uncertainty.

The energy of each configuration is minimized with respect to three parameters (the two oscillator parameters, and the external neutron-alpha particle distance), and we then solve the following equation to generate the minimized value of the energy of the system for a given angular momentum:

$$\begin{pmatrix}\langle\psi_1|H|\psi_1\rangle & \langle\psi_1|H|\psi_2\rangle \\ \langle\psi_2|H|\psi_1\rangle & \langle\psi_2|H|\psi_2\rangle\end{pmatrix}\begin{pmatrix}c_1 \\ c_2\end{pmatrix} = E\begin{pmatrix}1 & \langle\psi_1|\psi_2\rangle \\ \langle\psi_2|\psi_1\rangle & 1\end{pmatrix}\begin{pmatrix}c_1 \\ c_2\end{pmatrix}, \tag{5}$$

where c_1 and c_2 are the weighting coefficients of the two configurations. This equation generates the minimum value of energy and the two weighting coefficients, which are needed to calculate any other observable.

RESULTS AND DISCUSSION

In Figure 2, we show the results of the kinetic energy calculation in the alpha-dineutron configuration of ^{6}He as a function of d. At large values of d, the separated alpha-dineutron limit, we see rotational ordering of states, and all particles are in s-waves. As d becomes smaller, the kinetic energy increases until at $d = 0$, we see evenly-spaced,

TABLE 1. Results of calculations with the Volkov V2 potential (V) and the Minnesota potential (M). Experimental results are from [29] unless otherwise noted. Some additional results will be mentioned in the text.

	^{6}He			^{6}Li		
	V	**M**	**Exp**	**V**	**M**	**Exp**
$<E>$ (MeV)	-28.7	-22.1	-29.3	-37.4	-27.6	-32.0
c_1^2	.607	.441	—	.722	.741	—
c_2^2	.082	.168	—	.044	.028	—
$\sqrt{r_{ch}^2}$ (fm)	1.99	1.91	2.05[3]	2.27	2.09	2.52[30]
$\sqrt{r_m^2}$ (fm)	2.40	2.26	2.48[31]	2.27	2.09	2.52[30]
$B(E2;0^+ \rightarrow 2^+)$ (e^2 fm^4)	2.932	2.305	5.4[5]	—	—	—
Q_0 (e fm^2)	—	—	—	+1.04	+0.70	-.082
μ_0 (μ_N)	—	—	—	0.69	0.69	0.82

doubly degenerate levels. This is because at $d = 0$, we are left with a spherical harmonic oscillator. The increase in kinetic energy comes from the fact that two of the neutrons are automatically promoted to higher orbits.

Our Hamiltonian is

$$\hat{H} = \sum_i^A t_i - T_{cm} + \sum_{i<j}^A (V_c^{ij} + V_t^{ij}\hat{S}_{12} + V_{LS}^{ij}\mathbf{L}\cdot\mathbf{S}), \tag{6}$$

where T_{cm} is the center-of-mass correction to the kinetic energy, V_c^{ij} is the central potential, V_t^{ij} is the radial dependence of the tensor force, $\hat{S}_{12}$ is the spherical tensor operator, and V_{LS}^{ij} is the radial behavior of the spin-orbit force. These last two terms vanished in the case of ^{6}He, but are included in the ^{6}Li calculation. We used three nucleon-nucleon interactions: the Volkov V1 and V2 interactions[27], and the Minnesota potential[28]. The Volkov interactions contain a pure Wigner component and a Majorana exchange component, and were designed to fit features of the alpha particle. The Minnesota potential contains Wigner and Majorana components and explicit spin-spin dependence and was designed to fit two-nucleon scattering data. In Table 1 we have most of the numerical results of our calculations. For the Volkov potentials, we show only the results for V2, since the results are nearly identical between the two potentials, but the energy results are better for V2. The energy results for ^{6}He are good for the Volkov potentials, and less good for the Minnesota potential. The Minnesota potential binds singlet pairs appreciably less than triplet pairs, and since our ^{6}He is made of mostly singlet pairs, this accounts for the higher energy. In the case of ^{6}Li, we have a high amount of overbinding (with the Volkov potentials), but this can be put down to the tensor interaction. We used the Volkov radial dependence for the radial dependence of the tensor force, and the Volkov potentials were not originally designed with the use of a tensor force. The tensor force was not used with the Minnesota potential, and the ^{6}Li result is still bound with respect to the alpha particle in that potential (B.E.=-25.1 MeV for the alpha particle with the Minnesota potential).

The results for the weights show a dominance for the alpha-two particle configurations in both nuclei for both potentials. This is in agreement with some other theoretical findings [18, 15], but not all[16]. Experiments are currently being performed to further study the spatial correlation of the valence neutrons in ^{6}He.

Other results are mixed in agreement with experiment. The radii results for ^{6}He are a little smaller than experiment, but the halo behavior is clear, as the difference of the matter radius and charge radius is nearly the same as observed in experiment. The charge radius for ^{6}Li is less good. The spin-orbit interaction tends to pull the minimum to smaller d, while the tensor interaction pushes the minimum to higher d, at least in the dominant alpha-deuteron configuration. For the Minnesota potential, there is no tensor interaction, and thus the radius is comparatively very small. The electric quadrupole moment of ^{6}Li is quite large, and the magnetic moment slightly smaller than experiment. This model ignores many subtle contributions to the magnetic moment. The $B(E2)$ result agrees with most other theories that have calculated for this transition. The 2^+ excited state of ^{6}He is unbound to decay into an alpha and two neutrons, so the experiment is a difficult one. An excellent result, which is not in the table is the $\log ft$ result for the beta decay of ^{6}He. With the Volkov potentials, our result is 2.90, which compares very well with the experimental result of 2.91. This result is comparable with the results obtained by the *ab initio* models, and better than mean field results. The matching of correct exponential asymptotics can be performed but does not significantly change the results.

CONCLUSIONS

We have found that our simple, but quantum mechanically rigorous models captures essential features of the studied nuclei. ^{6}He is better described than ^{6}Li, but the description of ^{6}Li is still adequate. The problems with ^{6}Li seem to stem mostly from the potentials used, as the wave functions still produce a very good result for the β-decay. It should be noted that of the theoretical models mentioned in the introduction, only the *ab initio* theories have done calculations on ^{6}Li in recent years. To improve our results, we would like to try representing the systems with a continuous function of opening angle rather than just the two extreme configurations. This improvement allows the easy inclusion of more spatial information without sacrificing the essential simplicity of the model. Also, we would like to improve upon the potential models used, and use a potential designed with a spin-orbit and tensor interaction, such as the Argonne V8' interaction. The inclusion of the Coulomb interaction could also be an improvement for ^{6}Li. Also, for the proper description of excited states, we would need to include a larger set of basis functions. Other systems that would be interesting to use our model on are ^{8}He, ^{7}Li, and the beryllium isotopes.

ACKNOWLEDGMENTS

This work was supported by NSF grant PHY-0555366.

REFERENCES

1. A. N. Ostrowski *et. al.*, *Phys. Rev. C*,**63**, 024605, (2001).
2. L. R. Gasques *et. al.*, *Phys. Rev. C*, **67**, 024602, (2003).
3. L. -B. Wang *et al.*, *Phys. Rev. Lett.*, **93**, 142501, (2004).
4. L. Giot *et. al.*, *Phys. Rev.*, **71**, 064311, (2005).
5. J. J. Kolata *et al.*, *Phys. Rev. C*, **75**, 031302, (2007).
6. E. A. Benjamim *et al.*, *Phys. Lett. B*, **647**, 30, (2007).
7. P. Navrátil, J, P. Vary, W. E. Ormand, and B. R. Barrett, *Phys. Rev. Lett.*, **87**, 172502, (2001).
8. R. B. Wiringa and S. C. Pieper, *Phys. Rev. Lett*, **89**, 182501, (2002).
9. S. C. Pieper, R. B. Wiringa, and J. Carlson, *Phys. Rev. C*, **70**, 054325, (2004).
10. E. Caurier and P. Navrátil, *Phys. Rev. C*, **73**, 021302, (2006).
11. M. Pervin, S. C. Pieper, and R. B. Wiringa, arXiv:0710.1265v1, (2007).
12. B. V. Danilin *et al.*, *Phys. Rev. C*, **69**, 024609, (2004).; *Phys. Rev. C*, **73**, 054002, (2006).
13. R. Crespo, I. J. Thompson, and A. M. Moro, *Phys. Rev. C*, **74**, 044616, (2006).
14. J. S. Al-Khalili, R. Crespo, R. C. Johnson, A. M. Moro, and I. J. Thompson, *Phys. Rev. C*, **75**, 024608, (2007).
15. W. Horiuchi and Y. Suzuki, *Phys. Rev. C*, **76**, 024311, (2007).
16. C. A. Bertulani and M. S. Hussein, arXiv:0705.3998v5, (2007).
17. N. Itagaki, A. Kobayakawa, and S. Aoyama, *Phys. Rev. C*, **68**, 054302, (2003).
18. T. Neff *et al*, *Nucl. Phys.* **A738**, 357, (2004); *Nucl. Phys.* **A752**, 321, (2005).
19. S. Aoyama, N. Itagaki, and M. Oi, *Phys. Rev. C*, **74**, 017307, (2006).
20. J. A. Wheeler, *Phys. Rev.*, **52**, 1083, (1937).
21. K. Wildermuth and Y. C. Tang, *A Unified Theory of the Nucleus*. Academic Press, New York: 1977.
22. M. V. Zhukov, B. V. Danilin, D. V. Fedorov, J. M. Bang, I. J. Thompson, and J. S. Vaagen, *Phys. Rep.*, **231**, 151, (1993).
23. H. Feldmeier and J. Schnack, *Rev. Mod. Phys.*, **72**, 655, (2000).
24. Y. Kanada-En'yo, H. Horiuchi, and A. Ono, *Phys. Rev. C*, **52**, 628, (1995).
25. D. M. Brink, *International School of Physics "Enrico Fermi"*, XXXVI, 247, (1966).
26. P. Ring and P. Schuck, *The Nuclear Many-Body Problem*, Springer-Verlag, New York: 1980.
27. A. B. Volkov, *Nucl. Phys.*, **74**, 33, (1965).
28. D. R. Thompson, M. Lemere, and Y. C. Tang, *Nucl. Phys.* **A286**, 53, (1977).
29. D. R. Tilley *et al*, *Nucl. Phys.* **A708**, 3, (2002).
30. R. Sánchez *et al*, *Phys. Rev. Lett.*, **96**, 033002, (2006).
31. A. Ozawa, T. Suzuki, and I. Tanihata, *Nucl. Phys.* **A693**, 32, (2001).

Equation of State and Pairing Gaps in Cold Atoms and Low-Density Neutron Matter

J. Carlson, Alexandros Gezerlis and Sanjay Reddy

Theoretical Division, Los Alamos National Laboratory, Los Alamos, NM, 87545, USA

Abstract. Cold Atom experiments provide a direct experimental test of the properties of neutron matter at extremely low densities, as is expected to be found in the crusts of neutron stars. These systems are very unusual in that the pairing gap is of the order of the Fermi energy, rather than the very small fraction typical for superfluids or superconductors. We compare the equation of state and the pairing gaps in these systems.

Keywords: Superfluidity, Pairing, Neutron Matter, Cold Atoms
PACS: 21.65.+f, 03.75.Ss, 26.60.+c, 05.30.Fk

INTRODUCTION

There has recently been a great deal of interest in experiments on dilute systems of strongly-interacting cold Fermi atoms near the unitarity regime. These experiments are intriguing in that, in the limit of an infinite scattering length and zero effective range, the pairing gap and the ground-state energy are universal constants times the Fermi energy of the non-interacting system at the same density. Many experiments have been performed that measure the sound velocity [1], collective excitations [2], superfluidity [3] and critical temperature [4], among many other properties.

Part of the reason these experiments are so fascinating is that the cold atom experiments can be tuned to mimic the physics in other systems where the scattering length is much larger than the interparticle spacing. In particular they can mimic very low density neutron matter as found in the crust of neutron stars. While neutron matter has been investigated theoretically for more than twenty years, large discrepancies between various theoretical approaches, particularly for the pairing gap remain. Experiments in cold atoms hold the promise of resolving these issues.

EQUATION OF STATE

A very simple and fundamental problem in many-body physics is to determine the ground state properties of a Fermi system with a simple attractive two-body interaction:

$$H = \sum_i \frac{-\hbar^2}{2m} \nabla_i^2 + \sum_{i,j} v(r_{ij}), \tag{1}$$

where the interaction $v(r)$ is of short range compared to the interparticle spacing. We consider the transition over the region where the interaction (or equivalently the density) is varied from $-k_F a << 1$ to $-k_F a >> 1$, where k_F and a are the Fermi momentum $k_F = (3\pi^2\rho)^{1/3}$ and the scattering length, respectively. The former regime ($-k_F a << 1$) is the traditional weakly-coupled limit of typical superfluids and superconductors, the latter ($-k_F a >> 1$ is the so-called universal regime with very strong pairing. Much is known analytically about the weak-coupling regime, but the strongly-paired regime has been much less well-characterized until recently.

This simple Hamiltonian can be used to study neutron matter as long as the p-wave (and higher) interactions are relatively unimportant. In our studies, we conservatively consider densities only up to $k_F a = -10$, which corresponds in neutron matter to a $k_F = 0.54$ fm^{-1}. This is approximately 1/2 the Fermi momentum of neutrons at saturation density in nuclear matter. Even this limited regime is very interesting, though, because here the s-wave pairing gap in neutron matter is quite large in traditional BCS calculations. It is also the regime in which atomic experiments can directly confront many-body theories of superfluid Fermions.

At low densities the primary difference between cold atoms and neutron matter is the effective range of the interaction. For cold atoms it is extremely small, of the order of on ten-thousandth the interparticle spacing. For

CP995, *Nuclei and Mesoscopic Physics - WNMP 2007*, edited by P. Danielewicz, P. Piecuch, and V. Zelevinsky

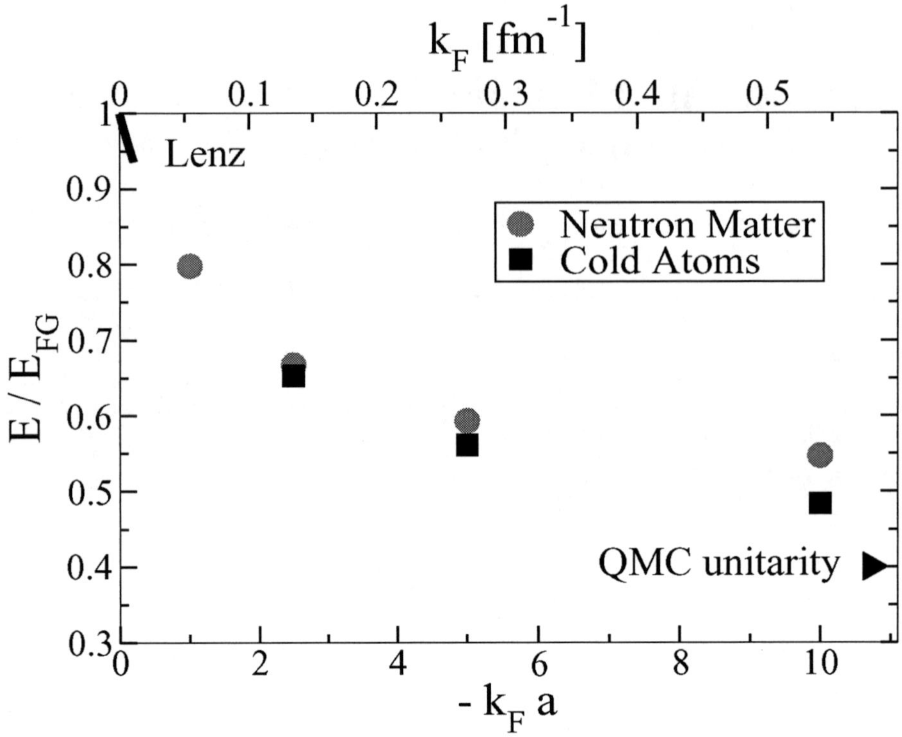

FIGURE 1. Ground-state energy as a function of $-k_F a$ for cold atoms and neutron matter. Cold atoms are assumed to have an effective range close to zero and neutron matter has an effective range over scattering length $(r_e/a) \approx 0.15$

neutrons the effective range in this density regime can be a significant fraction of the interparticle spacing. In our calculations we use two interactions $v(r)$, for simulations of cold atoms the interaction is taken as $v(r) = -v_0 \frac{2\hbar^2}{m} \frac{\mu^2}{\cosh^2(\mu r)}$, with $\mu = 24/r_0$, or an effective range of $r_0/12$, with $1/\rho = (4/3)\pi r_0^3$. The interaction's effective range is close enough to zero not to significantly differ from a true zero-range interaction.

For neutrons we use the s=0 s-wave part of the Argonne V18 (AV18)[5] interaction, which has been tuned to the nucleon nucleon scattering data and adjusted to give a good scattering length and effective range for neutrons. The repulsive core in the AV18 interaction prevents the collapse that would occur for a purely attractive interaction with a finite range. The real nucleon-nucleon interaction is quite complicated, requiring spin-spin, spin-orbit, and momentum-dependent terms to fully describe the proton-proton scattering data below pion production threshhold. At these low densities, though, we can replace it by a spin- and momentum-independent interaction acting only between spin-up and spin-down neutrons. Such an interaction incorrectly invokes p-wave interactions in the $S = 1$ m_S=0 spin state, corrections for this interaction are made perturbatively. At the highest densities considered, these corrections are of order 10% for the ground-state energy. Corrections to the pairing gap are much smaller than for the ground-state energy, and corrections at lower densities are also much smaller.

The results for the equation-of-state are shown in Figure 1 [6]. At very low densities the ground-state energy ignoring superfluid corrections can be obtained analytically in an expansion in $(k_F a)$. As the superfluid corrections are exponentially small for $-k_F a << 1$, these corrections can be safely ignored. The analytic expression for normal matter is[7] $E/E_{FG} = 1 + \frac{10}{9\pi} a k_F + \frac{4}{21\pi^2}(11 - 2\ln 2)(a k_F)^2$.

At higher densities, though, near $k_F a = -1$, the pairing contributions start to become imporant, and a more complete treatment is required. As $-k_F a \rightarrow \infty$, the ground state energy should be a universal constant ξ times the energy of the non-interacting system. Recent theoretical and experimental results yield $\xi \approx 0.40$[8, 9, 10, 11], as shown by the triangle in the figure.

Quantum Monte Carlo results for the equation-of-state are shown for both cold atoms and neutron matter. The results for the two cases are very similar to each other and to the analytic results for normal matter at $-k_F a = 1$, beyond the region where the analytic result clearly applies. Even at $-k_F a = 10$, where $k_F r_e$ for neutron matter is larger than one, the ground-state energy differs by only about 10 percent in going from cold atoms with zero effective range to neutron matter with $a/r_e \approx 0.15$.

We have performed calculations of the equation of state using from 14 to 90 particles in periodic boundary conditions. The estimated error in the ground-state energy from finite size effects is quite small, of the order of a few percent. These errors are particularly small near the unitary regime, as the system more nearly resembles fermions paired into effective bosons, and the shell effects traditionally associated with Fermion simulations are essentially absent.

From these calculations we find a small repulsive effect associated with the transition from a zero-range interaction to neutron matter. Beyond the densities considered here, the equation of state can be more complicated as it will involve contributions from p-wave parts of the interaction and beyond. However, for low densities cold atom experiments can provide a direct test of the equation of state for neutron matter to rather high precision.

These experimental and theoretical constraints could provide significant constraints on nuclear energy density functionals. Many current models diverge significantly from this low-density behavior, and constraints from improved theory and experiment could, for example, significantly reduce uncertainties in calculations of neutron radii in atomic nuclei.

PAIRING GAP: THEORY

The pairing gap, or the energy difference between even and odd-particle systems, is another fundamental ground-state property of Fermi systems. In nuclei these differences are a couple MeV, or a few precent of the Fermi energy of nuclear matter at saturation density. The pairing gap we find for bulk neutron matter is similar in magnitude, at most a couple of MeV, but a much larger fraction of the Fermi energy, up to about 25 per cent. These pairing gaps are much larger than traditional superfluids or superconductors, where the pairing gaps are many orders of magnitude smaller than the Fermi energy.

As for the equation of state, analytic results have been obtained in the weak-coupling, or small $-k_F a$ limit. Original results at zero and finite temperature were obtained in BCS theory, this theory has often been used to calculate pairing gaps in neutron matter as well. Gorkov calculated polarization corrections which go beyond the simple BCS theory, and found a reduction $\Delta/\Delta_{BCS} = (1/4e)^{1/3}$ in the weak coupling limit.

Calculations of neutron matter pairing gaps have diverged wildly over the years, primarily because of the difficulty in going reliably beyond the BCS theory in the strong-coupling ($-k_F a > 1$) region. Indeed calculations of the pairing gap have varied by an order of magnitude in the region of densities we are considering ($1 \leq -k_F a \leq 10$). Most of the difficulties arise because many theories start with an assumption of a well-defined Fermi surface. In fact this is far from the case in the unitary regime.

The divergence can most clearly be seen by looking at neutron matter as an intermediate point in the BCS-BEC transition. This transition between weakly paired systems in the BCS regime to fully bound dimers in the BEC regime is clearly seen in cold atom experiments. In the BCS regime the pairing gap is governed by BCS theory with the Gorkov correction. However in the extreme BEC regime the pairing gap is just given by 1/2 the binding energy of a molecule, a result given correctly by just solving the BCS equations. In this extreme case the polarization corrections are negligible. One would naturally expect then, in the fairly smooth BCS-BEC transition that the polarization effects (essentially anything beyond the BCS approximation) become smaller as one approaches stronger coupling. Theories that can address the entire BCS-BEC transition are therefore quite valuable for examining strongly-paired fermions.

Calculations of the pairing gap in many-body simulations are significantly more difficult than the ground-state energy, essentially because the pairing gap is much more susceptible to finite-size truncation errors than the ground-state energy. Indeed, Quantum Monte Carlo (QMC) calculations of the ground-state energy of liquid Helium have been performed for nearly 40 years. The pairing gaps have not been computed using these techniques, though, because of the vast difference in scale between the energy of the entire system (of the order of eVs per particle) and the pairing gap (of the order of meV).

Pairing gaps in the strongly paired regime are more amenable to QMC calculations in finite volumes, however. For example, calculations in the BEC limit would be accurate as long as the simulation volume is significantly larger than the size of a molecule. In the unitary regime this requires that the simulation volume be larger than the size of a pair, which is comparable to the interparticle spacing.

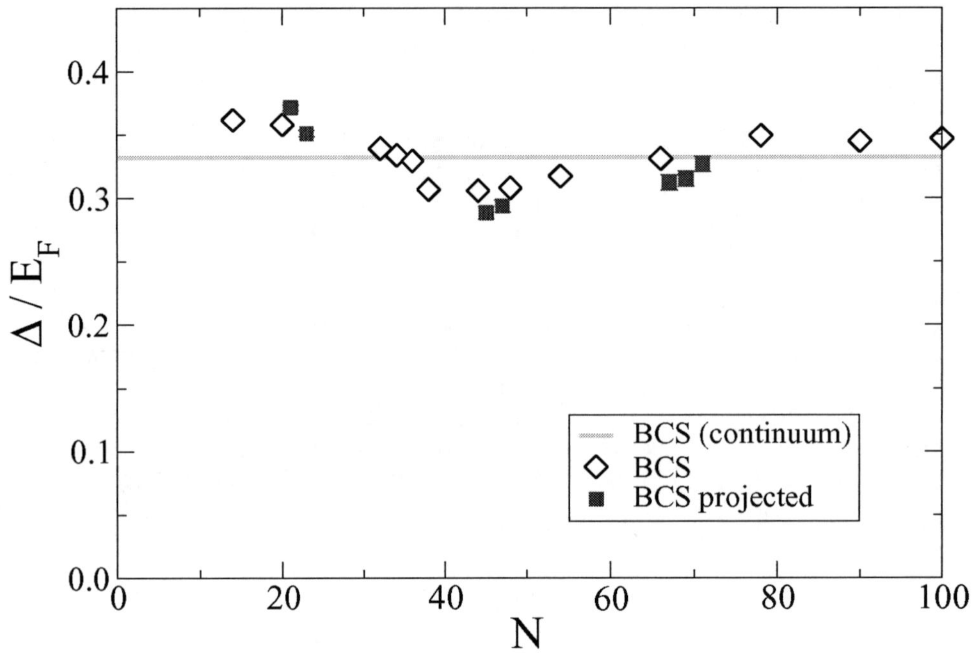

FIGURE 2. Neutron matter pairing gap at $k_F a = -10$ versus particle number in periodic boundary conditions, BCS and QMC calculations.

To examine this more concretely, we have compared calculations of the BCS pairing gap in finite volumes with periodic boundary conditions to the infinite volume limit. The finite volume calculations essentially require replacing the integral over all momenta in the BCS equations with finite sums over the momenta in periodic boundary conditions. A comparison of results is shown in Figure 2 for the case of $-k_F a = 10$.

The figure shows significant oscillations of the pairing gap with system size. Particularly at smaller system sizes, for N = 20 .. 44, the pairing gap can change by approximately 20 per cent. These oscillations are associated with the shells in the simulation cube, we note that N=14 corresponds to a closed shell with a cubic momentum structure and N=46 is a half-filled shell between N=38 and N=54. Beyond N=66, another closed shell, the oscillations in the pairing gap are somewhat smaller, of the order of 10 per cent. This uncertainty is dramatically smaller than the differences between previous calculations of the pairing gap, which can be much more than a factor of two.

In this figure we also show calculations with particle-projected BCS wave functions. In these calculations we simply project the BCS wave functions obtained to finite particle number, and use even-odd energy differences to calculate the pairing gap in a variational Monte Carlo calculation. These results are very similar to those of the original BCS calculations in periodic boundary conditions, indicating that particle number fluctuations are not an important effect for large $-k_F a$.

We have performed calculations of the pairing gap for cold atoms and neutron matter for N=66-70 particles for $-k_F a = 1, 2.5, 5, \text{and} 10$; the results are shown in Figure 3. For cold atoms the gap monotonically increases with $-k_F a$ as one would expect. The pairing gap at $-k_F a = 10$ for cold atoms is quite similar to the value of 0.5 (0.03) obtained previously. The pairing gap at unitarity is already an important ingredient for understanding cold atom experiments, as we discuss below.

For neutron matter, the ratio of pairing gap to Fermi energy reaches a maximum at around $-k_F a = 5$. Below this point the pairing is weak and the pairing is reduced. Above this point the effective range plays a significant role, again reducing the pairing gap. At high densities no s-wave pairing gap is expected. Typically this is attributed to the repulsive interaction at large relative momenta, but even in a simple model with only a scattering length and effective

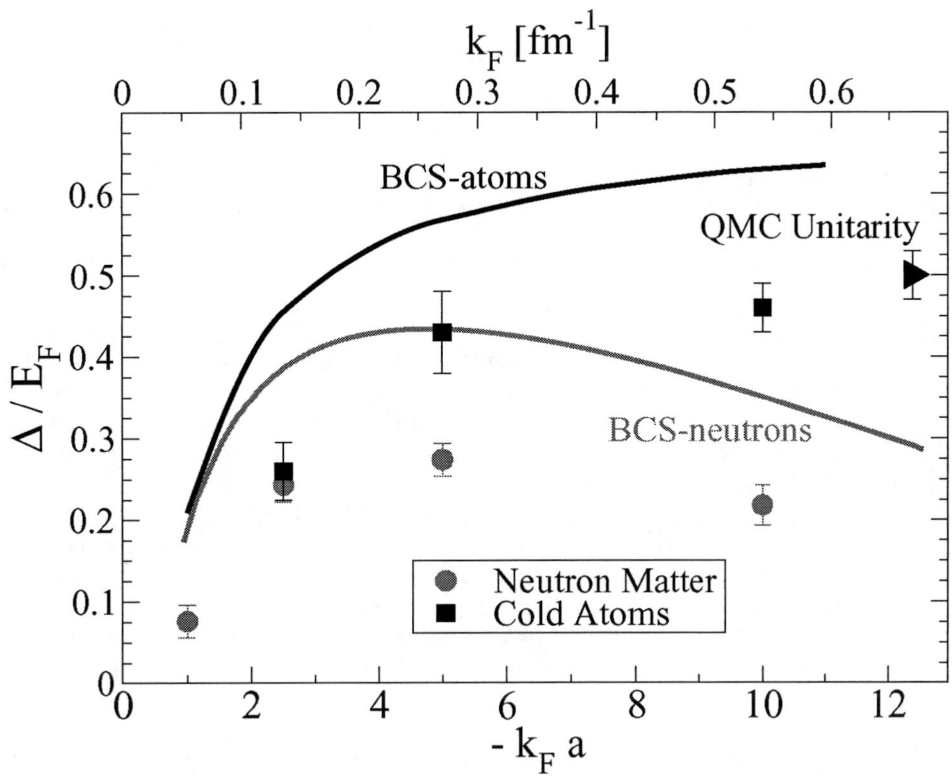

FIGURE 3. Neutron matter pairing gap vs. $-k_F a$ for cold atoms and neutron matter. Solid lines are solutions of the BCS equations, points with error bars are QMC results.

range the pairing gap will diminish as the range of the interaction overlaps more than one pair at a time.

These calculations produce much larger pairing gaps than those obtained in diagrammatic calculations. They are somewhat smaller, though, than those obtained in auxiliary field Difficusion Monte Carlo (AFDMC) calculations, which in some cases produced pairing gaps larger than the BCS results. AFDMC has the advantage of being able to treat a much more realistic interaction than considered here. On the other hand it does not have an upper bound property which makes it more difficult to choose the correct trial state. AFDMC calculations with larger system sizes and different trial wave functions are being pursued.

These simulations suggest that polarization corrections to the BCS pairing gap are much smaller than previously believed. This can have significant effects, for example, on the cooling of neutron stars. First, though, it is critical to understand if these results are indeed correct by comparing to experiment wherever possible.

PAIRING GAP: COLD ATOM EXPERIMENTS

To date the pairing gap has been probed in cold atom experiments through two beautiful kinds of experiments, RF response and local polarization measurements in polarized systems. The first experiments to probe the pairing gap was measurements of the RF response of the paired cold atom system. An RF field is tuned to an atomic transition to 'flip' the state (spin) of an initial minority spin atom into a third species, essentially without transferring any momentum. Clearly the response at different frequencies tells us something about the final states of the system, which will contain at least one unpaired particle. The other intriguing type of experiment that has been performed recently is to examine the local polarization in a trapped system of atoms with an overall net polarization. The polarization as a function of position in the trap at different temperatures yields rather direct estimates of the pairing gap, as discussed below. Below we discuss each of these experiments in turn.

The RF response of the trap essentially measures a response $R(\omega)$, which is essentially a measurement of the transition matrix elements and density of states at an excitation energy ω:

$$R(\omega) = \sum_f \langle i|O^\dagger|f\rangle\langle f|O|\rangle\delta(\omega-(E_f-E_i)), \tag{2}$$

with the transition operator proportional to $O=\sum_i a_3^\dagger(i)a_2(i)$ converting a particle of type 2 (say down spin) to type 3. The final states have at least one particle of type 3. Since there is no momentum transfer in the transition operator the dominant response at weak RF intensity comes from the transition of a single atom. Of course the transitions occur throughout the trap but to first approximation these are isolated transitions which can be treated in a local density approximation.

In an idealized case, the atoms transitioning to type 3 would not interact with the original atomic species. Starting from an initially fully paired state, the final states could then be characterized by the momentum of the particle undergoing the transition, and the energy of the final state could be separated into the kinetic energy of this free particle plus the energy of the residual system with one unpaired spin: $E_f(k) = k_3^2/(2m) + E_{res}(k)$. We've labeled the energy of the residual system by the momenta k which must be equal and opposite to the momenta of the transformed particle. The minimum of $E_{res}(k)$ then gives the pairing gap.

The real situation is more complicated because the atomic species in state 3 in general interacts with particles of type 1 and 2. An extreme case is when particles of type 3 interact the same way as the original particle. The original ground state then remains an eigenstate of the new Hamiltonian and the response is a delta function at zero energy. In the present experiments the situation lies between these two extremes, $k_F a_{13} \approx -1$. For general attractive short-range interactions between 1-2, 2-3, and 1-3 simultaneously the final states can contain collapsed or nearly-collapsed states and there is no lower bound to the response energy. These states contribute to significant evaporative loss seen in the experiments.[12]

One can derive interesting energy-weighted sum rules for the response,[13] but in general these yield only limited information about the pairing gap because of important contributions to the response from the high-energy regime. Indeed it is simple to show that the energy-squared sum rule diverges for short range interactions. A more complete theoretical treatment of this response is warranted, but may provide only limited information about the pairing gap.

In stark contrast, measurements of the local polarization in the trap provide direct and accurate information on the pairing gap. In such an experiment the spin-averaged local chemical potential is fixed by the trap potential: $\mu(r) = \mu_\uparrow(r) + \mu_\downarrow(r) = \lambda - V_{trap}(r)$, where λ is an overall constant chemical potential adjusted to give the total number of particles and $V_{trap}(r)$ is the trapping potential. The difference between the up and down chemical potentials $\delta\mu(r) = \mu_\uparrow(r) - \mu_\downarrow(r)$ is independent of the position in the trap. At zero temperature in a slowly-varying potential the system will separate into three regions: a superfluid region in the center which has been found to have zero net polarization, a normal polarized region in the middle with different numbers of spin up and spin down particles, and a fully polarized region in the exterior consisting of only one species. In this experiment the two exterior regions can be used to obtain λ and $\delta\mu$, and the fact that the interior region is unpolarized can be used to place a lower bound on the pairing gap. At finite temperature the interior superfluid will be polarized by thermal effects, and measurements of the polarization as a function of position in the trap provide a sensitive estimate of the pairing gap.

We have analyzed recent measurements by the MIT group[14] and obtained an estimate for the pairing gap $\Delta = 0.4-0.5\ E_F$ in the unitary regime.[15] This is to be compared to the quantum Monte Carlo calculations which predicted a gap of 0.5 (0.03) MeV.[8?] The experiments indicate that as the temperature is lowered the superfluid region is completely polarized and phase separated from the (polarized) normal state. This observation in itself provides a lower bound of approximately 0.4 E_F for the pairing gap.

One can go further and use calculations of the energy of a system with momentum k and one unpaired spin up particle: $E_{res}(k)$ to estimate the polarization as a function of position in the trap. The $E_{res}(k)$ is the energy of the unpolarized system plus the energy of a quasiparticle of momentum k. The calculated energy includes the full interactions of the unpaired particle with the medium, though in the fixed-node approximation. In the limit of low polarization, or equivalently dilute quasiparticles, one can then estimate the polarization as a function of radius, temperature, and the chemical potentials $\delta\mu$ and λ.[15] One comparison with experimental data is shown in Figure 4 for the lowest temperature experimental data where the comparison should be most accurate. The calculations yield the correct transition point between superfluid and normal matter, as well as accurate estimates for the polarization in each phase.

The experiment, combined with this analysis confirms at quantitative level the predictions of the QMC calculations for the pairing gap in the unitary regime. It will be very valuable to be able to extract the pairing gap across the whole BCS-BEC transition as a direct test of many-body theories over the whole strongly-paired regime.

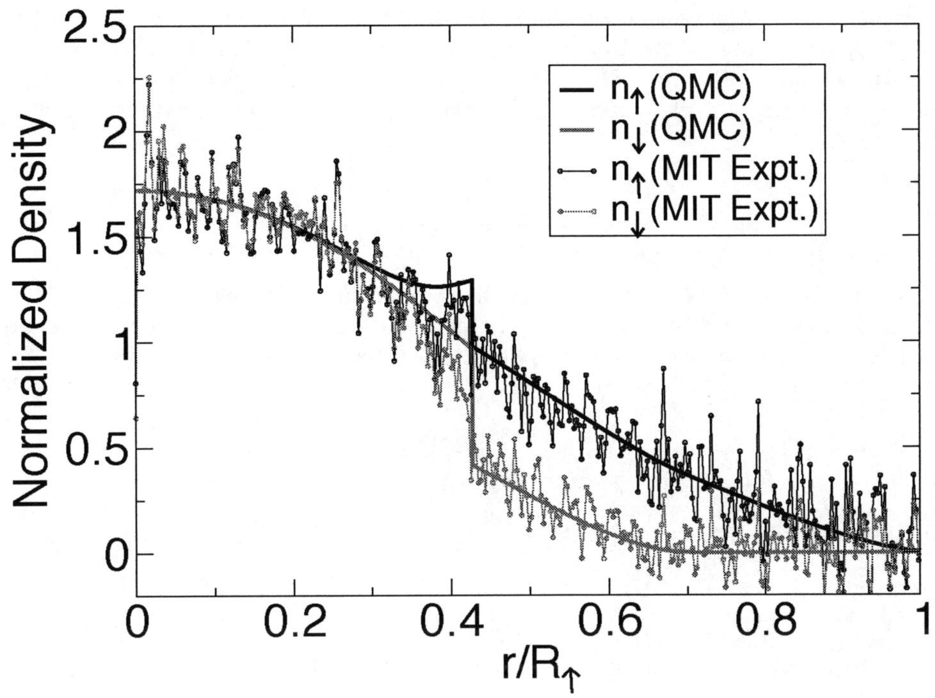

FIGURE 4. Spin up and Spin down densities as a function of distance from the center of the trap, theory (solid curves) and experiment (points).

UNANSWERED QUESTIONS AND THE TRANSITION FROM BULK TO MESOSCOPIC SYSTEMS

The comparisons between theory and experiment described above have all used the local density approximation to go from the calculations in a bulk medium to the observables in a trap where the density is varying. Such an approximation should be very good in cases where the variations in density are slow over many interparticle spacings. That is indeed the case for the MIT experiments described above, which contain of order 10^6 particles in a nearly spherical trap.

Intriguingly, experiments performed at Rice by Hulet et al.[? 17] show a somewhat different behavior for the local polarization than described above. Instead of the three regions described above, these experiments see only a nearly unpolarized central region and a fully polarized exterior. The clearest differerence between the MIT and Rice experiments is that the latter is in a very elongated geometry and about 10^4 particles. Thus the number of particles integrated along the short direction is much smaller than in the MIT experiment.

Calculations of the binding of a single spin down particle in a sea of spin up particles clearly show that this direct transition from fully polarized to fully unpolarized systems cannot happen in the bulk. It is always energetically favorable to produce an intermediate region which is not superfluid and which contains a finite polarization less than one.

Clearly, then, the observed transition has to do with the finite size of the system. This has been interpreted as the signature of a significant surface tension between the two phases, which could stabilize this direct transition in a comparatively small finite system.

Microscopically, the difference between the two experiments may be due to the difference between the ratio of single-particle energies to pairing gaps in the two experiments. In the bulk, the single-particle spacings go to zero, and the MIT experiment is much closer to this limit. In the Rice experiments the single-particle energy level spacings in the trap are comparatively larger, which will certainly affect the competition between phases.

Altering the trap geometry could then provide very important diagnostics on the transition of pairing gaps from the bulk to finite mesoscopic systems, a crucial question for nuclear physics. Significant theoretical effort has already been undertaken for the pairing gap in small finite systems, showing a significant reduction in the finite trapped systems. It would be very intriguing to study the evolution of the pairing gap as a function of system size and shape, both experimentally and theoretically.

ACKNOWLEDGMENTS

The work of J.C., A. G., and S. R., has been supported by the Department of Energy Office of Nuclear Physics, the UNEDF SCIDAC collaboration, and by the Los Alamos National Laboratory LDRD program. Computer time have been provided by the NERSC supercomputer Center and the LANL Open Supercomputing initiative.

REFERENCES

1. J. Joseph *et al.*, Phys. Rev. Lett. **98**, 170401 (2007).
2. M. Bartenstein *et al.*, Phys. Rev. Lett. **92**, 203201 (2004).
3. M. W. Zwierlein *et al.*, Nature **435**, 1047 (2005).
4. L. Luo *et al.*, Phys. Rev. Lett. **98**, 080402 (2007).
5. R. B. Wiringa, V. G. J. Stoks, and R. Schiavilla, Phys. Rev. C **51**, 38 (1995).
6. J. Carlson and Alexandros Gezerlis, arXiv:0711.3006, (2007).
7. W. Lenz, Z. Phys. **56**, 778 (1929); T. D. Lee and C. N. Yang, Phys. Rev. **105**, 1119 (1957).
8. J. Carlson *et al.*, Phys. Rev. Lett. **91**, 050401 (2003).
9. G. E. Astrakharchik *et al.*, Phys. Rev. Lett. **93**, 200404 (2004).
10. S. Y. Chang *et al.*, Phys. Rev. A **70**, 043602 (2004).
11. S. Zhang, K. E. Schmidt, and J. Carlson, private communication.
12. Y. Shin, C. H. Schunck, A. Schirotzek and W. Ketterle, Phys. Rev. Lett. **99**, 090403 (2007).
13. Gordon Baym, C. J. Pethick, Zhenhua Yu, Martin W. Zwierlein, arXiv:0707.0859 (2007).
14. Y. Shin, Christian H. Schunk. A. Schrotzek, and W. Ketterle, arXiv:0709.3027 (2007).
15. J. Carlson and S. Reddy, arXiv:0711.3006 (2007).
16. G. B. Partridge *et al.*, Science **311**, 503 (2006).
17. G. B. Partridge *et al.*, Phys. Rev. Lett. **97**, 190407 (2006).

Two-component Bose gases under rotation

S. Bargi*, K. Kärkkäinen*, J. Christensson*, G.M. Kavoulakis†, M. Manninen** and S.M. Reimann*

*Mathematical Physics, LTH, Lund University, SE-22100 Lund, Sweden
†Department of Sciences, TEI of Crete, P.O. Box 1939 Heraklion, 71004 Greece
**NanoScience Center, Department of Physics, FIN-40014 University of Jyväskylä, Finland

Abstract. We examine the formation of vortices in a one- and two-component gas of bosonic atoms in a harmonic trap that is set rotating. Both the mean-field Gross-Pitaevskii approach, and the numerical diagonalization method are employed. For a two-component Bose gas, we show that beside the well-known coreless vortices of single quantization, the interatomic interactions between the two species may lead to coreless vortices of multiple quantization. We furthermore comment on the geometries of the interlaced vortex patterns. In the limit of weak interactions, we finally demonstrate a number of exact results.

Keywords: Bose-Einstein condensation, vortices, quantum dots
PACS: 03.75.Lm, 05.30.Jp, 67.40.-w, 73.21.La, 71.10.-w, 85.35.Be

INTRODUCTION

Many of the properties of superfluids [1] such as liquid Helium, have more recently been reported in studies of dilute atomic quantum gases, where advances in cooling and trapping techniques made possible the observation of Bose-Einstein condensation [2, 3, 4, 5, 6]. One example is the existence of quantized vortex states, as seen in the early experimental work by Matthews *et al.* [7] or Madison *et al.* [8]. Following these experiments, much theoretical work was concerned with the problem of rotation, both in the Thomas-Fermi regime of strong interactions between the particles [9, 10, 11, 12], and in the limit of weak interactions between the bosonic atoms. Of much interest are the lowest-energy states at given angular momenta of the rotating bosonic cloud, forming the so-called "yrast" line (which borrows its name from the convention in nuclear physics [13]). Examining the problem of effective repulsion between the particles, Butts and Rokhsar [14] and Kavoulakis *et al.* [15] based their analysis on the Gross-Pitaevskii mean-field approach. They found that in a harmonic trap, as the angular momentum or the rotation frequency of the trap increase, successive transitions between stable patterns of singly-quantized vortices may occur. In analogy to finite-size superconductors [16], at moderate rotation rates where only a few vortices exist, they geometrically arrange themselves in stable "vortex molecules". At larger rotation rates, the well-known triangular Abrikosov lattice forms. The localization of the vortices at smaller angular momenta resembles that of *particles* [17, 18, 19] at extreme rotation, as Manninen *et al.* [20] have shown for the case of Coulomb interactions between the particles. Wilkin, Gunn, and Smith [21] and Mottelson [13] investigated the case of rotation for an attractive effective interaction between the atoms and showed that in a harmonic trap the lowest-energy state is the one where the angular momentum is carried by the center of mass. Numerical diagonalization of the many-body Hamiltonian for a rotating Bose gas was performed by Bertsch and Papenbrock [22], as well as Wilkin and Gunn [23]. Jackson *et al.* demonstrated that the mean-field Gross-Pitaevskii results emerge as the correct leading order approximation to exact calculations [24].

The above studies were carried out mainly for a single-component gas. Multi-component quantum gases, however, provide some surprises and are by far more than a trivial generalization of the single-component case. As the different components of the Bose gas interact and exchange angular momentum, the extra degree of freedom that is associated with the motion of the different species may lead to the occurrence of different and new phenomena [25, 26, 27]. In the past, this problem has been discussed extensively both experimentally and theoretically, see for example, Refs. [7, 25, 26, 27, 28, 29, 30, 31, 32, 33] and the review by Kasamatsu *et al.* [34].

In these Proceedings, we review the results of mean-field theory, and compare to a number of exact results in these many-body systems. Employing both the Gross-Pitaevskii approach and the numerical diagonalization method, we show that besides the well-known coreless vortices of single quantization [32, 33, 34], in a two-component gas the interatomic interactions between the two species may lead to vortex states of multiple quantization [35]. We furthermore comment on the interlaced vortex patterns [29] in two-component Bose gases.

CP995, *Nuclei and Mesoscopic Physics - WNMP 2007,* edited by P. Danielewicz, P. Piecuch, and V. Zelevinsky

ROTATING BOSE-EINSTEIN CONDENSATE IN THE GROSS-PITAEVSKII APPROACH

The many-body Hamiltonian for N particles of mass M with contact interactions of strength U_0 in a harmonic potential of frequency ω is

$$\hat{H} = \sum_{i=1}^{N} \left(-\frac{\hbar^2}{2M} \nabla_i^2 + \frac{1}{2} M \omega^2 r_i^2 \right) + \frac{1}{2} U_0 \sum_{i \neq j=1}^{N} \delta(|\mathbf{r}_i - \mathbf{r}_j|). \tag{1}$$

The corresponding Hamiltonian in a frame of reference that rotates with angular frequency Ω, is simply the one above, with the additional term $-\hbar\Omega\hat{L}_z$.

To obtain the exact quantum states, one needs to diagonalize the Hamiltonian Eq. (1). This procedure, which usually has to be done numerically, also provides information about the low-lying excitations of the system. However, it has a serious drawback: With increasing particle numbers or angular momenta, even at moderate interaction strengths (where one can effectively restrict the Hilbert space) the problem very soon becomes impossible to solve for all but the smallest system sizes.

Mean-field methods, such as the Gross-Pitaevskii approximation commonly used for the description of bosonic systems, here provide a suitable alternative. We apply this approach to a rotating one-component system of bosonic particles. For the rotational states, it is reasonable to assume a strongly oblate trap geometry along the rotational z-axis. The order parameter $\Psi(\mathbf{r})$, which is commonly referred to as the condensate wave function, then separates into the in-plane function $\Phi(x,y)$ and into the wave function along the rotation axis, which is assumed to be the harmonic oscillator ground state $\phi_0(z) = \exp(-z^2/2a_z^2)/(\pi a_z^2)^{1/4}$ with a_z being the oscillator length along this direction.

The energy functional to be minimized is given by

$$\mathscr{E}[\Phi,\Phi^*] = \int \mathrm{d}x\mathrm{d}y \left[\Phi^* \frac{1}{2M}(\hat{p}_x^2 + \hat{p}_y^2)\Phi + \frac{1}{2} M\omega^2 (x^2+y^2)|\Phi|^2 + \frac{g}{2}|\Phi(x,y)|^4 \right], \tag{2}$$

where $\hat{p}_x = -i\hbar\partial_x$ ($\hat{p}_y = -i\hbar\partial_y$) is the momentum in x (y)-direction, and $g = U_0 \int \mathrm{d}z |\phi_0(z)|^4 = \sqrt{8\pi}\hbar^2 a/Ma_z$ is the interaction strength.

The variational minimization of the functional Eq. 2 with respect to Φ^* yields the well-known Gross-Pitaevskii effective single-particle equation for the order parameter Φ,

$$\left(\frac{1}{2M}(\hat{p}_x^2 + \hat{p}_y^2) + \frac{1}{2} M\omega^2 (x^2+y^2) + g|\Phi(x,y)|^2 \right) \Phi(x,y) = \mu \Phi(x,y), \tag{3}$$

where the chemical potential μ arises from the normalization constraint of the order parameter, $\int \mathrm{d}x\mathrm{d}y |\Phi|^2 = N$. For clarity, in the following we normalize the order parameter to unity.

By minimizing $\mathscr{E} - L\hbar\Omega$, where $L = i\int \mathrm{d}x\mathrm{d}y \Phi^*(y\partial_x - x\partial_y)\Phi$ is the angular momentum in units of $\hbar$ (where we identify $L = \langle \hat{L}_z \rangle$), we obtain the dynamically stable states at angular momenta $L(\Omega)\hbar$ for a fixed Ω. However, in order to study the vortex dynamics in the condensate, one should be able to access the full dispersion relation $E(L)$ for any value of angular momentum. The full dispersion can be obtained in a similar way as it is done in the cranking model of nuclear physics. The angular momentum is fixed by imposing an additional functional to the Gross-Pitaevskii functional that "shifts" the minimum of the energy to the desired angular momentum $L_0\hbar$,

$$E_{L_0} = \mathscr{E} + \frac{A}{2}\hbar\omega(L - L_0)^2. \tag{4}$$

The constant A is positive and should be chosen to be large enough to ensure a positive curvature of E_{L_0}.

We solve the Gross-Pitaevskii equation using the imaginary-time evolution method, which is sometimes also called the relaxation method. The basic idea behind this method is to propagate the condensate wave function in imaginary time $t \to it$ with the time evolution given by the Schrödinger equation

$$-\hbar \frac{\partial \Phi(t)}{\partial t} = \hat{H}_{\mathrm{GP}} \Phi(t), \tag{5}$$

where the Hamiltonian $\hat{H}_{\mathrm{GP}}$ consists of the kinetic energy $\hat{T}$ and the confinement and interaction parts, $\hat{V}$. For a short time step Δt, the time-evolution is (approximately) given by the (non-unitary) time propagation operator,

$$\Phi(\Delta t) = \exp(-\Delta t(\hat{T} + \hat{V}))\Phi(0), \tag{6}$$

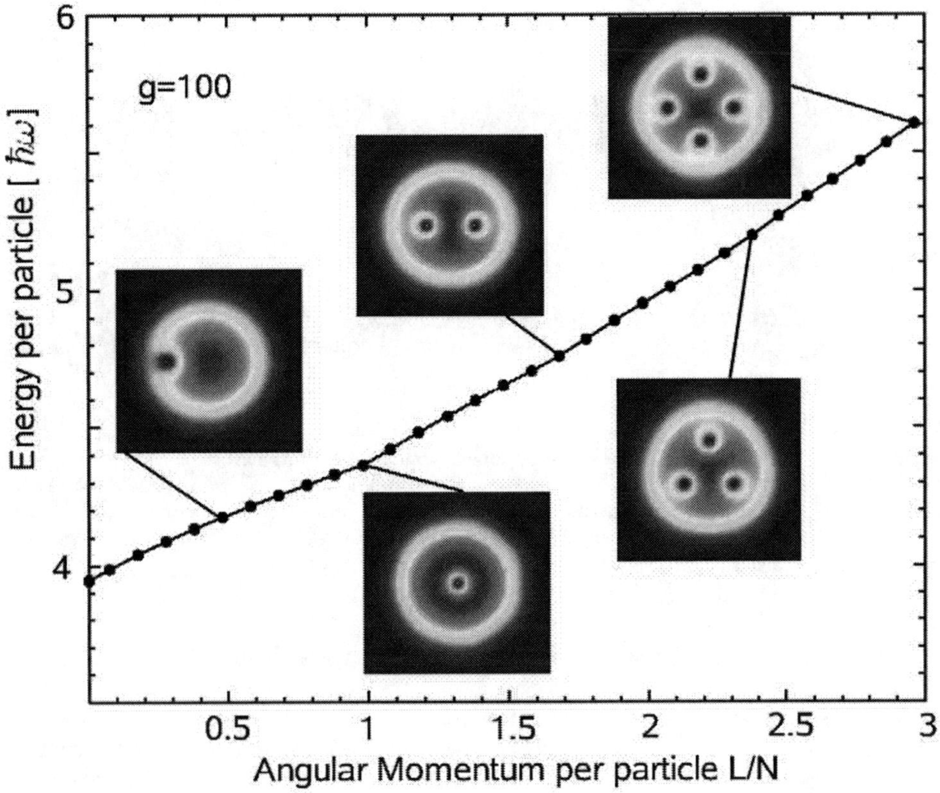

FIGURE 1. The Gross-Pitaevskii ground state energy per particle (in units of $\hbar\omega$) as a function of angular momentum per particle, for a coupling strength $g = 100$ atomic units (a. u.). The insets show the densities for certain angular momenta. The number of vortices in the condensate seen as holes in the densities, increases with increasing rotation.

which does not preserve the norm of the wave function (and thus has to be renormalized at each iteration). If we write the wave function Φ in terms of the eigenstates ψ_i of the Hamiltonian,

$$\Phi(t) = \sum_{i=0}^{\infty} c_i \psi_i \exp(-t\varepsilon_i), \tag{7}$$

we see that the ground state ψ_0 with the lowest energy ε_0 has the slowest decay rate. The ground state is thus obtained with long enough time evolution ($t \longrightarrow \infty$) so that the excited state contributions to the wave function decay away. In the end the ground state becomes projected out of the expansion.

In practice the propagator is evaluated using a "split-step" scheme, where $\exp(-\Delta t\,\hat{T})$ and $\exp(-\Delta t\hat{V})$ operate on the order parameter in a sequence, and the order parameter is Fourier transformed whenever the kinetic energy operator acts upon it. We use a fourth order splitting scheme described in refs. [36, 37]. For a one-component Bose gas, Fig. 1 shows the energies and densities obtained in the Gross-Pitaevskii mean field apporach as described above. The total energy per particle is plotted as a function of angular momentum (in contrast to Fig. 5 below, which shows only the interaction part of the energy). The coupling strength was $g = 100$ a.u., which is clearly beyond the limit of weak interactions.

Vortices appear as clear holes in the densities, with their number increasing, one by one, along the yrast line. The first vortex enters the bosonic cloud from the periphery, and it reaches the trap center at $L/N = 1$. The second and third singly-quantized vortices appear at higher angular momenta. The two-, three- and four-vortex solutions shown in Fig. 1 are calculated for $L/N = 1.7, 2.4$ and 3. (See also [15]). In Fig. 1 one observes small deviations from linearity in the yrast line [22] for $L < N$, which is due to the relatively strong interactions.

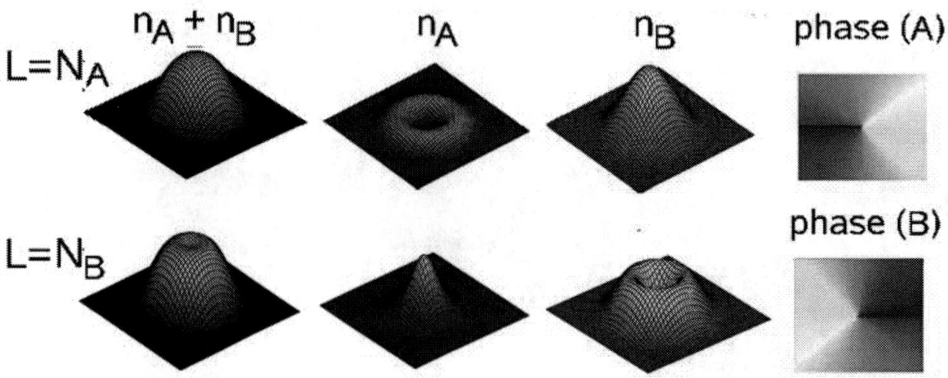

FIGURE 2. *Upper panel*: Coreless vortex state in a two-component condensate computed from the Gross-Pitaevskii equations, with $N_A/N_B = 0.36$ and equal coupling strengths $g = 50$ a.u. The angular momentum is $L = N_A$. Shown are the total density, the densities of the two components n_A and n_B, as well as the phase of the order parameter of the component A, changing from $-\pi$ (blue) to π (red). *Lower panel*: As above, but now for $L = N_B$, where a second single vortex forms in the larger component. Here, the phase is now shown for component B.

Two-component Bose gas in the Gross-Pitaevskii description

Let us now consider a bosonic gas of atoms that is a mixture of two species A and B with N_A and N_B atoms. For simplicity, we assume equal masses of the atoms, $M_A = M_B = M$, and equal coupling strengths between all particles, $g_{AA} = g_{AB} = g_{BB} = g$. (In experiments, the two species A and B could for example be different hyperfine states of the same atom, or different isotopes). We also choose the harmonic trapping potentials for the two components to be identical.

The two-component Bose-Einstein condensate is described by order parameters Φ_A and Φ_B for the two atom species. The Gross-Pitaevskii energy functional is then given by

$$\begin{aligned}\mathscr{E}[\Phi_A,\Phi_A^*,\Phi_B,\Phi_B^*] &= \int \mathrm{d}x\mathrm{d}y \left[\frac{1}{2M}\left(\Phi_A^*(\hat{p}_x^2+\hat{p}_y^2)\Phi_A + \Phi_B^*(\hat{p}_x^2+\hat{p}_y^2)\Phi_B\right)+\right.\\ &\quad + \frac{1}{2}M\omega^2(x^2+y^2)\left(|\Phi_A|^2+|\Phi_B|^2\right)+\\ &\quad \left. + \frac{g}{2}\left(|\Phi_A|^4+|\Phi_B|^4\right)+g|\Phi_A|^2|\Phi_B|^2\right]. \end{aligned} \tag{8}$$

Minimization with respect to the order parameters Φ_A^* and Φ_B^* yields the coupled GP-equations

$$\left(\frac{1}{2M}(\hat{p}_x^2+\hat{p}_y^2)+\frac{1}{2}M\omega^2(x^2+y^2)+g|\Phi_A|^2+g|\Phi_B|^2\right)\Phi_A(x,y) = \mu_A\Phi_A(x,y) \tag{9}$$

$$\left(\frac{1}{2M}(\hat{p}_x^2+\hat{p}_y^2)+\frac{1}{2}M\omega^2(x^2+y^2)+g|\Phi_B|^2+g|\Phi_A|^2\right)\Phi_B(x,y) = \mu_B\Phi_B(x,y). \tag{10}$$

The chemical potentials μ_A and μ_B are determined by the conservation requirement for the particle numbers $N_A = \int \mathrm{d}x\mathrm{d}y|\Phi_A|^2$ and $N_B = \int \mathrm{d}x\mathrm{d}y|\Phi_B|^2$. As above, we normalize the order parameter (here for component B) to unity. The ratio N_A/N_B then determines N_A. For the total angular momentum, we now have $L = \int \mathrm{d}x\mathrm{d}y(\Phi_A^*\hat{L}_z\Phi_A + \Phi_B^*\hat{L}_z\Phi_B) = L_A + L_B$. Figure 2 shows the densities and phases of the order parameters Φ_A and Φ_B computed from the Gross-Pitaevskii equations, for coupling strength $g = 50.0$ and a ratio $N_A/N_B = 0.36$. The rotational frequency is $\Omega = 0.45$.

In the upper panel, the angular momentum is $L = N_A$. The system has formed the "first" coreless single vortex in a two-component condensate, where the larger component rests at the origin with no vorticity (and correspondingly, no phase shift in the order parameter), while the smaller component forms a singly quantized vortex around the center

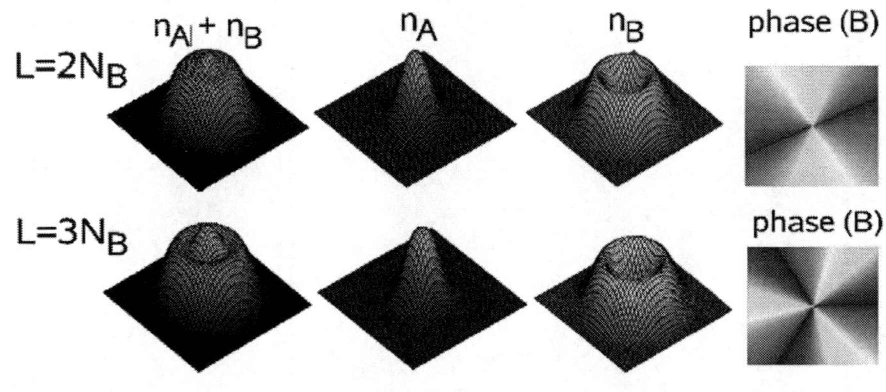

FIGURE 3. Densities and phases as in Fig. 2 but now for $L = 2N_B$ at $\Omega = 0.6$ (*upper panel*) and $L = 3N_B$ at $\Omega = 0.725$ (*lower panel*).

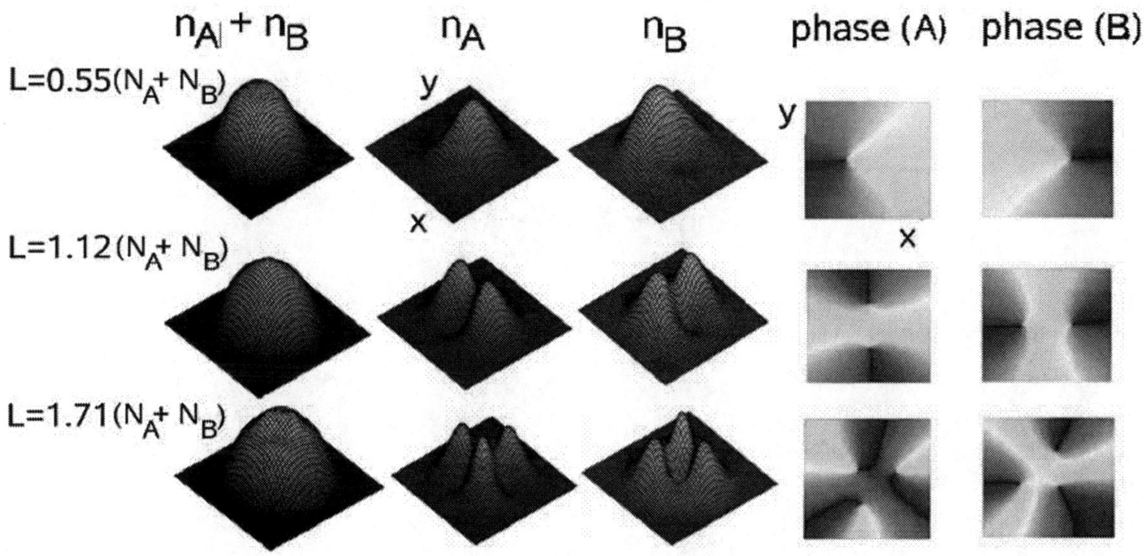

FIGURE 4. Densities and phases as in Fig. 2, but now for equally large components, $N_A = N_B = N/2$, and angular momenta $L = 0.55(N_A + N_B)$ and $\Omega = 0.35$ (upper panel), $L = 1.12(N_A + N_B)$ and $\Omega = 0.45$ (middle panel), as well as $L = 1.71(N_A + N_B)$ and $\Omega = 0.57$ (lower panel), forming interlaced vortices. Shown are also the phases for components A and B.

of the trap, with a clear minimum in the density. The phase for component A is plotted to the right, with the color changing from $-\pi$ (blue) to π (red). Clearly, the phase changes by 2π around the vortex at the center. When the angular momentum is $L = N_B$, a singly-quantized coreless vortex is formed in component B, while the component A becomes stationary (shown in the lower panel of Fig. 2).

At higher angular momenta, one obtains multiply quantized coreless vortices. We show in Fig. 3 the total densities and phases at angular momentum $L = 2N_B$ (upper panel) for $\Omega = 0.6$ (upper panel), and $L = 3N_B$ for $\Omega = 0.725$ (lower panel). Clearly, the central minimum in the density of component B expands with increasing angular momentum, and the phase changes by $2 \cdot 2\pi$ and $3 \cdot 2\pi$.

For equal populations, $N_A = N_B$, a different scenario emerges (see Fig. 4): The system is now symmetric with respect to the two components, in the density as well as in the distribution of angular momentum. A vortex enters each of the components, from "opposite" sides. These vortices reach a minimum distance from the center of the trap equal to a_0 (where a_0 is the oscillator length) when $L = N_A = N_B$ [35]. As the angular momentum is further increased, more vortices enter the system. The middle panel of Fig. 4 shows the case $L = 1.12(N_A + N_B)$ with two interlaced vortices, and the lower panel $L = 1.71(N_A + N_B)$ with three vortices in each component. For even higher angular momenta,

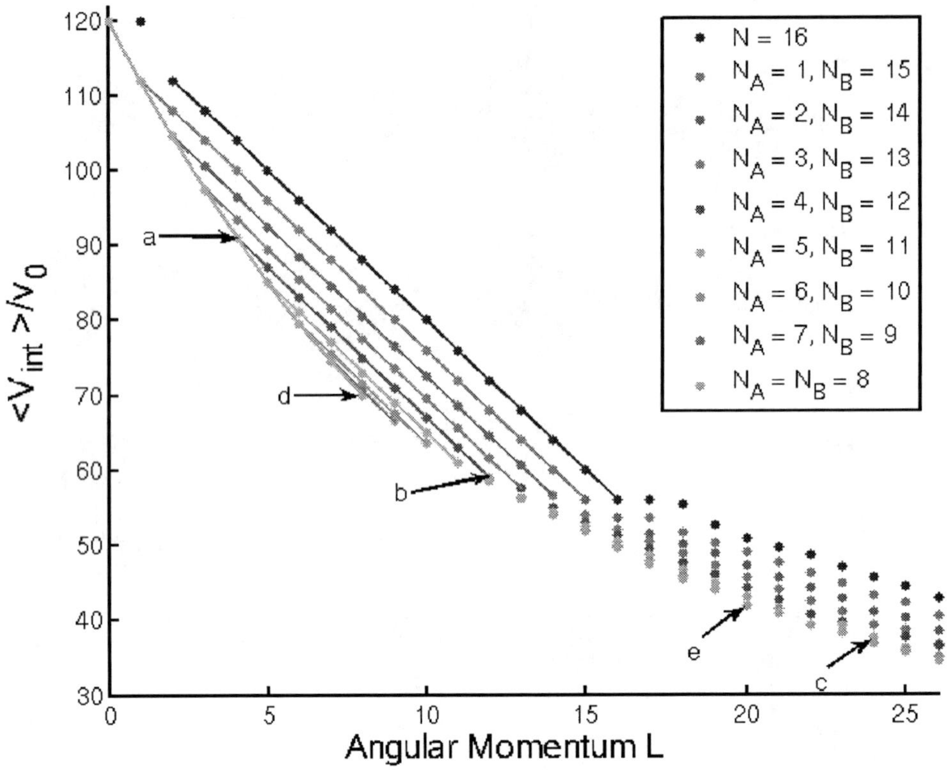

FIGURE 5. Yrast lines (interaction energies) for different compositions of the two-component Bose-Einstein condensate. (The letters a-e indicate the states for which the corresponding pair correlations are shown in Fig. (6) below.)

they will form a square lattice, as has been seen experimentally Schweikhard *et al.* [29], and theoretically by Mueller and Ho [31]. This is in contrast to the triangular Abrikosov lattice for the single-component case. (The interlaced two-vortex solutions are found also in the exact solutions described in the following section where they appear in the pair correlation densities as weak minima for each of the two components.)

MIXTURES OF BOSE GASES IN NUMERICAL DIAGONALIZATION

Let us now turn to study the exact solutions of the many-body Hamiltonian Eq. (1), where they are accessible. To ease the numerical effort, we consider the limit of weak interactions. The yrast line can then be well described by states formed within a restricted Hilbert space, where one assumes maximum alignment of (total) angular momentum $L = L_z$, and thus has as the only quantum number the angular momentum projection $m \geq 0$ for the harmonic oscillator single-particle basis states with $n = 0$ [13]. (This subspace is in fact identical to the so-called "lowest Landau level" at strong magnetic fields [38]). A non-interacting state with total angular momentum L then has the energy $\hbar\omega(N+L)$, where $N = N_A + N_B$. This state is highly degenerate for higher L-values, as there exists a large number of possible combinations that distribute the N_A (or N_B) particles on the orbitals with $m \geq 0$. The interactions split this degeneracy. To solve the many-body problem with interactions, one needs to consider all Fock states with N_A and N_B particles of each species that are eigenstates of the operator of total angular momentum $\hat{L}$.

In the defined subspace, the problem reduces to diagonalizing the interaction part of the Hamiltonian, $\hat{V}_{\text{int}}$. The total energy is then given by $E = \hbar\omega(N+L) + \langle\hat{V}_{\text{int}}\rangle$. In order for the resulting equations to be as simple as possible, here we set for the energy unit the interaction energy of two particles in their harmonic oscillator ground state, $v_0 = U_0 \int d^3r \mid \Phi_0(\mathbf{r}) \mid^4 = \sqrt{2/\pi}\hbar\omega(a/a_z)$. (This quantity is proportional to the factor g that was used earlier).

Figure 5 shows the yrast lines of interaction energies obtained for $N = 16$ bosonic particles in a harmonic trap, with different numbers of atoms N_A and N_B in the two components A and B. Clearly, for repulsive interactions, the interaction energies decrease with increasing angular momentum, as the particles on average are further apart.

Let us first consider only one kind of bosonic atoms in the trap, i.e. the case $N_A = 0$ and $N_B = 16$ as marked by the dark-blue bullets in Fig. 5. For weak interactions, the interaction energy is linear in L for $2 \leq L \leq N$, see Ref. [22]. As we have already seen in the mean-field results for the one-component gas, as the angular momentum increases, a vortex state enters the bosonic cloud and reaches the center of the trap at $L = N$ (see Refs. [14, 15]).

Considering now different numbers of atoms N_A and N_B, we see that the yrast energies are generally lowered when compared to the one-component case. Within a range of angular momenta, there are exact expressions for the energies: For angular momenta $0 \leq L \leq N_{\rm min}$, where $N_{\rm min} = \min\{N_A, N_B\}$, the interaction energy has a parabolic dependence on L, with

$$\langle \hat{V}_{\rm int} \rangle / v_0 = \frac{1}{2} N(N-1) - \frac{1}{2} NL + \frac{1}{4} L(L-1). \tag{11}$$

When L becomes larger, when $N_{\rm min} \leq L \leq N_{\rm max}$ with $N_{\rm max} = \max\{N_A, N_B\}$, the interaction energy varies linearly with total angular momentum L,

$$\langle \hat{V}_{\rm int} \rangle / v_0 = \frac{1}{2} N(N-1) - \frac{1}{4} N_{\rm min} N - \frac{1}{4} NL + \frac{1}{4} N_{\rm min}(N_{\rm min} - 1). \tag{12}$$

For equal populations, the parabolic expression for the interaction energy holds all the way between $0 \leq L \leq N_A = N_B$. These (to numerical precision exact) results were given by Bargi *et al.* in [39].

Let us now study the particle densities in order to obtain more insight into the underlying vortex patterns in the two-component system. Fig. 6 shows the particle densities n_A and n_B for $N = 16$ for different compositions of the two-component Bose gas. In the left column, the total density, $n_A + n_B$, is shown. For the upper three panels, the middle and right columns show the densities n_A and n_B separately. As described above in the context of the Gross-Pitaevskii method, in a single-component system a single vortex will form when $L = N$. In the two-component case, there are two such vortices, one at $L = N_A$ ($= 4$ in Fig. 6, panel (a)) and one at $L = N_B$ ($= 12$ in Fig. 6, panel (b)). At these angular momenta, one component is rotating around the other one, which is still located at the center of the trap, with its density still having the usual Gaussian profile of the lowest state in a harmonic potential. (This can be clearly seen from the single-particle occupation numbers, which are not shown here). The total density $n_A + n_B$ is thus "coreless", as seen to the left in Fig. 6.

Finally, at $L = 2 \cdot N_{\rm max}$, one obtains a doubly quantized vortex in the larger component (right panel; $N_B = N_{\rm max} = 12$, $L = 24$), while the smaller component (middle panel; $N_A = N_{\rm min} = 4$) is still located at the center of the trap. This is another example of a coreless vortex state, and it confirms the existence of the multiply-quantized vortices obtained within the Gross-Pitaevskii approach in the previous section .

When the two components are of equal population, there are no vortex states located at the center of the trap, which requires different measures to visualize the internal structure of the state. In contrast to the results obtained within the Gross-Pitaevskii mean-field method, any density plotted from the *exact* ground state retains the rotational symmetry of the Hamiltonian. While this allowed us to study the centered vortices, non-circular internal symmetries of the quantum states are more difficult to analyze. Thus, instead of densities, one often studies the pair correlations $\langle \Psi \mid \hat{\Psi}^\dagger(\mathbf{r}) \hat{\Psi}^\dagger(\mathbf{r}') \hat{\Psi}(\mathbf{r}) \hat{\Psi}(\mathbf{r}') \mid \Psi \rangle$ describing the probability of finding a particle at $\mathbf{r}$ given that one particle is fixed at the reference point $\mathbf{r}'$. These "conditional densities" are shown in Fig. 6 in panels (d) and (e) for $N_A = N_B = 8$ and angular momenta $L = 8$ and $L = 20$, respectively. The system is, again, symmetric with respect to the two components.

In the pair correlations for the few-particle system, we again see the same structure as for the Gross-Pitaevskii results in Fig. 4. We are restricted to lower angular momenta, but it is clear that there is an angular separation of the two components, and that each component has a "density" maximum where the other one has a minimum. Again, the total density does not reveal the complicated internal structure of the state.

CONCLUSIONS

In summary, we have seen that the addition of a second component in a Bose-Einstein condensate adds interesting new features to the rotational phase diagram. Mean-field calculations using the Gross-Pitaevskii equation were compared to the results obtained by numerical diagonalization of the many-body Hamiltonian (shown here for $N = 16$ bosons in the trap). The two approaches were found to agree remarkably well. In a single-component system in a harmonic trap,

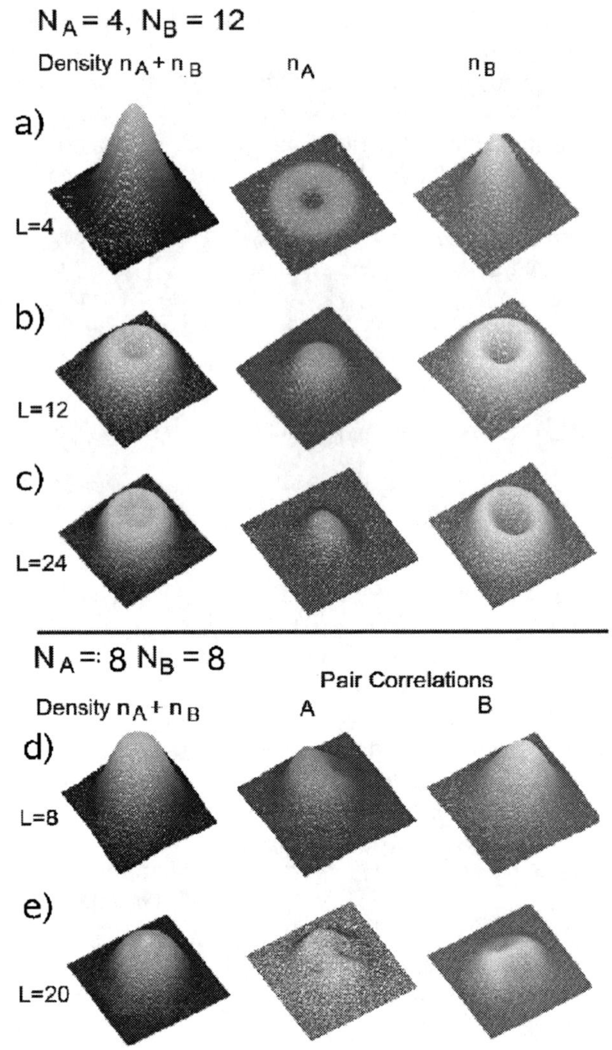

FIGURE 6. Total density $n_A + n_B$ (left column) and densities of the components A and B of a two-component Bose gas with a total of $N = 16$ atoms, set rotating at different angular momenta. Panels (a)-(c) are for $N_A = 4$ and $N_B = 12$ with $L = 4, 12$ and 24, respectively. Panels (d) and (e) are for equal populations, $N_A = N_B = 8$, at angular momenta $L = 8$ and $L = 20$. The (x,y)-range in all figures is from $-2.2a_0$ to $2.2a_0$, except in the lowest panel (e), where it is from -3 to 3.

only singly quantized vortices are mechanically stable. When two components are present, there are two circumstances that cooperate to make multiply quantized vortices stable: the repulsion between the different components make them separate in space, and the tendency of the bosons within each component to condense into one single-particle state then makes multiply quantized vortices energetically favorable. In general, these vortices will be "coreless", having one component in its non-rotating ground state at the center of the trap, with the other component forming a vortex around it. In the case where the number of particles is equal in the two components, they separate in a different way: Each component has a density minimum, corresponding to a vortex, where the other has a maximum, and the higher the angular momentum, the more vortices there are in the system. Finally, we obtained some exact analytical results for the lowest-energy spectrum in a rotating two-component Bose gas. This is a very interesting and surprising result for such a complicated many-body system.

ACKNOWLEDGMENTS

We thank Y. Yu for discussions. Financial support from the European Community project ULTRA-1D (NMP4-CT-2003-505457), the Swedish Research Council, the Swedish Foundation for Strategic Research, as well as NordForsk, is gratefully acknowledged.

REFERENCES

1. A. J. Leggett, Rev. Mod. Phys. **71**, S318 (1999).
2. M.H. Anderson, J.R. Ensher, M.R. Matthews, C.E. Wieman and S.A. Cornell, Science **269**, 198 (1995).
3. K.B. Davis, M.O. Mewes, M.A. Joffe, M.R. Andrews, and W. Ketterle, Phys. Rev. Lett. **74**, 5202 (1995).
4. K.B. Davis, M.O. Mewes, M.R. Andrews, N.J. Vandruten, D.S. Durfee, D.M. Kurn, and W. Ketterle, Phys. Rev. Lett. **75**, 3969 (1995).,
5. W. Ketterle, Rev. Mod. Phys. **74**, 1131 (2002).
6. E.A. Cornell, and C.E. Wieman, Rev. Mod. Phys. **74**, 875 (2002).
7. M. R. Matthews, B.P. Anderson, P.C. Haljan, D.S. Hall, C.E. Wieman and E.A. Cornell, Phys. Rev. Lett. **83**, 2498 (1999).
8. K.W. Madison, F. Chevy, W. Wohlleben, and J. Dalibard, Phys. Rev. Lett. **84**, 806 (2000).
9. A.A. Svdizinsky, and A.L. Fetter, Phys. Rev. Lett. **84**, 5919 (2000).
10. J.J. García-Ripoll and V.M. Pérez-García, Phys. Rev. A **60**, 4864 (1999).
11. D.L. Feder, C.W. Clark, and B.I. Schneider, Phys. Rev. A **60**, 011601(R) (2000).
12. D.L. Feder, C.W. Clark, and B.I. Schneider, Phys. Rev. Lett. **82**, 4956 (1999).
13. B. Mottelson, Phys. Rev. Lett. **83**, 2695 (1999).
14. D.A. Butts and D.S. Rokshar, Nature (London) **397**, 327 (1999).
15. G.M. Kavoulakis, B. Mottelson and C.J. Pethick, Phys. Rev. A **62**, 063605 (2000).
16. M.V. Milosevic, and F.M. Peeters, Phys. Rev. B **68**, 024509 (2003).
17. I. Romanovsky, C. Yannouleas, L.O.Baksmaty, and U. Landman, Phys. Rev. Lett. **97**, 090401 (2006).
18. S.M. Reimann, M. Koskinen, Y. Yu and M. Manninen, New J. Phys. **8**, 59 (2006).
19. N. Barberan, M. Lewenstein, K. Osterloh, and D. Dagnino, Phys. Rev. A **73**, 063623 (2006).
20. M. Manninen, S. M. Reimann, M. Koskinen, Y. Yu and M. Toreblad, Phys. Rev. Lett. **94**, 106405 (2005).
21. N.K. Wilkin, J.M.F. Gunn and R.A. Smith, Phys. Rev. Lett. **80**, 2265 (1998).
22. G.F. Bertsch and T. Papenbrock, Phys. Rev. Lett. **83**, 5412 (1999).
23. N.K. Wilkin and J.M.F. Gunn, Phys. Rev. Lett. **84**, 6 (2000).
24. A.D. Jackson, G.M. Kavoulakis, B. Mottelson and S.M. Reimann, Phys. Rev. Lett. **86**, 000945 (2001).
25. N. D. Mermin and T.-L. Ho, Phys. Rev. Lett. **36**, 594 (1976).
26. P. Bhattacharyya, T.-L. Ho, and N. D. Mermin, Phys. Rev. Lett. **39**, 1290 (1977).
27. T.-L. Ho, Phys. Rev. Lett. **49**, 1837 (1982).
28. A. E. Leanhardt, Y. Shin, D. Kielpinski, D. E. Pritchard, and W. Ketterle, Phys. Rev. Lett. **90**, 140403 (2003).
29. V. Schweikhard, I. Coddington, P. Engels, S. Tung, and E. A. Cornell, Phys. Rev. Lett. **93**, 210403 (2004).
30. J. E. Williams and M. J. Holland, Nature (London) **401**, 568 (1999).
31. E. J. Mueller and T.-L. Ho, Phys. Rev. Lett. **88**, 180403 (2002).
32. K. Kasamatsu, M. Tsubota, and M. Ueda Phys. Rev. Lett. **91**, 150406 (2003).
33. K. Kasamatsu, M. Tsubota, and M. Ueda Phys. Rev. A **71**, 043611 (2005).
34. K. Kasamatsu, M. Tsubota, and M. Ueda, Int. J. of Mod. Phys. B, **19**, 1835 (2005).
35. J. Christensson, S. Bargi, K. Kärkkäinen, Y. Yu, G.M. Kavoulakis, M. Manninen and S.M. Reimann, to be published (condmat/arXiv:0704.0447) (2007).
36. S.A. Chin, Phys. Lett. A **226**, 344-438 (1997).
37. S.A. Chin, and E. Krotscheck, Phys. Rev. E **72**, 036705 (2005).
38. S.M. Reimann, and M. Manninen, Rev. Mod. Phys. **74**, 1283 (2002).
39. S. Bargi, J. Christensson, G.M. Kavoulakis, and S.M. Reimann, Phys. Rev. Lett. **98**, 130403 (2007).

Cold atom ballistics by coherent control

Linda E. Reichl and Benjamin P. Holder

Center for Complex Quantum Systems, University of Texas at Austin, Austin, TX 78712

Abstract.
We use the technique of Stimulated Raman adiabatic passage (STIRAP) to affect transport of ultracold atoms between two optical lattices in relative motion. We show, using Floquet analysis and degenerate perturbation theory, that the dynamics of atoms in a particular time-dependent optical lattice system can be reduced to the 3-level STIRAP model, allowing for a simple description of their coherent acceleration.

Keywords: Stimulated Raman adiabatic passage (STIRAP), optical lattice, coherent control
PACS: 42.50.Vk, 03.75.Be, 05.45.Mt

INTRODUCTION TO THE STIRAP MODEL

Stimulated Raman adiabatic passage (STIRAP) is a model system introduced by Kuklinski *et. al.* [1] to describe the experimental results of Gaubatz *et al* [2] and others in which an atomic or molecular population was transferred using two laser beams. Given three energy levels (E_a, E_b, E_c) with corresponding states $(|a\rangle, |b\rangle, |c\rangle)$, it was observed that an initial population in $|a\rangle$ subjected to a counter-intuitive pulse sequence (the "pump" laser, coupling levels E_a and E_b, is applied *after* the "stokes" laser, coupling levels E_b and E_c) will undergo a complete transfer to state $|c\rangle$. The model which describes this phenomenon can be derived by assuming a three level system with nonzero dipole couplings d_{ab} and d_{bc}. Making the rotating wave approximation, the Hamiltonian for this model, in the basis $(|a\rangle, |b\rangle, |c\rangle)$, has the form

$$H(t') = -\frac{\hbar}{2}\begin{pmatrix} 0 & W_1(t') & 0 \\ W_1(t') & -2\Delta & W_2(t') \\ 0 & W_2(t') & 0 \end{pmatrix}, \tag{1}$$

where W_1 and W_2 are the Rabi frequencies associated to the two transitions, Δ is the equal detuning of the two lasers from their corresponding transition frequencies, and t' is a parameter. It is easy to see that this matrix has an eigenvector, with eigenvalue zero, of the form

$$|1(t')\rangle = \cos\theta(t')|a\rangle - \sin\theta(t')|c\rangle\,, \tag{2}$$

where $\tan\theta(t') \equiv \frac{W_1(t')}{W_2(t')}$. The characteristic STIRAP transition ($|a\rangle \to |c\rangle$ as t' goes from $-\infty$ to $+\infty$) is therefore guaranteed under the conditions

$$\lim_{t'\to-\infty}\frac{W_1(t')}{W_2(t')} \to 0 \quad \text{and} \quad \lim_{t'\to+\infty}\frac{W_2(t')}{W_1(t')} \to 0\,. \tag{3}$$

The adiabatic behavior of this model system, using Gaussian dependence of the W_i on t', is displayed in Figure 1.

Although STIRAP was derived for a three level system in the electronic energy spectrum of atoms and molecules, its usefulness as a model quantum system is much broader. For example, a novel application of the model was achieved by Eckert *et. al* [3], who considering states $|a\rangle$, $|b\rangle$ and $|c\rangle$ to be the ground states of three neighboring optical microtraps. By varying their relative spatial separations, atoms could be coherently transported between the two farthest separated traps. Here, we describe how the STIRAP model can be applied to cold atoms trapped in an optical lattice, using three Floquet states of an effective Hamiltonian (see next section) and couplings induced by the modulation of the lattice amplitudes.

CP995, *Nuclei and Mesoscopic Physics - WNMP 2007*, edited by P. Danielewicz, P. Piecuch, and V. Zelevinsky

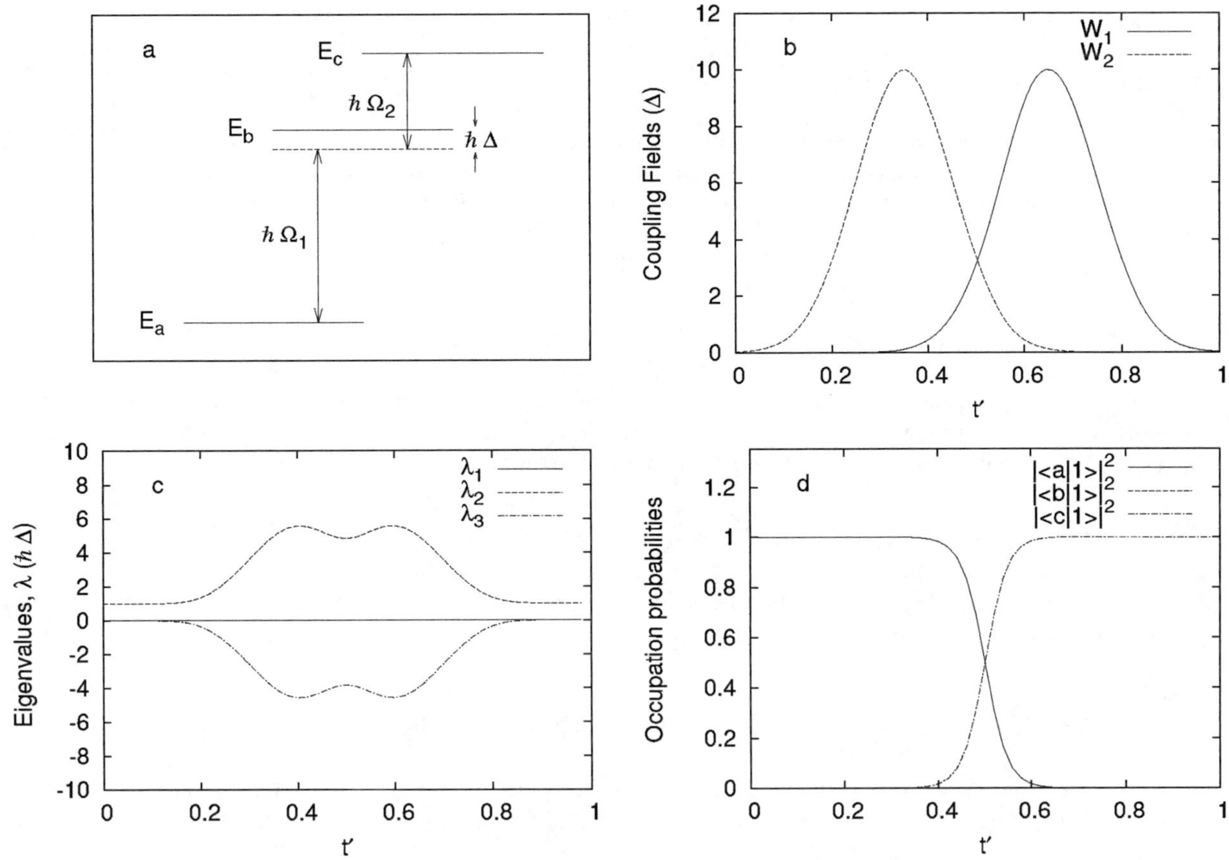

FIGURE 1. The 3-level "ladder" STIRAP system (Eq. (1)) with parameters $\Delta = 0.05$, $W_1 = W_2 = 1$ and $\hbar = 2$. Coupling fields are applied with frequencies equally detuned from the energy level spacings spacings by Δ (a). Adiabatic variation of the amplitudes of the coupling fields (b) in the manner described by Eq. (3) affects a transition within the $|1\rangle$ eigenvector from basis state $|a\rangle$ to $|c\rangle$ (d). The eigenvalue corresponding to this state remains unchanged at zero throughout the transition (c).

THE OPTICAL LATTICE EXPERIMENT AND ITS THEORETICAL DESCRIPTION

The cooling and trapping of neutral atoms with lasers relies primarily on the repeated interaction of laser photons with a single transition in the atomic energy structure. By manipulating this interaction with an atom's *internal* degrees of freedom, one can control the transfer of momentum to the atom and thus its *external* degrees of freedom, i.e. center of mass motion. Alkali atoms are some of the best candidates for cooling and trapping because the strong optical D_2 transition has a cycling transition in its hyper-fine structure, allowing for many photon absorptions without loss of population to other levels via spontaneous emission. In a typical experiment [4], millions of Cesium atoms are first cooled and trapped in a magneto-optical trap and then further cooled by the velocity-selective technique of Raman cooling [6] to obtain a dilute gas of atoms with sub-recoil velocities.

An optical lattice is created by subjecting the cloud of cold atoms to a pair of counter-propagating lasers, far-detuned from the atomic transition. The detuning of the lasers prohibits excitation of an individual atom but induces a spatially-periodic "light shift" of the ground-state energy. Theoretically, the Schrödinger equation for the dipole interaction of a classical standing wave of light (directed along the x-axis) with a two-level system can be reduced, via adiabatic elimination of the excited state dynamics, to a single equation for the center of mass dynamics of an atom in the ground state:

$$i\hbar\frac{\partial}{\partial t}|\psi_g(x,t)\rangle = \left[\frac{p_x^2}{2m} - \frac{d^2|E(x,t)|^2}{\hbar\Delta}\right]|\psi_g(x,t)\rangle\,. \tag{4}$$

The term in brackets on the right hand side of this equation is called the *effective Hamiltonian* where d is the dipole matrix element coupling the two levels, Δ is the detuning of the laser frequency from the transition and $E(x,t)$ is the complex electric field amplitude [7, 8]. To the extent that the cold atoms can be considered non-interacting, the dynamics of the entire cloud in an optical lattice can be reduced to this simple one-particle equation. This will be the object of our study in the following sections.

A wide variety of effective Hamiltonians can be constructed through the use of multiple pairs of laser beams, variation of the relative frequencies of a given pair and modulation of the laser amplitudes. The stationary optical lattice, in the laboratory frame, is created by a single pair of lasers with equal frequencies yielding a complex electric field amplitude

$$|E(x,t)|^2 \sim \frac{E_0^2}{2}\cos(2k_L x)\,, \tag{5}$$

where E_0 is the real electric field amplitude and $k_L = \omega_L/c$ the wave number of the laser. Therefore the effective potential in Eq. (4) is proportional to a stationary cosine wave with half the periodicity of the laser light. The harmonic modulation of this stationary lattice amplitude ($E_0 \rightarrow E_0(t)$), creating an effective potential with three cosine terms, was used in the experimental discovery of "chaos-assisted tunneling" [5]. A pair of lasers with slightly offset frequencies $\omega_L \pm \delta\omega$ yields a "traveling" optical lattice in the laboratory frame, where

$$|E(x,t)|^2 \sim \frac{E_0^2}{2}\cos(2k_L x - \delta\omega t)\,. \tag{6}$$

The sum of Eqs. (5) and (6) allows for the "two-resonance" effective potential, which is the focus of our analysis in the following sections [9].

THE TWO-RESONANCE SYSTEM AND A STIRAP SUBSYSTEM

We will consider a "two-resonance" system, constructed by two pairs of counter propagating lasers: one pair with equal frequencies the second pair with slightly offset frequencies. The effective Hamiltonian for this system has the form

$$\hat{H}_0 = \hat{p}^2 + \kappa_0\left[\cos(\hat{x}) + \cos(\hat{x} - \omega_0 t)\right]\,, \tag{7}$$

where all variables are in dimensionless units [from the units of the previous section: $p_x \rightarrow 2\hbar k_L p$, $x \rightarrow x/2k_L$, $t \rightarrow t/4\omega_r$, $\delta\omega \rightarrow 4\omega_r\omega_0$ and $(d^2E^2/2\hbar\Delta) \rightarrow -4\hbar\omega_r\kappa_0$ with recoil frequency $\omega_r \equiv \hbar k_L^2/2m \approx 13$kHz for cesium] and the effective Planck's constant is unity. Solving the Schrödinger Equation with a time-periodic Hamiltonian is accomplished by Floquet Theory, which guarantees solutions of the form

$$|\psi_\alpha(t)\rangle = \exp^{-i\bar{\varepsilon}_\alpha^0 t}|\bar{\phi}_\alpha^0(t)\rangle \quad \text{with} \quad |\bar{\phi}_\alpha^0(t+T_0)\rangle = |\bar{\phi}_\alpha^0(t)\rangle\,, \tag{8}$$

where $|\bar{\phi}_\alpha^0\rangle$ are called *Floquet eigenstates*, $\bar{\varepsilon}_\alpha^0$ are called the *Floquet eigenvalues* and $T_0 = 2\pi/\omega_0$. In the following, we will also assume that the solutions are spatially periodic with the period of the optical lattice. This can be approximately satisfied experimentally by the preparation of a cloud of atoms with sub-recoil velocities. Ten of the Floquet eigenvalues for this Hamiltonian, with $\omega_0 = 10/\gamma\,(\gamma = 1.6180\ldots$ is the golden mean), are plotted as a function of κ_0 in Figure 2a. Associated Floquet eigenstates, at $\kappa_0 = 1$, are shown in their Husimi representation in Figure 3.

In order to accelerate trapped atoms, we will consider three Floquet states of the two-resonance system as a basis on which to perform a STIRAP-like transfer. We assume that atoms are initially prepared in the stationary cosine potential and that the state of an individual atom is well approximated by the Floquet state $|\bar{a}\rangle$ $(\equiv |\bar{\phi}_a^0(t)\rangle)$ shown in Figure 3. In order to transfer the atomic population to the state $|\bar{c}\rangle$, trapped in the traveling potential, a harmonic coupling of the two eigenvalues $\bar{\varepsilon}_a^0$ and $\bar{\varepsilon}_c^0$, through an intermediate eigenvalue $\bar{\varepsilon}_b^0$ is performed. By varying the strength of these couplings, a STIRAP transfer may be achieved. While classic STIRAP involves the dipole coupling of an atom's electronic eigenstates using a laser, the Floquet eigenvalues here can be coupled [8] by applying a small perturbation to the Hamiltonian in Eq. (7) of the form

$$\lambda\hat{V} = \lambda\cos(\hat{x})\left[\kappa_1(t')\cos(\Omega_1 t) + \kappa_2(t')\cos(\Omega_2 t)\right]\,, \tag{9}$$

where the frequencies Ω_1 and Ω_2 are chosen to be equal (or nearly equal) to the eigenvalue spacings $|\bar{\varepsilon}_b^0 - \bar{\varepsilon}_a^0|$ and $|\bar{\varepsilon}_c^0 - \bar{\varepsilon}_b^0|$, respectively, at a particular value of κ_0; κ_1 and κ_2 will usually be Gaussian functions of the parameter t'.

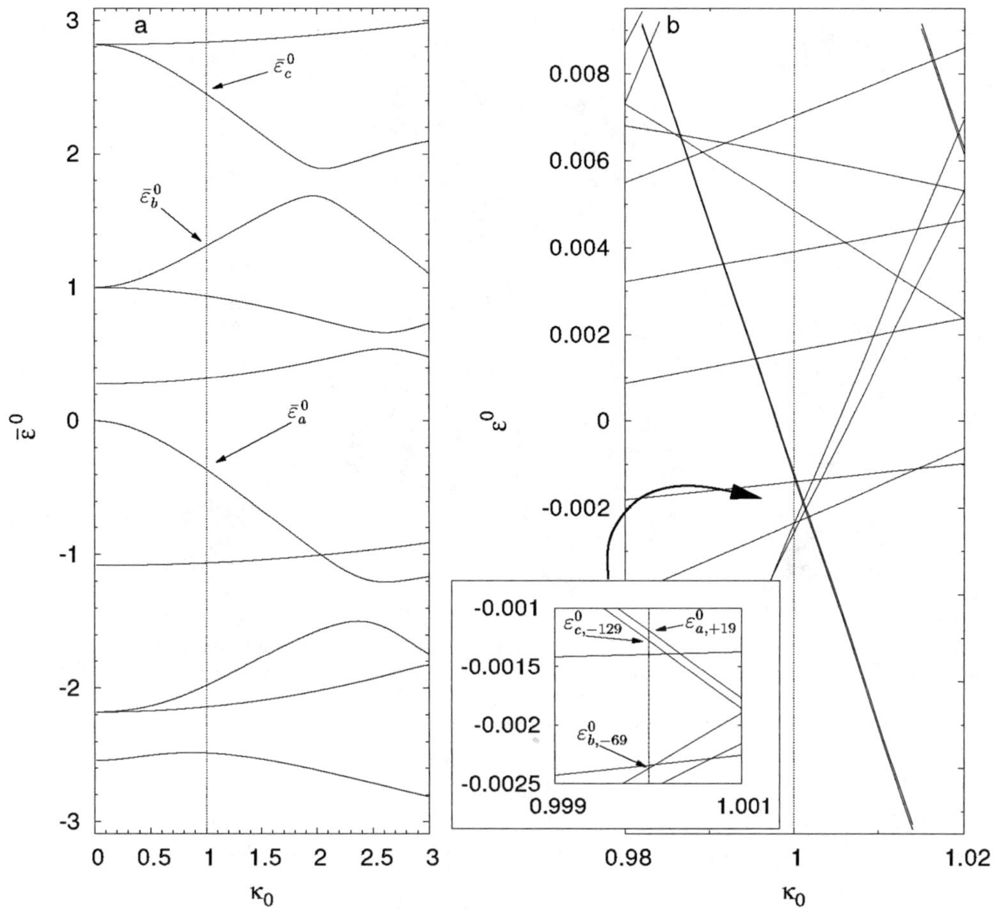

FIGURE 2. A few of the Floquet eigenvalues for the two-resonance Hamiltonian with $\omega_0 = 10/\gamma$ as a function of κ_0 (a). Figure (b) shows the corresponding "unbarred" eigenvalues in zone $-\omega/2 < \varepsilon^0 \leq \omega/2$ for the choice $(m_0, m_1, m_2) = (325, 88, 60)$ where $\omega_0 = m_0\omega$, $\Omega_1 = m_1\omega$, $\Omega_2 = m_2\omega$ and (Ω_1, Ω_2) have been chosen to approximately couple $(\bar{\varepsilon}_a^0, \bar{\varepsilon}_b^0)$ and $(\bar{\varepsilon}_b^0, \bar{\varepsilon}_c^0)$, respectively, at $\kappa_0 = 1$. The Q-indices of the unbarred eigenvalues (labeled in the inset) indicate the number of zones of width ω from the barred eigenvalue and satisfy $|Q_b - Q_a| = m_1$ and $|Q_c - Q_b| = m_2$.

Physically, this perturbation is achieved by modulating the stationary cosine potential amplitude about its κ_0 value (see Figure 4).

In order to see how the perturbation in Eq. (9) implies the existence of a STIRAP three-state model, we must consider a new Floquet analysis which applies to the full Hamiltonian $\hat{H} = \hat{H}_0 + \lambda\hat{V}$. To use Floquet theory, the frequencies ω_0, Ω_1 and Ω_2 must be chosen to be commensurate. Since ω_0 is assumed to be given, call it $\omega_0 = m_0\omega$, we must choose $\Omega_1 = m_1\omega$ and $\Omega_2 = m_2\omega$, with $m_i \in \mathbb{Z}$. Given a particular choice, the full Hamiltonian is periodic in time with period $T = 2\pi/\omega$ and solutions to the corresponding Schrödinger equation same have the form as in Eq. (8) except that the Floquet states now have period T, i.e.

$$|\psi_\alpha(t)\rangle = \exp^{-i\varepsilon_\alpha t}|\phi_\alpha(t)\rangle \quad \text{with} \quad |\phi_\alpha(t+T)\rangle = |\phi_\alpha(t)\rangle\,. \tag{10}$$

Note that we have written the Floquet solutions of the unperturbed system with an overbar and those of the full system without. When $\lambda = 0$, the two sets of physical solutions $(\{|\psi_\alpha(t)\rangle\})$ must be identical, however the Floquet solutions for $\hat{H}_0$ when considered a T-periodic system represent an infinite multiplicity of those when it is considered a T_0-periodic system. More precisely, for each barred Floquet state $|\bar{\phi}_\alpha^0\rangle$, there is a $(\lambda = 0)$ unbarred state $|\phi_{\alpha,Q_\alpha}^0\rangle$ for every $Q_\alpha \in \mathbb{Z}$ corresponding to the same physical solution $|\psi_\alpha(t)\rangle$ [8]. The eigenvalues of these unbarred states occur in "zones" of width ω such that, in every zone, each barred eigenvalue $\bar{\varepsilon}_\alpha^0$ has a corresponding unbarred eigenvalue $\varepsilon_{\alpha,Q_\alpha}^0 = \bar{\varepsilon}_\alpha^0 + Q_\alpha\omega$. Since ω was chosen as the common factor in the frequencies Ω_1 and Ω_2 and those frequencies

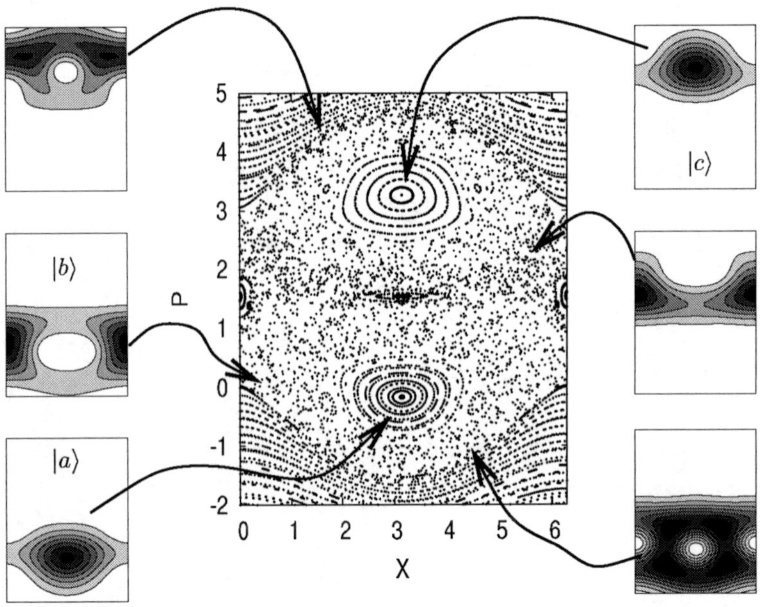

FIGURE 3. Husimi plots of the Floquet eigenstates (taken at $t = 0$) for the two-resonance Hamiltonian with $\omega_0 = 10/\gamma$ and $\kappa_0 = 1$. The association of each state to the classical phase space of the same Hamiltonian (strobed at the period $2\pi/\omega_0$) is shown with arrows (axes on the Husimi plots are the same as the classical phase space plot). Since these are time-independent representations of the Floquet states, the barred representation is identical to the unbarred (e.g. $|a\rangle = |\bar{a}\rangle$).

were chosen to couple the eigenvalues $(\bar{\varepsilon}_a^0, \bar{\varepsilon}_b^0, \bar{\varepsilon}_c^0)$, the unbarred eigenvalues $(\varepsilon_a^0, \varepsilon_b^0, \varepsilon_c^0)$ in a particular zone will be nearly degenerate. Figure 2b shows the near-degeneracy of the three unbarred eigenvalues corresponding to the states considered in the example above.

The perturbation analysis of a time-periodic system is identical to that of Rayleigh-Schrödinger (time-independent) perturbation theory when time is considered an additional *coordinate* in an extended Hilbert space [10, 11]. Therefore, following the usual degenerate perturbation procedure, the near-degeneracy of the three $\lambda = 0$ eigenvalues ε^0_{a,Q_a}, ε^0_{b,Q_b} and ε^0_{c,Q_c} requires that we write the zeroth-order solution $|\phi^{(0)}\rangle$ as a superposition of $|\phi^0_{a,Q_a}\rangle$, $|\phi^0_{b,Q_b}\rangle$ and $|\phi^0_{c,Q_c}\rangle$. Keeping terms to first order in λ, it can be shown that the perturbed system, in the limit of $\lambda \to 0$, reduces to an

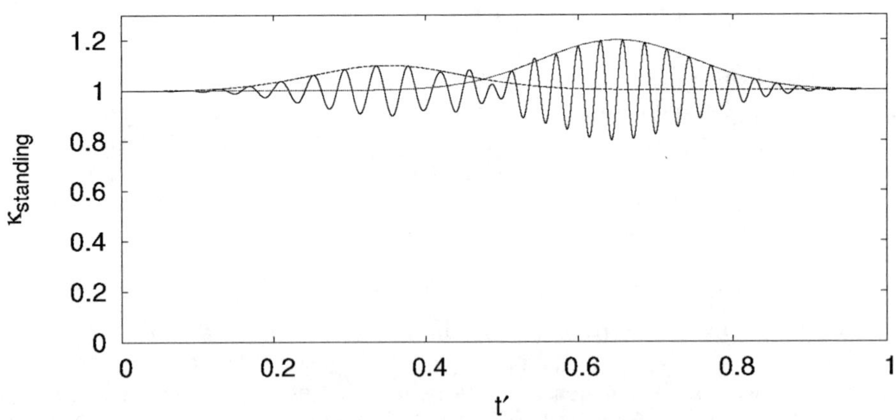

FIGURE 4. STIRAP transfer is performed by modulation of the stationary lattice amplitude (square of the electric field for one pair of lasers). Coupling of the two "upper" levels before the two lower levels, requires that the earlier perturbation has frequency Ω_2 and the later has frequency Ω_1. Here $\Omega_1 > \Omega_2$.

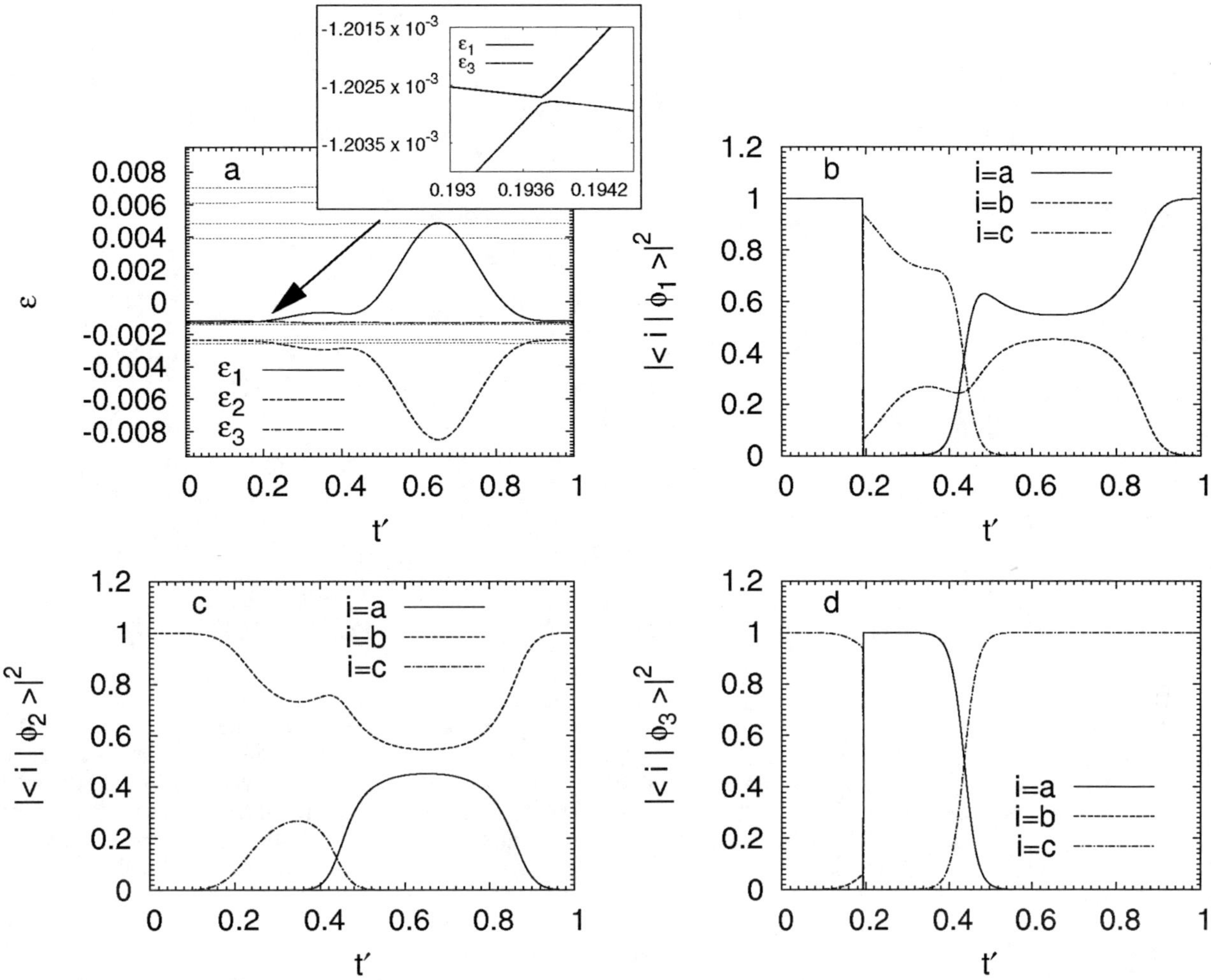

FIGURE 5. Floquet eigenvalues for the two-resonance system with applied perturbation $\hat{V}$ and $\lambda = 0.03$ are plotted as a function of the adiabatic time parameter t' (a). The three bold lines are eigenvalues ε_1, ε_2 and ε_3 (solid, dashed and dot-dashed, respectively) while the light dashed lines are the other eigenvalues shown in Figure 2 (avoided crossings between bold and light lines are vanishingly small and ignored). At $t' = 0$ and $t' = 1$, the eigenvalues are identical to those in Figure 2b at $\kappa_0 = 1$. The occupation probabilities in the basis $\{|a\rangle, |b\rangle, |c\rangle\}$ are plotted in (b)-(d), for states $|\phi_1\rangle$, $|\phi_2\rangle$ and $|\phi_3\rangle$ (initially equal to $|a\rangle, |b\rangle$ and $|c\rangle$, respectively). A narrow avoided crossing at $t' \approx 0.2$ (inset) prohibits the STIRAP transfer in the adiabatic limit [compare (b) and (d) to Figure 1(d)].

isolated three-level eigen-equation

$$\begin{pmatrix} 0 & \lambda\kappa_1(t')v_{ab} & 0 \\ \lambda\kappa_1(t')v_{b,a} & \Delta_1 & \lambda\kappa_2(t')v_{bc} \\ 0 & \lambda\kappa_2(t')v_{c,b} & \Delta_1 - \Delta_2 \end{pmatrix} \begin{pmatrix} C_a \\ C_b \\ C_c \end{pmatrix} = \varepsilon \begin{pmatrix} C_a \\ C_b \\ C_c \end{pmatrix}, \tag{11}$$

where $C_i = \langle \phi^0_{i,Q_i} | \phi^{(0)} \rangle$, $v_{ij} = \langle \bar{\phi}^0_i | \cos\hat{x} | \bar{\phi}^0_j \rangle / 2$, $\Delta_1 = \left(\bar{\varepsilon}^0_b - \bar{\varepsilon}^0_a\right) - \Omega_1$ and $\Delta_2 = \Omega_2 - \left(\bar{\varepsilon}^0_c - \bar{\varepsilon}^0_b\right)$. Apart from the unequal detunings Δ_1 and Δ_2 – which are a result to the requirement that the Hamiltonian $\hat{H}_0 + \lambda\hat{V}$ be time-periodic given an arbitrary ω_0 – this three level system is identical to the STIRAP model presented in Eq. (1). However, this slight difference destroys the adiabatic STIRAP transition when $\Delta_1, \Delta_2 \neq 0$ [8]. Figure 5 shows the adiabatic behavior of the example introduced in Figure 2 under parametrization

$$\kappa_i(t') = \exp\left[-\frac{(t' - t'_i)^2}{2(\sigma'_i)^2}\right], \tag{12}$$

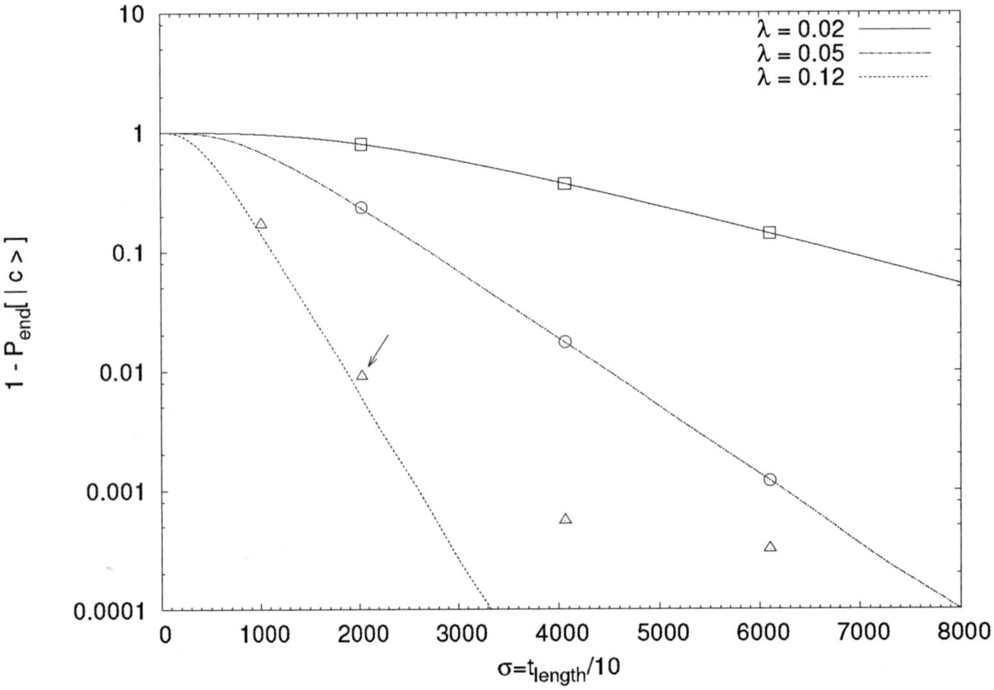

FIGURE 6. The approach to adiabaticity for the two-resonance system with resonant STIRAP coupling ($\Delta_1 = \Delta_2 = 0$). The probability of a non-adiabatic transition (transition from $|a\rangle$ to any state other than $|c\rangle$) is plotted as a function of σ, measured in dimensionless time. Lines are calculated using a three-level STIRAP model with matrix elements v_{ij} calculated from the Floquet eigenvectors associated to $\hat{H}_0$ with $\kappa_0 = 1$ and $\omega_0 = 10/\gamma$. The square, circle and triangle points are calculated by evolution of the Schrödinger equation for the full perturbed system, $\hat{H}_0 + \lambda\hat{V}$, for $\lambda = 0.02, 0.05$ and 0.12, respectively (Gaussian parameters here are those used in the adiabatic case of Figure 5, multiplied by t_{length}). The full time evolution of the point marked with an arrow is shown in Figure 7.

where $t_1' = 0.65$, $t_2' = 0.35$, $\sigma_1' = \sigma_2' = 0.1$ and the parameter t' varies between 0 and 1. A narrow avoided crossing between eigenvalues ε_1 and ε_3 (equal to ε_a and ε_c, respectively, at $t' = 0$) occurs at $t' \approx 0.2$ which induces an exchange of character between the two corresponding states $|\phi_1\rangle$ and $|\phi_3\rangle$. Therefore, in the truly adiabatic limit, atoms prepared in $|a\rangle$ at $t' = 0$ will remain in that state at $t' = 1$.

In the physical or numerical implementation of STIRAP transport for the two-resonance system, this problem can be avoided in two ways: one can apply resonant coupling frequencies ($\Delta_1 = \Delta_2 = 0$) or the evolution can proceed at a rate fast enough to pass through the avoided crossing. Considering the latter option, it is instructive to look at the Landau-Zener tunneling probability for the avoided crossing at $t' \approx 0.2$. This probability has the form

$$P_{\text{L-Z}} = \exp\left[\frac{-\pi\Delta_{\text{ac}}^2}{2\frac{dt'}{dt}\left|\frac{d}{dt'}(\varepsilon_1 - \varepsilon_3)\right|}\right], \tag{13}$$

where Δ_{ac} is the closest approach of eigenvalues ε_1 and ε_3 and $\frac{dt'}{dt} \equiv t_{\text{length}}^{-1}$ is the inverse of the non-adiabatic evolution time [12]. For the parameters used in Figure 5, the avoided crossing in the inset has $\Delta_{\text{ac}} = 1.05 \times 10^{-7}$ and $\frac{d}{dt'}(\varepsilon_1 - \varepsilon_3) = 2.58 \times 10^{-3}$ which gives an evolution time of $t_{\text{length}} = 1.5 \times 10^9$ for a 99% diabatic transition. Since this implies a physical time for Cesium atoms of about 8 hours (using ω_r and the non-dimensionalization given above), there is no danger of passing adiabatically through this avoided crossing in a real experiment.

ACCELERATION OF COLD ATOMS

In this section we consider the real-time evolution (numerically or in a laboratory) of the two-resonance system under the STIRAP coupling described above. An effective transfer of atoms involves a competition between two

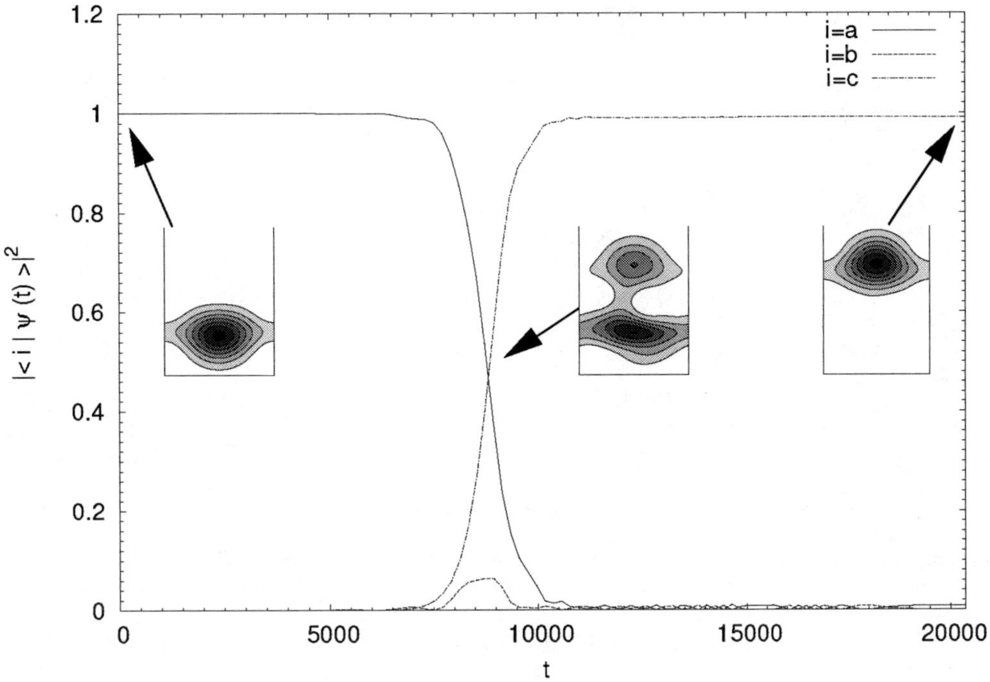

FIGURE 7. Time evolution of a state $|\psi(t)\rangle$ initially prepared in $|a\rangle$, under STIRAP coupling of the two resonance system with parameters $\kappa_0 = 1$, $\omega_0 = 10/\gamma$, $\lambda = 0.12$ (Gaussian parameters of the coupling fields are the same as in Figure 5). Husimi plots, with axes identical to those in Figure 3, are shown at three times.

factors. First, the STIRAP-like model in Eq. (11) is only an appropriate description of the system in the limit of small coupling, i.e. small λ. In the adiabatic analysis we have seen that there may be avoided crossings between the three relevant Floquet eigenvalues, which can impede the desired transfer. As λ is increased, these avoided crossings will grow larger – and others may appear – which will invalidate the simple STIRAP interpretation of the perturbed two-resonance system. The second factor is the necessity of a reasonable transfer time in the laboratory: population transfer must take place in less than a second (preferably quicker) in order for a given technique to be useful. This requirement is in competition with the first in that it requires that the applied couplings be relatively large, i.e. large values of λ.

The interplay of these two factors is evident in the graph shown in Figure 6. For smaller values of λ, the approach to adiabaticity in the full two-resonance system follows very closely that of the three-level STIRAP model, whereas it can be seen that the $\lambda = 0.12$ data points do not, particularly for long evolution times. As is expected, the transition of a state from $|a\rangle$ to $|c\rangle$ is not guaranteed for larger values of λ. The approach to adiabaticity itself is very similar to a Landau-Zener transition: the lines in Figure 6, for the evolution of the STIRAP model, clearly show an exponential approach. We have considered this evolution of the STIRAP model in Eq. (11) (with resonant, $\Delta_1 = \Delta_2 = 0$, coupling) over a wider range of parameters and found very good agreement with the empirical relationship

$$P_{\text{na}} = \exp[-a\sigma\lambda\bar{v}f + 1.1] \tag{14}$$

where P_{na} is non-adiabatic transition probability; σ is the Gaussian width (in time) of both coupling fields κ_1 and κ_2; a is a constant of order unity, dependent on the overlap of the two Gaussian fields; $\bar{v} \equiv \left[v_{ab}^2 v_{bc}^2 / \sqrt{v_{ab}^2 + v_{bc}^2}\right]^{1/3}$; and $f \equiv \Gamma/(X - 1 + \Gamma)$ with $X \equiv \max\left(\frac{v_{ab}}{v_{bc}}, \frac{v_{bc}}{v_{ab}}\right)$ and $\Gamma \approx 100$. Therefore, we see that the desired transfer occurs more quickly (smaller σ) when the coupling strength λ is large and the coefficients v_{ab} and v_{bc} are nearly equal.

An example of the full STIRAP transition for the two-resonance system is shown in Figure 7. The evolution of the occupation probabilities is very similar to that of the STIRAP model (see Figure 1d) and the transfer efficiency here is better than 99%. Converting back to physical units, using the non-dimensionalization given above, the transfer time of $\sim 2 \times 10^4$ corresponds to about 0.4 seconds. Our chosen value of ω_0 implies that the momentum difference between the two optical lattices is $\Delta p \approx 3.3 = 3.3 \times 2\hbar k_L$, which for Cesium implies a change of velocity $\Delta v \approx 2.3\frac{\text{cm}}{\text{s}}$.

ACKNOWLEDGMENTS

The authors thank the Robert A. Welch Foundation (Grant No. F-1051) for support of this work. The authors thank the Texas Advanced Computing Center (TACC) for the use of their computing facilities in performing calculations for this paper.

REFERENCES

1. J.R. Kuklinski, U. Gaubatz, F.T. Hioe, and K. Bergmann. "Adiabatic population transfer in a three-level system driven by delayed laser pulses," Phys. Rev. A **40**, R6741 (1989).
2. U. Gaubatz, P. Rudecki, M. Becker, S. Schiemann, M. Külz, and K. Bergmann. "Population switching between vibrational levels in molecular beams," Chem. Phys. Lett. **149**, 463 (1988).
3. K. Eckert, M. Lewenstein, R. Corbalán, G. Birkl, W. Ertmer and J. Mompart. "Three-level atom optics via the tunneling interaction," Phys. Rev. A **70**, 023606 (2004).
4. D.A. Steck, W.H. Oskay, M.G. Raizen. "Observation of chaos-assisted tunneling between islands of stability," Science **293**, 274 (2001).
5. D.A. Steck, W.H. Oskay and M.G. Raizen. "Fluctuations and decoherence in chaos-assisted tunneling," Phys. Rev. Lett. **88** , 120406 (2002).
6. M. Kasevich, and S. Chu. "Laser Cooling below a Photon Recoil with Three-Level Atoms," Phys. Rev. Lett. **69**, 1741 (1992).
7. R. Graham, M. Schlautmann, P. Zoller. "Dynamical localization of atomic-beam deflection by a modulated standing light wave," Phys. Rev. A **45**, R19 (1992).
8. B.P. Holder and L.E. Reichl. "Stimulated-Raman-adiabatic-passage–like transitions in a harmonically modulated optical lattice," Phys Rev. A **76**, 013420 (2007).
9. Since the light shift potential is proportional to the square of the electric field amplitude, the two-resonance potential cannot be exactly achieved with two pairs of lasers (one pair having equal frequencies and the other with offset frequencies). However, the effect of unwanted cross terms created by the modulus squared of $E(x,t)$ for the four laser system can be minimized by an offset of the carrier frequency for the two pairs [8].
10. B.P. Holder and L.E. Reichl. "Avoided crossings in driven systems," Phys. Rev. A **72**, 043408 (2005).
11. H. Sambe. "Steady States and Quasienergies of a Quantum-Mechanical System in an Oscillating Field," Phys. Rev. A **7**, 2203 (1973).
12. M. Wilkinson and M.A. Morgan. "Nonadiabatic transitions in multilevel systems," Phys. Rev. A **61**, 62104 (2000).

Nondynamic Correlation and Coupled-Cluster Methods

Josef Paldus and Xiangzhu Li

Department of Applied Mathematics, University of Waterloo, Waterloo, Ontario N2L 3G1 Canada

Abstract.
The coupled-cluster (CC) methods that are based on the exponential Ansatz for the wave operator proved to be extremely valuable in quantum chemical computations of the molecular electronic structure and represent nowadays most accurate and often used post-Hartree-Fock approaches that are capable to attain chemical accuracy for many molecular properties of interest. Nonetheless, while the widely exploited single-reference (SR) CC approaches, such as the CCSD method accounting for one- and two-body cluster amplitudes or its CCSD(T) version perturbatively corrected for three-body clusters – all available in numerous software packages – are remarkably efficient in handling of dynamical correlation effects, their performance rapidly deteriorates in the presence of quasidegeneracy of the reference configuration when the nondynamic correlations become important. A possible, computationally affordable, remedy for this failure is the so-called reduced multireference (RMR) CCSD method, as well as its triple-corrected version RMR CCSD(T). These methods exploit the complementarity of the configuration interaction (CI) and CC approaches and represent the topic of this communication.

Keywords: Molecular electronic structure, Coupled-cluster methods, Ab initio calculations, Quasidegeneracy, Static, nondynamic, and dynamic correlation effects
PACS: 31.10.+z, 31.15.-p, 31.50.Bc, 33.15.-e

INTRODUCTION

In view of an enormous mathematical complexity that we face when trying to solve Schrödinger equation even for the simplest molecular systems, the main goal of quantum chemical methodology has been the design of computationally manageable, yet reliable, approximation schemes that are based on various model Hamiltonians. Even when we ignore relativistic effects and freeze the nuclear framework by relying on the Born-Oppenheimer approximation, the problem remains too formidable to be solved exactly for any system having two or more electrons. For this very reason, almost all molecular applications are based on finite dimensional models.

In this regard, both valence bond (VB) and molecular orbital (MO) formalisms proved to be of an enormous value in the interpretation, understanding, and prediction of molecular spectra, of various electric and magnetic properties, chemical reactivity and related topics. Although the VB approach, first introduced by Heitler and London [1], has many desirable features from the chemical point of view, it was nonetheless soon eclipsed by a computationally more amenable Hund-Mulliken MO formalism [2, 3] that relies on the independent particle model (IPM), whose typical representative is the Hartree-Fock (HF) approximation [4, 5]. The numerical solution of HF equations being restricted to atoms, or at most to diatomic (or linear) molecules, one relies on a discretized algebraic form of these equations that is based on the *linear combination of atomic orbitals* (LCAO) *self-consistent field* (SCF) formalism [6]. The choice of an AO basis set thus defines the employed *ab initio* model. Generation of SCF solutions, even for systems with several hundreds of electrons, is nowadays routine, although the handling of general open-shell states can still be frustrating due to the ambiguity in the choice of the trial wave functions.

With large enough AO basis sets the HF approximation yields well over 99% of the total energy. This fact led initially to a belief that the SCF solutions are sufficiently accurate for chemical purposes. However, already in the late sixties it became abundantly clear that in spite of the conceptual importance of HF solutions, and their utility in supplying at least a semiquantitative information about some molecular properties, their ability to provide a reliable description of the electronic structure in general was rather limited. The first dramatic demonstration of the inadequacy of SCF solutions emerged from Wahl's study of the F_2 molecule [7], which showed that at the SCF level of approximation F_2 is not bound by as much as 1.6 eV! The shortcomings of IPM are nowadays well recognized, since the energy differences of chemical interest are too small in comparison with total energies of the systems involved. For this very reason, all present day quantitative studies account, in one way or another, for the many-electron correlation effects that are lacking in the IPM description.

CP995, *Nuclei and Mesoscopic Physics - WNMP 2007*, edited by P. Danielewicz, P. Piecuch, and V. Zelevinsky

The post-Hartree-Fock methods that are currently widely used are basically of two types: variational and perturbative. The former ones are typified by CI (or shell model) methods, employing a linear Ansatz for the wave function in the spirit of the Ritz variation principle, while the latter ones are based on the many-body perturbation theory (MBPT) of either finite or an infinite order variety, and are in many respects complementary to the variational ones, being size-extensive but, at least in their single reference (SR) version, unable to properly account for nondynamic correlation as soon as the state considered becomes quasidegenerate with some higher-lying state(s). In contrast, the CI methods, which invariably rely on truncated expansions in view of a rapidly increasing dimension of the CI problem with the increasing size of the system and the basis set employed, lack the size-extensive property. The finite order SR MBPT methods are also limited to at most the fourth order in view of the rapidly increasing computational demands. Since the higher-order terms are responsible for a proper account of dynamic correlations, it is thus advantageous to employ the coupled-cluster (CC) approaches, which are based on the exponential Ansatz for the wave operator, and in which certain classes of the MBPT diagrams are automatically summed to an infinite order by virtue of solving the energy-independent CC equations [8, 9, 10] (for reviews see Refs. [11] – [21], for a historical viewpoint, see [22, 23]). These equations may be viewed as recursion formulas for the generation of higher order MBPT contributions of a certain type on the basis of the lower order ones.

Even here, however, one has to truncate the cluster operator at a relatively low excitation level in order to achieve a computationally viable algorithm. In most cases only the one- and two-body [or singly- (S) and doubly- (D) excited] connected clusters are retained, resulting in the CCSD method, while all higher-level terms are approximated by the disconnected clusters thanks to the exponential Ansatz. For nondegenerate closed-shell ground states, or even for high-spin open-shell states, the CC approach at the SD level of truncation (SR CCSD method) generally provides very reliable and precise results that are fully size-extensive. Yet, to achieve 'the chemical accuracy' of $\sim$1 kcal/mol, at least triexcited clusters should be accounted for as well. This is usually done perturbatively via the CCSD(T) method [24], which has been often referred to as the 'gold standard' of quantum chemistry [25] – [30]. Indeed, CCSD and CCSD(T) approaches are presently very popular in view of their high accuracy and reliability (se, e.g., Ref. [31]) and yield excellent results for closed-shell ground states (or, generally, for the lowest state of a given symmetry species) of systems near their equilibrium geometry.

Unfortunately, away from the equilibrium structure, as when breaking genuine chemical bonds or, generally, for various open-shell systems, the perturbative nature of the (T)-correction for triples breaks down, further enhancing an already inadequate performance of CCSD itself, arising due to the lack of size-consistency associated with the SR Ansatz that is based on a restricted Hartree-Fock (RHF) single-determinantal reference (see, e.g., Refs. [14] – [21]). Moreover, with the increasing quasidegeneracy of the reference configuration increases the importance of higher-than-pair clusters. These shortcommings can be, in principle, overcome by relying on appropriate multireference (MR) CC approaches. Unfortunately, these methods are not only computationaly very demanding, but also suffer from intruder state and other problems (see below). For these reasons the recent emphasis focussed on the so-called state-selective or state-specific (SS) approaches, in particular those that are essentially of a SR nature.

A great progress in rectifying the inadequacy of the standard CCSD(T) approach in the presence of quasidegeneracy has been achieved by introducing the renormalized (R) and completely-renormalized (CR) versions of CCSD(T) (see Refs. [20, 32]) and, most importantly, by extending them to a completely size-extensive CR-CC(2,n), $n = 3$, 4, etc. methods [27]. These ideas were also implemented in the equations-of-motion (EOM) methods [20, 33]. Of course, with the increasing value of n these approaches become again computationally demanding but, happily, the CR-CC(2,3) method yields in most cases perfectly adequate results, unless connected quadruples come into play.

An alternative useful idea of accounting for higher-than-pair clusters in an efficient and economic way is represented by the so-called *externally-corrected* (ec) CCSD approaches [34] – [40]. These are essentially the standard SR CCSD approaches in which, however, higher-than-pair connected clusters are approximately accounted for via a small subset of important three- and four-body cluster amplitudes, which are in turn obtained from some independent, external source. In this way we achieve a physically more meaningful decoupling of the CCSD equations from the rest of the CC chain. In fact, using the FCI three- and four-body amplitudes for this purpose, the resulting ecCCSD gives back the FCI result. It is of course desirable that the wave function serving as an external source is capable to efficiently describe nondynamic correlation effects.

The most useful external source turned out to be a modest size MR CISD wave function thanks to the complementarity of CC and CI approaches in their description of the dynamic and nondynamic correlations, leading to the reduced MR (RMR) CCSD method [41, 42], which proved to be very effective in handling of various MR situations, as will be pointed out below. Our objective here is to describe this approach and to report on its performance in various quasidegenerate situations. First, however, we briefly outline the standard SR CC method.

SINGLE-REFERENCE COUPLED-CLUSTER APPROACH

In the standard SR CC approach, the exact (nonrelativistic) N-electron wave function $|\Psi\rangle$ for the state of interest (energetically the lowest state of a given symmetry species) is represented via the exponential cluster Ansatz

$$|\Psi\rangle = e^{T}|\Phi_0\rangle, \quad \langle\Phi_0|\Phi_0\rangle = \langle\Psi|\Phi_0\rangle = 1\,, \tag{1}$$

where $|\Phi_0\rangle$ is some nondegenerate IPM wave function and the cluster operator T is given by the sum of its i-body components T_i,

$$T = \sum_{i=1}^{N} T_i\,, \qquad T_i = \sum_{j} t_J^{(i)} G_J^{(i)}\,, \tag{2}$$

where $t_J^{(i)}$ are the unknown cluster amplitudes and $G_J^{(i)}$ are excitation operators generating i-fold excited configurations $|\Phi_J^{(i)}\rangle$ when acting on the IPM (usually HF) reference $|\Phi_0\rangle$, $|\Phi_J^{(i)}\rangle = G_J^{(i)}|\Phi_0\rangle$. The configurations $|\Phi_J^{(i)}\rangle$ span the i-times excited subspace of the N-electron space considered. The subscript J represents an index set of occupied and virtual (relative to $|\Phi_0\rangle$) molecular spin orbitals (MSOs) that fully specify the excited configuration $|\Phi_J^{(i)}\rangle$, $J \equiv \{A_1, A_2, \cdots, A_i \rightarrow R_1, R_2, \cdots, R_i\}$, i.e.,

$$G_J^{(i)} \equiv G_{A_1,A_2,\cdots,A_i}^{R_1,R_2,\cdots,R_i} = \prod_{k=1}^{i} X_{R_k}^{\dagger} X_{A_k}\,, \tag{3}$$

where $X_M^{\dagger}$ and X_M designate, respectively, the creation and annihilation operators associated with an orthonormal basis $\{\,|M\rangle\}$ of MSOs.

The exponential character of the cluster expansion (1) warrants the size-extensivity of the resulting formalism regardless the truncation scheme employed, as implied by a comparison with the standard linear CI expansion of (intermediately normalized) $|\Psi\rangle$,

$$|\Psi\rangle = C|\Phi_0\rangle = \sum_{i=0}^{N} C_i|\Phi_0\rangle, \quad C_0 = 1\,, \tag{4}$$

which yields

$$\begin{aligned} C_1 &= T_1, \\ C_2 &= T_2 + \tfrac{1}{2}T_1^2, \\ C_3 &= T_3 + T_1T_2 + \tfrac{1}{6}T_1^3, \\ C_4 &= T_4 + \tfrac{1}{2}T_2^2 + \tfrac{1}{2}T_1^2T_2 + T_1T_3 + \tfrac{1}{24}T_1^4, \quad \text{etc.} \end{aligned} \tag{5}$$

Setting, for simplicity, $T_1 = 0$ (i.e., using Brueckner's maximum overlap orbitals), we have that $C_1 = 0, C_2 = T_2, C_3 = T_3$ and $C_4 = T_4 + \frac{1}{2}T_2^2$. Thus, even when we approximate T solely by its pair cluster component, $T \approx T_2$ (i.e., $T_i = 0, i \neq 2$), we obtain contributions from all relevant even-number-of-times excited configurations, since $C_{2n} \approx T_2^n/n!$, the most important one being $C_4 \approx \frac{1}{2}T_2^2$. Since the contribution of the connected four-body cluster T_4 is usually negligible in comparison with that of its disconnected counterpart $\frac{1}{2}T_2^2$, we see that the CCD approach, employing the same number of unknowns as CID, also accounts for a large part of quadruples that are entirely missed by CID.

The unknown cluster amplitudes $t_J^{(i)}$ are determined by the energy-independent CC equations, obtained by projecting Schrödinger equation, premultiplied with e^{-T}, onto the excited state manifold, i.e.,

$$\langle\Phi_J^{(i)}|e^{-T}He^{T}|\Phi_0\rangle = 0\,, \tag{6}$$

while the energy results by projecting onto the reference $|\Phi_0\rangle$,

$$E = \langle\Phi_0|e^{-T}He^{T}|\Phi_0\rangle = \langle\Phi_0|He^{T}|\Phi_0\rangle, \tag{7}$$

so that

$$E - \langle\Phi_0|H|\Phi_0\rangle = \langle\Phi_0|H(T_1 + T_2 + \tfrac{1}{2}T_1^2)|\Phi_0\rangle\,. \tag{8}$$

The standard CCSD approximation involving single and double excitations uses T truncated to the sum of T_1 and T_2, and CCSD(T) then adds a perturbative correction for T_3.

Explicit CC equations (6) in terms of the unknown cluster amplitudes t_I [for simplicity we drop the superscript (i) indicating the excitation order] take the form of nonlinear algebraic equations,

$$a_I + \sum_J b_{IJ} t_J + \sum_{J<K} c_{IJK} t_J t_K + \cdots = 0\,, \tag{9}$$

where at most quartic terms appear in view of the two-body nature of the Coulomb potential and the coefficients a_I, b_{IJ}, c_{IJK}, etc. are determined by one- and two-body matrix elements of the Hamiltonian (see, e.g., Refs. [11, 16, 21] for details).

EXTERNALLY-CORRECTED CCSD METHODS

In view of the fact that a molecular electronic Hamiltonian involves at most two-body interactions, the full chain of CC equations has the following general structure

$$\begin{aligned}
\langle\Phi_J^{(1)}|H[1+T_1+(T_2+\tfrac{1}{2}T_1^2) &+ (T_3+T_1T_2+\tfrac{1}{6}T_1^3)]|\Phi_0\rangle_C = 0\,,\\
\langle\Phi_J^{(2)}|H[1+T_1+(T_2+\tfrac{1}{2}T_1^2) &+ (T_3+T_1T_2+\tfrac{1}{6}T_1^3)\\
&+ (T_4+T_1T_3+\tfrac{1}{2}T_1^2T_2+\tfrac{1}{24}T_1^4)]|\Phi_0\rangle_C = 0\,,\\
\langle\Phi_J^{(3)}|H[(T_2+\tfrac{1}{2}T_1^2) &+ (T_3+T_1T_2+\tfrac{1}{6}T_1^3)\\
&+ (T_4+T_1T_3+\tfrac{1}{2}T_1^2T_2+\tfrac{1}{24}T_1^4)\\
&+ (T_5+T_1T_4+\cdots)]|\Phi_0\rangle_C = 0, \text{ etc.}\,,
\end{aligned} \tag{10}$$

where the subscript C indicates that only connected components are to be retained. This structure reflects the block diagonal structure of the corresponding CI chain, namely

$$\langle\Phi_J^{(i)}|H(C_{i-2}+C_{i-1}+C_i+C_{i+1}+C_{i+2})|\Phi_0\rangle = E\langle\Phi_J^{(i)}|C_i|\Phi_0\rangle,\ (i=0,1,2,\cdots,N) \tag{11}$$

where $C_i = 0$ if $i < 0$ or $i > N$.

Thus, formally, we can obtain the CC chain by setting to zero the left hand side of Eqs. (11) and by replacing the C_i operators by their cluster analogues, Eq. (5), while retaining only the connected terms, the disconnected ones being cancelled by the right hand side energy term in CI Eqs. (11) (cf., e.g., Ref. [43] for details). Without truncation, both CI and CC chains are equivalent, providing the full CI (FCI) or full CC (FCC) result, representing the exact solution for a given finite dimensional *ab initio* model (as defined by the chosen AO basis set). However, FCI or FCC can only be carried out for relatively small model systems, while in all practical applications one has to employ truncated schemes. The standard CCSD method thus arises by setting $T_3 = T_4 = 0$, which decouples the second equation in (10) from the rest of the CC chain. However, as already mentioned, T_3 and T_4 are no longer negligible when we deal with quasidegenerate situations and cannot be treated via MBPT. Thus, if we are able to obtain at least approximate values of T_3 and T_4 clusters from some independent source, we could achieve a physically more meaningful truncation of the CC chain and extend the validity of the CCSD-like approximation into the region of significantly stretched geometries. This is precisely the idea of the *externally corrected* CCSD (ecCCSD). In terms of the explicit algebraic form of CC equations, Eq. (9), we see that once we have found a suitable approximation $T_3 \approx T_3^{(0)}$ and $T_4 \approx T_4^{(0)}$ for the three- and four-body clusters, we can evaluate the T_3 and T_4 dependent terms in the first two equations (11) and correct, once and for all, the absolute terms a_I in Eq. (10). Only the T_1T_3 term contributes to the linear b_{IJ} terms that are associated with monoexcited $t_J^{(1)}$ cluster amplitudes. However, even this latter term can be treated noniteratively using the approximate $T_1^{(0)}$ clusters provided by the external source [38].

The basic requirement for the choice of approximate $T_3^{(0)}$ and $T_4^{(0)}$ clusters is their ability to account for the missing nondynamic correlation, while the practical aspect requires their affordability. The first attempt in this direction exploited an unrestricted Hartree Fock (UHF) wave function [of different orbitals for different spins (DODS) type]. Its implicit exploitation lead to the so-called ACPQ (approximate coupled pairs with quadruples) method [44, 45]. Later on, its explicit version was also developed and implemented [46]. Although in many cases this source enables one to reach the correct dissociation channel, its main shortcoming is the fact that for the closed-shell systems it can only provide T_4 clusters, since the T_3 contribution vanishes due to the spin symmetry of the DODS wave function. Nonetheless, the ACPQ method enabled an effective handling of extended linear systems (at the semiempirical level), which are very demanding, since the standard CCSD method completely breaks down in this case [45]. Another suitable source, at least in principle, is represented by simple VB-type wave functions, which again are capable to correctly describe bond breaking situations. This possibility was tested at the semiempirical level with considerable success [34, 36].

At the *ab-initio* level, the most obvious possibility is offered by the complete active space (CAS) SCF or CAS FCI wave functions, assuming that we employ a sufficiently large active space which is flexible enough to correctly describe the desired dissociation channel (cf. also Ref. [35]). Our tests proved to be very encouraging [38, 39], particularly for open-shell systems, in which case we employed the spin-adapted CCSD method that is based on the unitary group approach (UGA) [16, 47]. Even in the case of a triple bond breaking, the use of the ecCCSD methods can significantly extend the range of applicability of the SR-type CC approaches, as will be mentioned later on. However, the most promissing external source turned out to be a modest-size MR CISD wave function, as described in the next section.

REDUCED MULTIREFERENCE CCSD METHOD

In view of the above mentioned complementarity of the CC and CI approaches in their handling of, repectively, the dynamic and nondynamic or static correlations, it seems particularly attractive to employ the latter as an external source for the approximate T_3 and T_4 clusters. A small reference space MR CISD will involve only *a very small subset* of triple and quadruple excitations relative to the leading configuration $|\Phi_0\rangle$, namely those that are primarily responsible for a proper account of a nondynamic correlation and guarantee the correct description of the desired dissociation channel.

For example, considering two-reference (2R) CISD involving two active (or frontier) orbitals $|i\rangle$ and $|j\rangle$ belonging to different symmetry species, i.e., the highest occupied (HOMO) and the lowest unoccupied (LUMO) MOs [i.e., four active MSOs $|i\alpha\rangle$, $|i\beta\rangle$, $|j\alpha\rangle$, and $|j\beta\rangle$], we have in addition to the IPM reference $|\Phi_0\rangle = |(\text{core})i\alpha i\beta\rangle$ an additional reference $|\Phi_1\rangle = |(\text{core})j\alpha j\beta\rangle$. Considering now (core) $\rightarrow$ (virtual) doubles, we see that doubles relative to $|\Phi_1\rangle$ are qudruples with respect to $|\Phi_0\rangle$, and their number is about the same as the number of doubles relative to $|\Phi_0\rangle$. Thus, 2R CISD involves the same number of quadruples as there are (core) $\rightarrow$ (virtual) doubles, both proportional to $n_c^2 n_v^2$ in number (where n_c and n_v designate the number of core and virtual orbitals, respectively; the number of the remaining doubles (core) $\rightarrow$ (active) and (active) $\rightarrow$ (virtual) is much smaller, being proportional to n_c^2 and n_v^2, respectively).

Thus, although MR CISD that is based on a small reference space involves only a small subset of quadruples relative to the reference $|\Phi_0\rangle$, these quadruples are clearly those that are required to account for the quasidegeneracy of $|\Phi_0\rangle$ and $|\Phi_1\rangle$. Indeed, the number of all (core) $\rightarrow$ (virtual) quadruples is proportional to $n_c^4 n_v^4$. The majority of these will be, of course, primarily responsible for the dynamic correlation and thus effectively accounted for through the CCSD cluster Ansatz. We can thus expect the low-dimensional MR CISD to supply us with the most important connected triples and quadruples, while relying on the CC Ansatz to represent the remaining higher than pair clusters.

In general, the MR CISD wave function based on an M-dimensional reference space has the form

$$|\Psi_0\rangle = \sum_{p=0}^{M-1} C^{[p]}|\Phi_p\rangle\,, \tag{12}$$

where $C^{[p]} = C_1^{[p]} + C_2^{[p]}$ designates the CI-type SD excitation operator relative to the reference $|\Phi_p\rangle$. (The fact that the same configuration may arise via excitations from different references must be properly taken into account.) We can rewrite this expansion in the SR CI form, using $|\Phi_0\rangle$ as a reference,

$$|\Psi_0\rangle = \left(1 + \sum_{i=1}^{N} C_i^{(0)}\right)|\Phi_0\rangle\,, \tag{13}$$

distinguishing the relevant i-fold excitation operators relative to $|\Phi_0\rangle$ by the superscript '(0)'. Applying a standard cluster analysis [48], it is straightforward to rewrite this wave function in the SR CC form, namely

$$|\Psi_0\rangle = \exp(T_1^{(0)} + T_2^{(0)} + \Delta)|\Phi_0\rangle = \exp(T_1^{(0)} + T_2^{(0)} + \{T_3^{(0)} + T_4^{(0)} + \cdots\}_{\text{subset}})|\Phi_0\rangle\,, \tag{14}$$

where Δ represents a subset of higher-than-pair cluster operators arising through the MR CISD Ansatz.

Assuming, next, that $T_3^{(0)}$ and $T_4^{(0)}$ amplitudes so obtained represent a reasonable approximation of actual T_3 and T_4 clusters, we can employ the RMR CCSD Ansatz

$$|\Psi\rangle = e^{T_1+T_2+T_3^{(0)}+T_4^{(0)}}|\Phi_0\rangle\,, \tag{15}$$

leading to a special case of the general ecCCSD approach. Clearly, the decoupling of the SR CC chain of equations, Eq.(10), using approximate values of T_3 and T_4, should provide a superior approximation to the standard one, which simply assumes that T_3 and T_4 vanish. This is particularly the case in quasidegenerate situations, when T_3 and T_4 are no longer negligible, in fact not even accessible to a perturbative treatment.

To summarize, the RMR CCSD method involves the following three steps: (i) We choose a suitable reference space and compute the corresponding MR CISD. Next, (ii) we find $T_3^{(0)}$ and $T_4^{(0)}$ clusters by cluster analyzing the MR CISD wave function of step (i) and, finally, (iii) we use these amplitudes to generate and solve ecCCSD equations. The details of actual implementations of the RMR CCSD method for various types of reference spaces and applications to actual problems can be found in our earlier papers ([41] and [49] – [54]).

Finally, we can further improve the performance of RMR CCSD by accounting for the remaining connected triples that have not been taken care of via the external corrections. These triples are very large in number and individually contribute very little, being primarily responsible for the dynamic correlation. Yet in their totality they can improve the RMR CCSD result just as CCSD(T) improves CCSD in the absence of quasidegeneracy. We thus account for these remaining triples perturbatively in the same way as in the standard CCSD(T), obtaining the RMR CCSD(T) approach [55]. For larger systems, we have also introduced truncated versions [56, 57] of both RMR CCSD and RMR CCSD(T).

APPLICATIONS

Initial exploitation of the ecCCSD-type methods [36] for π-electron systems, described via semiempirical Pariser-Parr-Pople Hamiltonian, was based on VB wave functions as an external source and yielded very encouraging results. In view of the lack of general VB codes for *ab initio* model Hamiltonians, the attention was turned towards the CI-type wave functions. An obvious choice seemed to be the CAS SCF and CAS FCI wave functions (see also [35]), as well as truncated CAS FCI (see [37] – [40]). However, since these approches account only for the three- and four-body clusters that arise through the excitations within the active (spin) orbitals, one has to use large enough active spaces in order to obtain a meaningful subset of the desired higher-than-pair cluster amplitudes. For example, the above mentioned minimal 2-electron/2-orbital CAS will not provide any such amplitudes. Moreover, a generation of CAS SCF wave functions with a reasonably large active space is computationally very demanding. We have ultimately decided to rely on a modest-size MR CISD as an external source. The resulting RMR CCSD method proved to be very successful even in very challenging situations.

In general, we require a model (or reference) space for MR CISD that involves at least one bonding and a corresponding antibonding active MO (or four MSOs) in order to describe the breaking of a single bond. For example, this is the case when we compute the potential energy curves (PECs) or harmonic frequencies for molecular or radical species, such as HF, F_2, OH, or the H4 model [58], or more demanding NO or F_2^+ species [55]. Since the absolute energies are of little importance for a comparison with the experimental data, one often characterized the quality of the PEC or PES by the so-called 'non-parallelism error' (NPE), defined as the maximal deviation of the approximate potential from the exact FCI one, both shifted to coincide at the equilibrium geometry, for a given range of geometries. Thus, should both potentials be separated by the same energy difference for all geometries, i.e., be 'parallel', we would have NPE = 0.

Considering, for example, the X $^2\Pi$ ground state of the OH radical over the range of internuclear separations ranging from $R_e = 1.832$ bohr to $2R_e$ and using AO basis set of a DZP (double-zeta plus polarization) quality, which still makes it possible to generate the exact FCI energies, we find that the NPE amounts to 10.26 mhartree for SR CCSD, but only 3.43 mhartree for 2R RMR CCSD and 0.44 mhartree for 5R RMR CCSD. The FCI result involves 441,792 configuration state functions (CSFs), while the 4R and 5R RMR methods require only 2,031 and 3,705 Slater determinants, respectively, for the MR CISD wave function. Note that a single-determinantal SCF PEC has NPE of 75.90 mhartree.

In all these cases the CR-CC(2,3) approach, that is designed to mimic the exact SR CCSDT, also yields very satisfactory results. Even when simultaneously breaking two single bonds as, e.g., in a symmetric dissociation of water, such methods work even for significantly stretched geometries. However, when connected quadruples become important, as seems to be the case, for example, when computing singlet-triplet splitting in the BN radical [59], one has to revert to computationally more demanding CR-CC(2,4), while 4R RMR CCSD still provides accurate and reliable results [60]. Nonetheless, even higher than quadruples may be of essence when we wish to access the most general dissociation channels, as we shall now indicate.

Let us illustrate the difficulties that are encountered when dissociating multiple bonds on a very challenging example of a triple-bond breaking in the nitrogen molecule N_2. We shall again use a basis set of a double-zeta quality, in which case we can still generate the exact FCI energies (this involves 2,388,528 Slater determinants with $S_z = 0$ and adapted to the A_g symmetry species). Needless to say that both SR CCSD and the 'gold standard' SR CCSD(T) break down when we encounter a higly stretched regime (see Fig. 1). Clearly, SR CCSD(T) improves the potential in the vicinity of the equilibrium geometry, but then utterly fails in the quasidegenerate region due to the failure of the SR MBPT. The model space that we have to consider here will involves three occupied and corresponding three virtual MOs, i.e., a 6-orbital active space. The minimal reference space will require one bonding-antibonding pair for each bond, i.e., $2^3 = 8$ references corresponding to double excitations from the σ, π_x, and π_y bonding MOs to the corresponding antibonding ones. Such an 8R space indeed greatly improves the computed PEC even for sizeably stretched geometries (say, up to about $2R_e$). In fact, such a potential enabled us to obtain excellent results precisely reproducing the available ro-vibronic spectra [50, 61, 62], as well as to extend them to higher vibrational and rotational quantum numbers. Even SR CCSD works rather well in the neighborhood of the equilibrium geometries (up to $\sim 1.5R_e$), enabling computations of property surfaces and of the related rovibrational dependence for the quadrupole moment and transition quadrupole moments [63, 64].

Yet, if we wish to generate the entire PEC up to the fully dissociated limit ($R \to \infty$), and especially in the intermediate range of geometries, even the 8R RMR CCSD will fail, just as all the CC or even CI methods that consider at most triple and quadruple excitations [see, e.g., the CR-CC(2,3) result in Fig. 1, which will be close to the SR CCSDT one]. The same problem is encountered when we approach the intermediate region of geometries from the dissociated limit, where we can employ localized or partially localized MOs and corresponding suitable reference(s). In order to bridge the intermediate range of internuclear separation it is indeed essential to employ full 6-electron/6-orbital active space. This space involves generally 175 spin-adapted singlet CSFs and can be reduced to 56R space and spin ($S_z = 0$) adapted determinants. Such a space then involves even octuply-excited configurations relative to $|\Phi_0\rangle$. We have shown recently [65] that indeed the corresponding 56R RMR CCSD(T) PEC practically reproduces the FCI PEC in the entire range of internuclear separations, as shown in Fig. 1. Deleting even one hexuply-excited configuration at $R \sim 4$ a.u. already deteriorates the result by couple of mhartree and using 28R RMR CCSD(T) leads to errors exceeding 30 mhartree!

In closing, let us mention that the RMR-type CCSD and CCSD(T) methods are not only capable to handle effectively the nondynamic correlation in the presence of quasidegeneracy, but are also computationally affordable, as witnessed by recent applications to transition metal componds, such as nickel carbonyls [57] $Ni(CO)_n$, $n = 1, 2,$ and 4, or transition metal ion – methylene complexes [66] MCH_2^+ (M = Sc to Cu). They were also successfully used to describe forward and reverse barrier heights for numerous heavy atom transfer, nucleophilic substitution, association, and unimolecular reactions [67], as well as in very challenging problems of small singlet–triplet splittings in the C_2 and BN radicals [59], or in the description of the spontaneous symmetry breaking in the N_3 and BNB radicals [68, 69].

CLOSING REMARKS

We have outlined the essence of the RMR CCSD and CCSD(T) approaches and described their ability to overcome the quasidegeneracy problems, while only moderately increasing the required computational effort over that of the SR CCSD and CCSD(T) methods. We have seen that even in the very challenging case of the triple-bond breaking the RMR approaches can generate a highly accurate and realistic potentials. The only weakness of these methods is their slight violation of the size-extensivity [70], even though in all practical applications and tests the deviations from size-extensivity (arising from the use of variational three- and four-body amplitudes) are very small (offen in the range of μhartrees).

To render this approach fully size-extensive, we have recently developed the so-called partially-linearized (pl) MR CC method, in which higher-than-pair cluster amplitudes are generated by solving linearized higher-level CC equations, namely those projected onto the higher-than-doubly-excited configurations that span the first-order interacting space of the corresponding MR CISD that is used in the RMR approaches [65]. This method is indeed fully size-extensive but, unfortunately, in our preliminary tests has always underperformed the RMR results, not to mention the convergence and other difficulties.

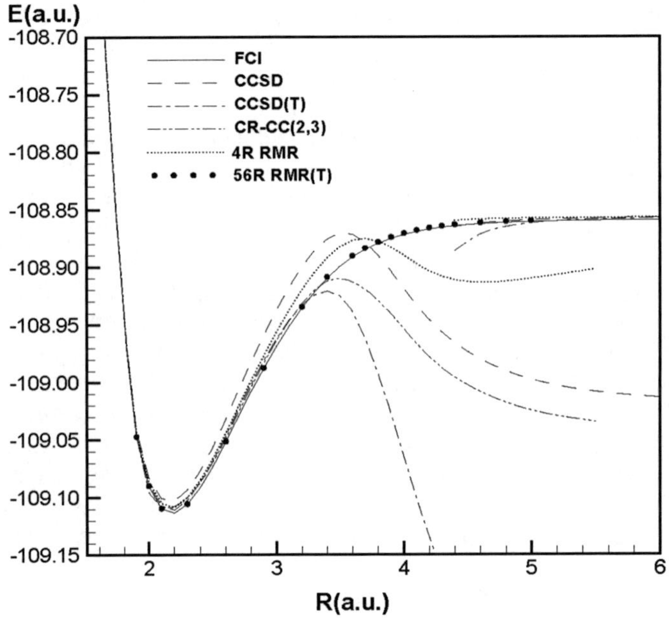

FIGURE 1. Computed potential energy curves for a double-zeta *ab initio* model of the nitrogen molecule as obtained with the full configuration interaction (FCI) and various coupled-cluster (CC) methods. See the text for details.

On the whole, while the size-extensive property is higly desirable (and absolutely essential for extended systems), it is the size-consistency that plays the crucial role when generating potential energy surfaces. For example, the full size-extensivity of the SR CCSD method is of a little benefit when breaking the triple bond, as amply demonstrated in Fig. 1, in which case the size-consistency is crucial for reaching the correct dissociation channel.

In closing let us mention that the idea of external corrections was also exploited in the state-universal (SU) MR CCSD method [71]. In order to overcome the problem of intruder states [72] and the necessity of using CAS to warrant the size-extensivity, we have formulated the so-called general model space (GMS) SU CCSD method [73, 74] that can employ an arbitrary model space while preserving the intermediate normalization property. This was accomplished by introducing the so-called 'C-conditions' (constraint or connectivity conditions). Nonetheless, for highly quasidegenerate systems, the use of external corrections proved again to be useful and was realized in the so-called (M,N)-CCSD method [75] that employs a M-reference MR CISD as the external source for a N-reference SU CCSD ($M \geq N$). This enabled us to generate very precise results for the excited states of the same symmetry species (see, e.g., Refs. [76, 77]) in the case of the ozone and water molecules.

ACKNOWLEDGMENTS

Continued support by the Natural Sciences and Engienering Research Council of Canada (J.P.) is acknowledged. The senior author also gratefully acknowledges kind invitation to take part in the WNMP07 Workshop and the hospitality extended to him by the MSU faculty.

REFERENCES

1. W. Heitler and F. London, *Z. Phys.* **44**, 455 (1927).
2. F. Hund, *Z. Phys.* **36**, 657 (1926).
3. R. S. Mulliken, *Phys. Rev.* **32**, 186 (1927).
4. D. R. Hartree, *Proc. Cambridge Phil. Soc.* **24**, 89 (1927); **45**, 230 (1948).
5. V. Fock, *Z. Phys.* **61**, 126 (1930); **75**, 622 (1932).

6. C. C. J. Roothaan, *Rev. Mod. Phys.* **23**, 69 (1951).
7. A. C. Wahl, *J. Chem. Phys.* **41**, 2600 (1964).
8. J. Čížek, *J. Chem. Phys.* **45**, 4256 (1996); *Adv. Chem. Phys.* **14**, 35 (1969).
9. J. Čížek and J. Paldus, *Int. J. Quantum Chem.* **5**, 359 (1971).
10. J. Paldus, J. Čížek, and I. Shavitt, *Phys. Rev. A* **5**, 50 (1972).
11. J. Paldus, in *Methods in Computational Molecular Physics*, edited by S. Wilson and G. H. F. Diercksen, NATO ASI Series B: Physics, Vol. 293, Plenum, NY, 1992, pp. 99–194.
12. R. J. Bartlett and J. F. Stanton, in *Reviews in Computational Chemistry*, Vol. 5, edited by K. B. Lipkowitz and D. B. Boyd, VCH Publishers, NY, 1994, pp. 65–169.
13. J. Paldus, in *Relativistic and Correlation Effects in Molecules and Solids*, edited by G. L. Malli, NATO ASI Series B: Physics, Vol. 318, Plenum, NY, 1994, pp. 207–282.
14. R. J. Bartlett, in *Modern Electronic Structure Theory*, Part I; edited by D. R. Yarkony, World Scientific, Singapore, 1995; pp. 1047–1131.
15. J. Gauss, in *Encyclopedia of Computational Chemistry*, Vol. 1; edited by P. von R. Schleyer, Wiley, NY, 1998; pp. 615–636.
16. J. Paldus and X. Li, *Adv. Chem. Phys.* **110**, 1 (1999).
17. J. Paldus, in *Handbook of Molecular Physics and Quantum Chemistry*, Vol. 2, Part 3, Chap. 19; edited by S. Wilson, Wiley, Chichester, 2003; pp. 272–313.
18. T. D. Crawford and H. F. Schaefer III, in *Reviews in Computational Chemistry*, Vol. 14, Chap. 2, edited by K. B. Lipkowitz and D. B. Boyd, Wiley, NY, 2000; pp. 33–136.
19. J. F. Stanton and J. Gauss, *Int. Rev. Phys. Chem.* **19**, 61 (2000).
20. P. Piecuch, K. Kowalski, I. S. O. Pimienta, and M. J. Guire, *Int. Rev. Phys. Chem.* **21**, 527–655 (2002); P. Piecuch, K. Kowalski, I. S. O. Pimienta, P.-D. Fan, M. Lodriguito, M. J. Guire, S. A. Kucharski, T. Kuś, and M. Musiał, *Theor. Chem. Acc.* **112**, 349–393 (2004); and references therein.
21. R. J. Bartlett and M. Musiał, *Rev. Mod. Phys.* **79**, 291 (2007).
22. J. Paldus, in *Theory and Applications of Computational Chemistry: The First Forty Years*, edited by C. E. Dykstra, G. Frenking, K. S. Kim, and G. E. Scuseria (Elsevier, Amsterdam, 2005), Chap. 7, pp. 115–147.
23. R. J. Bartlett, in *Theory and Applications of Computational Chemistry: The First Forty Years*, edited by C. E. Dykstra, G. Frenking, K. S. Kim, and G. E. Scuseria (Elsevier, Amsterdam, 2005), Chap. 42, pp. 1191–1221.
24. K. Raghavachari, G. W. Trucks, J. A. Pople, and M. Head-Gordon, *Chem. Phys. Lett.* **157**, 479 (1989).
25. C. J. Cramer, *Essentials of Computational Chemistry: Theories and Models*, 2nd ed., Wiley, NY, 2004, p. 226.
26. S.-I. Lu, *J. Chem. Phys.* **121**, 10495 (2004).
27. P. Piecuch and M. Włoch, *J. Chem. Phys.* **123**, 224105 (2005); P. Piecuch, M. Włoch, J. R. Gour, and A. Kinal, *Chem. Phys. Lett.* **418**, 467 (2006).
28. J. Song, E. Aprá, Y. G. Khait, M. R. Hoffmann, and K. Kowalski, *Chem. Phys. Lett.* **428**, 277 (2006).
29. A. Kinal and P. Piecuch, *J. Phys. Chem. A* **110**, 367 (2006).
30. See also *Wikipedia* (`http://en.wikipedia.org/wiki/Coupled_cluster`)
31. T. J. Lee and G. E. Scuseria, in *Quantum Mechanical Electronic Structure Calculations with Chemical Accuracy*, edited by S. R. Langhoff, Kluwer, Dordrecht, 1995, pp. 47-108
32. K. Kowalski and P. Piecuch, *J. Chem. Phys.* **113**, 18 (2000); **122**, 074107 (2005), and references therein.
33. K. Kowalski and P. Piecuch, *J. Chem. Phys.* **120**, 1715 (2004); M. Włoch, J. R. Gour, K. Kowalski, and P. Piecuch, *J. Chem. Phys.* **122**, 214107 (2005).
34. J. Paldus and J. Planelles, *Theor. Chim. Acta* **89**, 13 (1994).
35. L. Z. Stolarczyk, *Chem. Phys. Lett.* **217**, 1 (1994).
36. J. Planelles, J. Paldus, and X. Li, *Theor. Chim. Acta* **89**, 33 (1994); **89**, 59 (1994).
37. G. Peris, J. Planelles, and J. Paldus, *Int. J. Quantum Chem.* **62**, 137 (1997).
38. X. Li, G. Peris, J. Planelles, F. Rajadell, and J. Paldus, *J. Chem. Phys.* **107**, 90 (1997).
39. G. Peris, F. Rajadell, X. Li, J. Planelles, and J. Paldus, *Mol. Phys.* **94**, 235 (1998).
40. J. Planelles, G. Peris, and J. Paldus, *Int. J. Quantum Chem.* **77**, 693 (2000).
41. X. Li and J. Paldus, *J. Chem. Phys.* **107**, 6257 (1997).
42. J. Paldus and X. Li, in *Correlation and Localization*, Topics in Current Chemistry, Vol. 203, edited by P. R. Surján, Springer, Berlin, 1999, pp. 1–20; in *Advances in Quantum Many-Body Theory*, Vol. 5, edited by R. F. Bishop et al., World Scientific, Singapore, 2002, pp. 393–404; X. Li and J. Paldus, in *Low-Lying Potential-Energy Surfaces*, ACS Symposium Series No. 828, edited by M. R. Hoffmann and K. G. Dyall, ACS Books, Washington, 2002, pp. 10–30.
43. J. Čížek and J. Paldus, *Int. J. Quantum Chem.* **5**, 359 (1971); J. Paldus and B. Jeziorski, *Theor. Chim. Acta* **73**, 81 (1988); J. Paldus, J. Čížek, and B. Jeziorski, *J. Chem. Phys.* **90**, 4356 (1989).
44. J. Paldus, J. Čížek, and M. Takahashi, *Phys. Rev. A* **30**, 2193 (1984).
45. J. Paldus, M. Takahashi, and R. W. H. Cho, *Phys. Rev. B* **30**, 4267 (1984).
46. P. Piecuch, R. Toboła, and J. Paldus, *Phys. Rev. A* **54**, 1210 (1996).
47. X. Li and J. Paldus, *J. Chem. Phys.* **101**, 8812 (1994).
48. J. Paldus and X. Li, *J. Chem. Phys.* **118**, 6769 (2003).
49. X. Li and J. Paldus, *J. Chem. Phys.* **108**, 637 (1998).
50. X. Li and J. Paldus, *Chem. Phys. Lett.* **286**, 145 (1998).
51. X. Li and J. Paldus, *Coll. Czech. Chem. Commun.* **63**, 1381 (1998).

52. X. Li and J. Paldus, *J. Chem. Phys.* **110**, 2844 (1999).
53. X. Li and J. Paldus, *Mol. Phys.* **98**, 1185 (2000).
54. X. Li, *J. Mol. Struct.* (THEOCHEM) **547**, 69 (2001).
55. X. Li and J. Paldus, *J. Chem. Phys.* **124**, 174101 (2006).
56. X. Li and J. Paldus, *Int. J. Quantum Chem.* **80**, 743 (2000).
57. X. Li and J. Paldus, *J. Chem. Phys.* **125**, 164107 (2006).
58. K. Jankowski and J. Paldus, *Int. J. Quantum Chem.* **18**, 1243 (1980).
59. X. Li and J. Paldus, *Chem. Phys. Lett.* **431**, 179 (2006).
60. X. Li, J. R. Gour, P. Piecuch, and J. Paldus, to be published.
61. X. Li and J. Paldus, *J. Chem. Phys.* **113**, 9966 (2000).
62. X. Li and J. Paldus, *J. Chem. Phys.* **115**, 5774 (2001).
63. A. E. Kondo, P. Piecuch, and J. Paldus, *J. Chem. Phys.* **102**, 6511 (1995).
64. P. Piecuch, A. E. Kondo, V. Špirko, and J. Paldus, *J. Chem. Phys.* **104**, 4699 (1996).
65. X. Li and J. Paldus, to be published.
66. X. Li and J. Paldus, *J. Chem. Phys.* **126**, 234303 (2007).
67. X. Li and J. Paldus, *J. Phys. Chem. A* **111**, 11189 (2007).
68. X. Li and J. Paldus, *J. Chem. Phys.* **126**, 224304 (2007).
69. X. Li and J. Paldus, *Int. J. Quantum Chem.* (in press).
70. X. Li and J. Paldus, *Int. J. Quantum Chem.* **99**, 914 (2004).
71. B. Jeziorski and H. J. Monkhorst, *Phys. Rev. A* **24**, 1668 (1981).
72. J. Paldus and X. Li, *Coll. Czech. Chem. Commun.* **69**, 90 (2004).
73. X. Li and J. Paldus, *J. Chem. Phys.* **119**, 5320, 5346 (2003); J. Paldus, X. Li, and N. D. K. Petraco, *J. Math. Chem.* **35**, 213 (2004).
74. X. Li and J. Paldus, *J. Chem. Phys.* **120**, 5890 (2004); **124**, 034112 (2006); *Mol. Phys.* **104**, 2047 (2006).
75. X. Li and J. Paldus, *J. Chem. Phys.* **119**, 5334 (2003);
76. X. Li, *Coll. Czech. Chem. Commun.* **70**, 755 (2005).
77. X. Li and J. Paldus, *Mol. Phys.* **104**, 661 (2006).

A compact spin-free coupled-cluster theory for open-shell systems

Dipayan Datta* and Debashis Mukherjee*,†

*Department of Physical Chemistry, Indian Association for the Cultivation of Science, Calcutta 700 032, INDIA.
†Raman Centre for Atomic, Molecular and Optical Sciences, Indian Association for the Cultivation of Science

Abstract. We present in this Paper a brief account of a novel spin-free compact coupled cluster (CC) theory for simple open-shell configurations, e.g. doublet and biradicals, which are not necessarily single determinants. A new cluster Ansatz for the wave-operator is introduced, in which the spin-free cluster operators are either of the type of closed-shell-like n hole-n particle excitations or contain valence excitations, which may involve exchange spectator scatterings. The cluster operators with exchange valence spectator scatterings and the pure valence excitation operators are allowed to contract among themselves through the spectator orbitals. The novelty of the Ansatz is in the choice of a suitable automorphic factor accompanying each composite of non-commuting operators, ensuring that each such composite appears only once. This leads to CC equations which terminate exactly at the quartic power of the cluster amplitudes, reminiscent of the closed-shell CC theory. As example applications, we compute the state energies of the ground state of OH and NH_2 radicals with cc-pVDZ basis set and assess the performance of the theory by comparing the results with the benchmark full CI results in the same basis set. The results show the power and the efficacy of the method.

Keywords: coupled-cluster theory, open-shell system
PACS: 31.15.bw

INTRODUCTION

Although the standard single-reference coupled cluster (SRCC) theory [1, 2, 3] has been proved to be very succesful in providing reasonably accurate description of the molecules in their ground state, a generalization of the CC formalisms to open-shell systems leading to an explicitly spin-free correlation model has neither been straightforward nor easy. Several methodologies have emerged during the last two decades to treat states with open-shell or multi-reference character of arbitrary complexity. Most of these methods are based on effective Hamiltonian strategy [4] using reference functions spanning a model space. The more widely used among these are the Valence-Universal Multi-reference Coupled Cluster (VU-MRCC) theories [5, 6, 7], which are widely used to compute energy differences of states with varying degrees of ionization relative to a closed-shell ground state. These formalisms are explicitly spin-free, but require hierarchical generation of cluster amplitudes of various valence ranks. Valence-Specific (VS) (or equivalently State-Universal (SU)) [8] theories, on the other hand, use reference functions spanning a Hilbert space of a fixed number of valence electrons and therefore generate the state energies *per se*, without solving the hierarchy. At a lower level of complexity, there are simple open-shell reference configurations about which a cluster expansion can be effected. Such exercises have been attempted using unrestricted Hartree-Fock (UHF) or restricted open-shell Hartree-Fock (ROHF) determinants as references and defining the cluster operators in terms of spin-orbitals [9, 10, 11]. The major disadvantages of these latter methods are their relatively high computational cost, since the number of cluster amplitudes is roughly three times more in the spin-orbital basis than in the orbital basis and also they lead generally to spin-broken solutions. These methods, however, retain the simplicity and compactness of the single-reference coupled cluster (SRCC) methods and generate CC equations which are at most quartic in cluster amplitudes.

In the symmetry-adapted-cluster (SAC) expansion formalism, developed by Nakatsuji *et. al.* [12], the cluster operator generates only those excitations from the reference, which belong to the same symmetry. The symmetry-adapted CC expansion, therefore, generates a pure spin-state, but the spin-adapted expressions are unwieldy. Janssen and Schaefer [13] first pointed out that utilization of the full flexibility of the spin-space requires the use of spin-free cluster operators, and in addition to excitation operators, one must generally include operators with spectator scatterings involving the indices of the open-shell (valence) orbitals. Unlike in a closed-shell theory or spin-orbital based open-shell theory, these operators are non-commuting and therefore can be joined among themselves in all powers. As a result, the Baker-Hausdorff expansion does not terminate after the quartic term. In their work, Janssen

CP995, *Nuclei and Mesoscopic Physics - WNMP 2007,* edited by P. Danielewicz, P. Piecuch, and V. Zelevinsky

and Schaefer [13] came across terms involving up to eight-fold commutators. An orthogonally spin-adapted open-shell CC formalism based on unitary group approach (UGA) to many-electron correlation problem has been discussed by Li and Paldus [14]. The chief difficulty arising out of the Janssen and Schaefer method [13] is present in this method too and terms with up to eight-fold commutators are present.

Recently, Szalay and Gauss [15] have developed a spin-restricted CC theory which imposes spin constraints on the wave function to reduce the number of independent cluster amplitudes. In this approach, there are a set of spin equations in addition to the usual projected Schrödinger equations. In the spin-restricted approach, both the spin and the amplitude equations are truncated at the same excitation level. The CC wave function is not spin-adapted, but its spin expectation value corresponds to the exact value. More recently, they have extended the spin-restricted approach to formulate a spin-adapted CC theory [16], which requires full inclusion of the spin equations and therefore truncations are allowed only in the projection space and not in the excitation rank.

A spin-free formalism proposed recently from our group [17], uses a cluster Ansatz which is intermediate between a simple exponential and a normal ordered exponential Ansatz. This formalism allows contractions among the various cluster operators via spectator orbitals. The combinatoric factors associated with the contracted composites formed by these operators are taken to be the inverse of the automorphic factors (the number of ways n operators can be contracted among themselves generating the same composite). This ensures that *each distinct excitation* generated by these composites appears *only once* in the wave-operator. Applications to core and valence IP's [18] and electron affinities of neutral radicals [19] demonstrated that the Ansatz leads to CC equations which are not only spin-free but also compact in the sense of having terms which are at most quartic in cluster amplitudes. A valence-specific theory using the same Ansatz has also been developed [20].

In this Paper, we present a succinct account of a spin-free compact cluster expansion formalism for simple open-shell configurations e.g. doublets which can be described either by a single non-singlet determinant or by just one CSF, where the combining coefficients are determined entirely by spin-symmetry. The cluster operators are defined in terms of spin-free generators of the unitary group with respect to a closed-shell vacuum. Pure closed-shell-like n hole-n particle excitation operators without any spectator scatterings involving the open-shell (valence/active) orbitals are included. In our cluster Ansatz, we do not allow these cluster operators to contract with any other operator. This allows us to incorporate a significant portion of the orbital relaxation and correlation effects in a simple way through an ordinary exponential parametrization of these operators. For the cluster operators with 'exchange' spectator scatterings and the other excitation operators, we invoke a combinatoric cluster expansion as in Ref. [20]. The new formalism is much more compact and simpler owing to the absence of a large number of contracted terms involving the operators with 'direct' spectator scatterings. We present here the preliminary results, leaving out the details of the theory and the proof of connectivity for a comprehensive communication to follow [21]. We test the numerical performance of our method by computing the ground state energies of the neutral doublet radicals OH and NH_2 and compairing the results with the benchmark full CI (FCI) results and the results of the theory presented in Ref. [20].

DEVELOPMENT OF A VALENCE-SPECIFIC THEORY FOR OPEN-SHELL STATES

In what follows, we shall illustrate the essence of our formalism by taking as an example the case of a single open-shell reference doublet. Using the terminology of the effective Hamiltonian formalism, we write the Bloch equation for the wave-operator Ω

$$H\Omega P = \Omega P H \Omega P \tag{1}$$

where P is the projector onto the model space. The effective Hamiltonian H_{eff} is defined by the equation

$$PH_{eff}P = PH\Omega P = E_\alpha \tag{2}$$

using intermediate normalization condition $P\Omega P = P$ for Ω. E_α is the energy of the ionized state. In our current development, the model space is spanned by a single spin-adapted reference determinant Φ_α, obtained by deleting one electron in the orbital α from a closed-shell N-electron determinant Φ_0. We emphasize here once again that our theory is valid for an arbitrary configurational state function (CSF) as the reference.

The various components of Ω are obtained by projecting by Q, the projector for the virtual space

$$QH\Omega P = Q\Omega P H \Omega P \tag{3}$$

If we confine ourselves to just the model space spanned by Φ_α and its virtual components, then the above equations refer to a specific Hilbert space and is a VS theory.

We formulate our spin-free CC theory using Φ_0 as vacuum. The cluster operators can then be defined in terms of the spin-free generators of the unitary group with respect to Φ_0 as vacuum. In all our applications we will use the orbitals of Φ_0. This necessitates a compact and efficient modeling of the orbital relaxation effects especially when the closed-shell state described by Φ_0 and the target open-shell state are of widely different nature. Core-ionized states or neutral radicals, for which Φ_0 is the closed-shell anion, are the examples where electron correlation and orbital relaxation effects are both strong. We could, instead, use the orbitals of the ionized states themselves. We want to emphasize at this point that our motivation for the development of a new cluster Ansatz is to use a common set of orbitals for all the core, inner and outer valence-ionized states and avoid orbital optimization for the individual doublet states. Thus, our choice of the ground state Hartree-Fock orbitals (i.e., those of Φ_0) as the starting orbitals is somewhat deliberate.

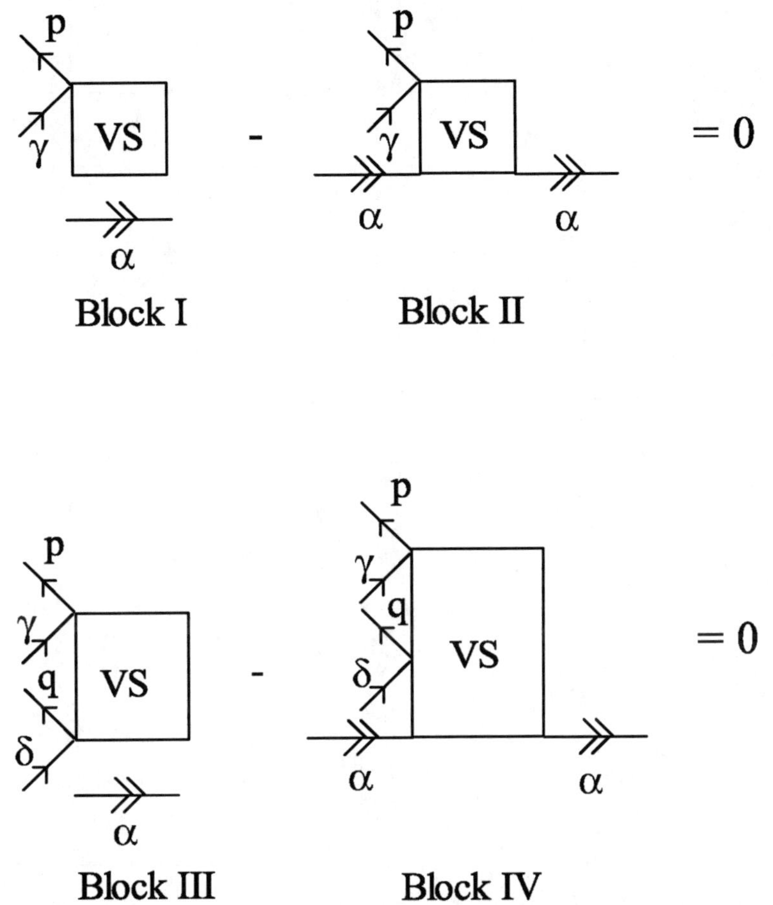

FIGURE 1. Block structure of the equations of zero-valence operators: (a) equation for $\overline{S}_1^{(0,0)}$, (b) equation for $\overline{S}_2^{(0,0)}$

As first pointed out by Janssen and Schaefer [13], spectator scatterings must have to be generally included in the spin-free choice of the cluster operators to exhaust the configurations in the Q space which have the same orbitals but which differ in spin functions. For example, the spin-orbital based single excitation amplitude from a hole γ to a particle p should be replaced by two pseudo two-body operators: one is spin-conserving $\langle p\alpha|s|\gamma\alpha\rangle\{E^{p\alpha}_{\gamma\alpha}\}$ and the other is spin-flipping $\langle p\alpha|s|\alpha\gamma\rangle\{E^{p\alpha}_{\alpha\gamma}\}$, with the curly braces indicating normal ordering with respect to Φ_0, corresponding to the 'direct' and 'exchange' spectator scatterings, respectively. $\{E\}$s are spin-free unitary generators.

Since we are interested in the energy of a specific open-shell state, we would not bring in any ground state correlation operator in Ω, corresponding to n hole-n particle excitations out of Φ_0. Therefore, the various n hole-n particle excitation blocks of a similarity-transformed Hamiltonian are not zero. Let us illustrate this point by taking a concrete example of the equations for the single excitation operators from a hole γ to a particle p. This would consist of the sum of two blocks equated to zero, as shown in Fig. 1a. The minus sign on the terms of Block II relative to Block I arises because of an 'extra' internal hole line in these terms. The single excitation amplitude will be the sum of two terms: $\langle p\alpha|s_2|\gamma\alpha\rangle - \langle p|t_1|\gamma\rangle$, with Φ_0 taken as the vacuum. Since there is only one equation to solve them, these amplitudes are linearly dependent. In order to resolve the linear dependency problem, we define a shifted one-body amplitude

$\langle p|\bar{s}_1^{(0,0)}|\gamma\rangle = \langle p|t_1|\gamma\rangle - \langle p\alpha|s_2|\gamma\alpha\rangle$ to replace the one-body component in the above sum, where $A^{(m,n)}$ represents a Fock space operator with m particle and n hole destructions, respectively. The operator $\bar{s}_\gamma^p\{E_\gamma^p\}$ ($\bar{S}_1^{(0,0)}$) resembles structurally the closed-shell single excitation operator. The VS theory described in Ref. [20], henceforth termed as VSCC, retained the two-body amplitude. Likewise, we define a closed-shell-like 2 hole-2 particle excitation operator $\bar{s}_{\gamma\delta}^{pq}\{E_{\gamma\delta}^{pq}\}$ ($\bar{S}_2^{(0,0)}$) as the only unknown in the equation of the double excitation operator shown in Fig.2.

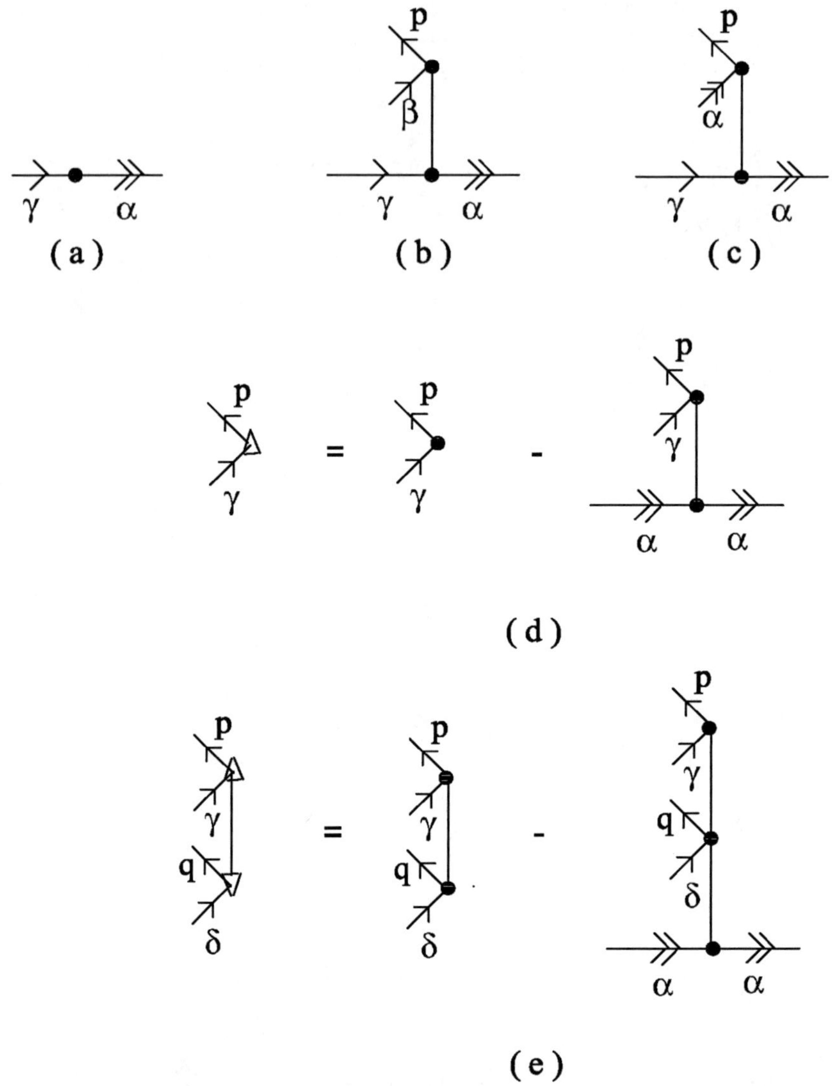

FIGURE 2. Operators for single-reference doublets, $\phi_\alpha = a_\alpha\phi_0$: (a) single excitation, $S_{1e}^{(0,1)}$, (b) double excitation, $S_{2e}^{(0,1)}$, (c) pseudo two-body excitation with exchange spectator scattering, $S_{re}^{(0,1)}$, (d) closed-shell-like one-body ($\bar{S}_1^{(0,0)}$) excitation operator as the sum of one-body and pseudo two-body operator with spectator α, (e) zero-valence two-body ($\bar{S}_2^{(0,0)}$) excitation operator as the sum of two-body and pseudo three-body operator with spectator scattering.

The operators with exchange spectator scattering $\langle p\alpha|s|\alpha\gamma\rangle\{E_{\alpha\gamma}^{p\alpha}\}$ ($S_{re}^{(0,1)}$) of VSCC theory are retained. Along with these operators, the cluster operators with rank up to two will include one-body operators with scattering between the non-valence holes and the valence hole α, $s_\gamma^\alpha\{E_\gamma^\alpha\}$ ($S_{1e}^{(0,1)}$) and true two-body excitations involving the valence hole $s_{\beta\gamma}^{p\alpha}\{E_{\beta\gamma}^{p\alpha}\}$ ($S_{2e}^{(0,1)}$). All these operators are shown in Fig. 2.

Our specific choice of $\bar{S}_1^{(0,0)}$ as discussed above, has removed the operators with direct spectator scatterings from Block II of Fig. 1a. This block would then consist of terms constructed from H, $\bar{S}^{(0,0)}$ and the $S^{(0,1)}$ operators. Similarly,

Block IV would not explicitly include the two hole-two particle excitation operator with the spectator scattering α. The equations for zero-valence and one-valence operators are, therefore, all coupled together and have to be solved simultaneously.

DEVELOPMENT OF A NEW CLUSTER EXPANSION ANSATZ AND EMERGENCE OF THE COMPACT SPIN-FREE CC EQUATIONS

In this section, we give a succinct description of the choice of our new cluster Ansatz and the generation of the spin-free CC equations. The details of the proof will appear in a forthcoming comprehensive paper [21]. As discussed in Sec., we have both $\overline{S}^{(0,0)}$ and $S^{(0,1)}$ operators in our formalism. Use of a normal ordered Ansatz for Ω would include the $S^{(0,1)}$ only up to the linear term, since the action of any power of these operators would try to destroy the single hole vacancy in Φ_α more than once and hence, would annihilate it. The $\overline{S}^{(0,0)}$ operators can still be included to all powers. This is a specific feature of our new formalism. Use of this Ansatz would lead to CC equations which essentially have a terminating structure. However, as we are intending to use the Hartree-Fock orbitals of the closed-shell vacuum Φ_0, a linear parametrization of the $S^{(0,1)}$ operators would not suffice for the states where orbital relaxation effects are severe.

An ordinary exponential Ansatz, on the other hand, would allow all the cluster operators to contract among themselves in all possible manner and all powers and thus would lead to a non-terminating structure for the CC equations. Our motivation for the present work is to provide an adequate description of the orbital relaxation effects in a compact manner through a Thouless-like exponential parametrization of the relaxation-inducing cluster operators. Thus, we allow the $S_{re}^{(0,1)}$ operators to contract among themselves via their spectator lines, in addition, we should also allow the $S_{1e}^{(0,1)}$ and $S_{2e}^{(0,1)}$ operators (henceforth, they are collectively called as $S_{e}^{(0,1)}$) to contract with the $S_{re}^{(0,1)}$ operators via the spectator lines in $S_{re}^{(0,1)}$. *Unlike in an ordinary exponential Ansatz, we do not allow the $\overline{S}^{(0,0)}$ operators to contract with any other operator.*

If we substitute this Ansatz for Ω in Eq. (1) and write both sides of this equation in normal order with respect to Φ_0 as vacuum, we will have all the $\overline{S}^{(0,0)}$ operators connected to H. In addition, there will be a number of $S^{(0,1)}$ operators joined to H through at least one non-spectator line. We call such composites as 'strongly connected' [18] composites. Also, there will be a set of $S^{(0,1)}$ operators contracted among themselves in a chain-like fashion. These operators may either be connected to the operators which are strongly joined to H or they may have connection to H via spectator lines only. These operators are said to be 'weakly connected' [18]. We denote them by X_i, again i distinguishing the various terms. On the right-hand side of the Bloch equation, $\{\widehat{\Omega H_{eff}}\}P$, there will be a set of $S_{e}^{(0,1)}$ operators contracted directly to H in H_{eff} from the left through the valence line α. They are also strongly connected. We denote them by $\{(S_{e}^{(0,1)})\}_h$. Rest of the $S_{e}^{(0,1)}$ will be weakly joined from the left to $S_{re}^{(0,1)}$ operators either in Ω or in H_{eff}.

It might appear that this Ansatz, while incorporating the orbital relaxation effects to all orders, would lead to a non-terminating series structure for the CC amplitude equations,since we are allowing certain S operators to contract through the spectator orbitals. However, if one insists that *each composite, obtained upon contractions of $S^{(0,1)}$ operators, should appear only once in the wave-operator*, then the factors associated with these contracted composites would not be the same as those in an ordinary exponential Ansatz. The factor associated with a composite consisting of k $S_{e}^{(0,1)}$ operators and l $S_{re}^{(0,1)}$ operators would be $(1/k!)(1/f_l)$, where f_l is the number of distinct ways of joining l $S_{re}^{(0,1)}$ operators leading to the same excitation. In combinatoric language, f_l is the the *automorphic* factor for a composite containing l non-commuting operators. Fig. 3 illustrates the choice of the combinatoric factors for contracted composites formed by $S_{1e}^{(0,1)}$ and $S_{re}^{(0,1)}$ operators. Such a choice of the combinatoric factors leads to complete factorization of the strong and weakly connected terms, contrary to a pure exponential Ansatz. We will discuss the implications of choosing these factors in details in the forthcoming paper [21]. This combinatoric series is symbolically represented as $\{\{\exp(S)\}\}$. It differs from a pure exponential in two ways: (a) it allows contractions among the $S^{(0,1)}$ operators only via the spectator orbitals and (b) the factors are not $1/n!$ for S^n. We will call this formalism as the *'Combinatoric Open-shell Coupled Cluster'* (COS-CC) formalism.

The new Ansatz for the wave-operator Ω can be compactly written as

$$\Omega = \{\{\exp(\overline{S}^{(0,0)} + S^{(0,1)})\}\} \tag{4}$$

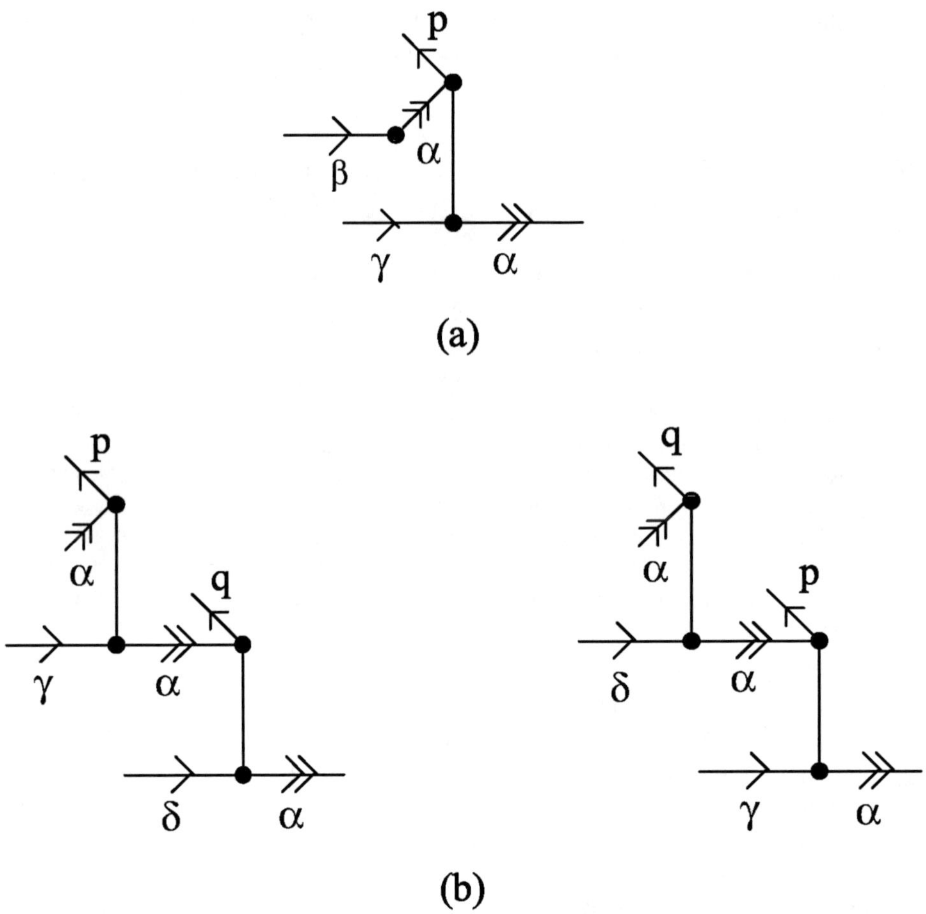

FIGURE 3. Contractions of one-valence cluster operators. (a) $S_{1e}^{(0,1)} - S_{re}^{(0,1)}$ contraction. The $S_{1e}^{(0,1)}$ operator can contract with $S_{re}^{(0,1)}$ only from the left. The product excitation, therefore, should appear only once with a factor of 1. (b) Non-commuting exchange mode contraction of two $S_{re}^{(0,1)}$ operators. The $S_{re}^{(0,1)}$ operators in the contracted composites lead to different excitations depending on their order in the composite. Each composite should have a factor of 1.

Since the $\overline{S}^{(0,0)}$ operators cannot be contracted among themselves or with any other operator, we do not need a combinatoric cluster expansion for them. The wave-operator, therefore, factorizes as

$$\Omega = \exp(\overline{S}^{(0,0)})\{\{\exp(S^{(0,1)})\}\} \tag{5}$$

Now we discuss in brief, how the new cluster Ansatz leads to a terminating series structure for the CC amplitude equations. The strongly connected composites have one set of terms Z which are 'external' or 'open' in the sense of inducing excitations out of the model space and another set W which is 'closed'. Only W will contribute to H_{eff}. The expressions for Z and W can be compactly written as

$$Z = [\overbrace{H\exp(\overline{S}^{(0,0)})\{\{\exp(S^{(0,1)})\}\}}]_{strong,open}, W = [\overbrace{H\exp(\overline{S}^{(0,0)})\{\{\exp(S^{(0,1)})\}\}}]_{strong,closed} \tag{6}$$

Unlike in the ordinary exponential Ansatz, in the COS-CC formalism, the strongly and weakly connected composites in each term have only a *local* automorphic factor. This stems from the fact that in our new cluster Ansatz, the combinatoric factors accompanying a composite of cluster operators is decided not by the number of the cluster operators in the composite, they are rather decided by the number of ways the cluster operators can be contracted among themselves generating the same excitation, respecting the non-commuting nature of the cluster operators. Weakly connected composites of the same structure, therefore, appear on either side of the Bloch equation with the same combinatoric factor. This leads to complete factorization of the weakly connected terms. Using the above definitions of Z, W and X, we can write the Bloch equation as

$$[(\Sigma_i(Z_i+W_i))(\Sigma_j X_j)]_w = [(\Sigma_j X_j)\overbrace{\{\{\exp(S_e^{(0,1)})\}_h(\Sigma_i W_i)\}_{strong}}]_w \tag{7}$$

where $[(A)(B)]_w$ indicates that A and B are joined weakly, i, j distinguishing the various terms. The various weakly joined composites X_i can again be grouped into $\{\{\exp(S^{(0,1)}\}\}$. The use of appropriate automorphic factors therefore leads to cancellation of the weakly connected terms from the Bloch equation. Because of the strong connectivity of all the terms in the resulting factored out equations, the termination at the quartic power of the cluster amplitudes is automatic and natural.

Cancelling the weakly connected terms X_i, we get from the Q projection of Eq. (7)

$$Z - [\{\exp(S_e^{(0,1)})\}_h W]_{ex} = 0 \tag{8}$$

The suffix 'ex' denotes the external operators which lead to excitations out of the model space. Eq. (8) represents the spin-free CC equations for the cluster amplitudes. The subscript h in Eq. (8) indicates that $S_e^{(0,1)}$ operators in $\{\exp(S_e^{(0,1)})\}_h$ are joined only to H from the left and not to any $S^{(0,1)}$ operators in H_{eff}. So these operators also are strongly joined.

We will now discuss the essentials of the construction of working equations Eq. (8) by applying it to one-valence problems, viz. to a single-reference open-shell doublet Φ_α. The various sets of equations can be formed by constructing the strongly connected composites Z and W from H, $\overline{S}^{(0,0)}$ and $S^{(0,1)}$ vertices, and then connecting W from the left with the $S_{1e}^{(0,1)}$ or $S_{2e}^{(0,1)}$ operators via the H vertices only. The CC equations for the $S^{(0,1)}$ operators are compactly written as

$$s_\gamma^\alpha\{E_\gamma^\alpha\} : Z_\gamma^\alpha + [(S_{1e}^{(0,1)})_\gamma^\alpha]_h W_\alpha^\alpha = 0 \tag{9}$$

$$s_{\beta\gamma}^{p\alpha}\{E_{\beta\gamma}^{p\alpha}\} : Z_{\beta\gamma}^{p\alpha} + [(S_{1e}^{(0,1)})_\gamma^\alpha]_h Z_{\beta\alpha}^{p\alpha} + [(S_{1e}^{(0,1)})_\beta^\alpha]_h Z_{\alpha\gamma}^{p\alpha} + [\{(S_{1e}^{(0,1)})_\gamma^\alpha (S_{1e}^{(0,1)})_\beta^\alpha\}]_h Z_{\alpha\alpha}^{p\alpha} + [(S_{2e}^{(0,1)})_{\beta\gamma}^{p\alpha}]_h W_\alpha^\alpha = 0 \tag{10}$$

$$s_{\alpha\gamma}^{p\alpha}\{E_{\alpha\gamma}^{p\alpha}\} : Z_{\alpha\gamma}^{p\alpha} + [(S_{1e}^{(0,1)})_\gamma^\alpha]_h Z_{\alpha\alpha}^{p\alpha} = 0 \tag{11}$$

The positive signs in the $[S_e^{(0,1)}]_h Z$ terms originate from two (-1) factors, one comes from $\{\exp(-S_e^{(0,1)})_h\}$ of equation Eq. (8) and the other comes from one internal hole line joining $S_e^{(0,1)}$ and Z.

Following the arguments of the previous section and introducing the definition $Y = \{\overbrace{H\overline{S}^{(0,0)}}\}$, the equations for the $\overline{S}^{(0,0)}$ operators are written as

$$\overline{s}_\gamma^p\{E_\gamma^p\} : f_\gamma^p - V_{\gamma\alpha}^{p\alpha} + Y_\gamma^p - Y_{\gamma\alpha}^{p\alpha} - Z_{\gamma\alpha}^{p\alpha} - [(S_{1e}^{(0,1)})_\gamma^\alpha]_h Z_{\alpha\alpha}^{p\alpha} = 0 \tag{12}$$

$$\overline{s}_{\gamma\delta}^{pq}\{E_{\gamma\delta}^{pq}\} : V_{\gamma\delta}^{pq} + Y_{\gamma\delta}^{pq} - Y_{\gamma\delta\alpha}^{pq\alpha} - Z_{\gamma\delta\alpha}^{pq\alpha} - [(S_{1e}^{(0,1)})_\gamma^\alpha]_h Z_{\alpha\delta\alpha}^{pq\alpha} + [(S_{2e}^{(0,1)})_{\gamma\delta}^{p\alpha}]_h Z_{\alpha\alpha}^{q\alpha} = 0 \tag{13}$$

The cluster operators $\overline{s}_{\alpha\alpha}^{pq}\{E_{\alpha\alpha}^{pq}\}$ which correspond to double occupancy of the valence orbital α are not present. Not only these operators, the two-body blocks in Fig. 2 with both the holes labeled by α are not present as these blocks do not have any hole destruction. The three-body blocks with all the three holes carrying the label α on the right are Pauli principle violating and hence, are absent.

The idea that one may generate open-shell CC equations for state energies *per se* by considering only operators for the open-shell state itself and clumping together in the CC equations blocks of various valence ranks leading to the same excitation with different number of direct spectator excitations, has been considered earlier by Mukherjee and Zaitsevskii [22] using a normal ordered wave-operator. The present development is built on a similar idea, but it contains the orbital relaxation effects to all order.

An attractive feature of the new formalism is that it allows the effect of higher-body operators to be included without explicitly taking them in the cluster Ansatz. As for example, inclusion of two hole-two particle excitation operators with direct spectator scatterings, which bring in the dominant open-shell correlation effects, would be required if one wishes to work with operators with direct spectators. This is a genuine three-body operator. The valence-specific theory described in Ref. [20] used singles-doubles truncation scheme and therefore did not include this three-body operator, instead it was approximated by the closed-shell two-body excitation operator without any reference to the spectator, i.e., the same operator was used for both the closed-shell state and the ionized state. This is somewhat awkward. The present formulation, on the other hand, replaces the three-body operator by an effective two-body excitation operator without spectator excitation, as discussed above.

NUMERICAL APPLICATIONS

The efficacy and performance of our method is tested by computing the ground state energy of the OH and NH_2 radicals using the spherical cc-pVDZ basis. We have performed our calculations at the equilibrium geometries optimized by the FCI method presented in Ref. [23]. The O-H distance in the OH radical is taken to be 0.97953 Å and the geomertical paramters for the NH_2 radical are: the N-H distance 1.04069 Å and $\langle(HNH)$=100.732 degrees. We carry out our calculations using the Hartree-Fock orbitals of the closed-shell OH^- and NH_2^- anions. This is deliberately done, so that we can assess the full import of the orbital relaxation accompanying ionization from an anionic species. We compare our results with the benchmark full CI results and the results of the VSCC method of Ref. [20]. In order asses the efficacy of the COS-CC Ansatz in incorporating the orbital relaxation effects to all orders, we compare our results with the Valence-specific method using the traditional normal ordered cluster Ansatz. The results are shown in Table I.

TABLE 1. Ground-state energy of neutral doublet radicals (in Hartree) using cc-pVDZ basis

OH radical		
	COS-CC	-75.560034
	VSCC	-75.558111
	VS theory using $\Omega = \{\exp(S)\}$	-75.566592
	FCI	-75.561655
NH_2 radical		
	COS-CC	-55.734031
	VSCC	-55.730051
	VS theory using $\Omega = \{\exp(S)\}$	-55.737928
	FCI	-55.735654

[a] Reference [23]

The results clearly indicate that the COS-CC energies are much closer to the FCI energies compared to the VSCC energies. The comparison with the valence-specific method using a traditional normal ordered exponential Ansatz indicates the superiority of our Ansatz in providing a balanced description of the orbital relaxation and correlation effects.

SUMMARY

In this Paper , we have presented the formulation and numerical applications of a compact spin-free coupled cluster theory for simple open-shell configurations e.g. single-reference doublets. In our formalism, the n hole-n particle excitation operators with the diagonal spectators are replaced by closed-shell-like one- and two-body excitation operators, although the equations determining these cluster amplitudes are quite different from the SRCC amplitude equations. Such a choice of the cluster operators reduces the number of contracted composites of operators and simplifies the theory greatly. For the operators with exchange spectator scattering and other valence cluster operators, we use a combinatoric cluster expansion with the factor associated with a composite of non-commuting operators being the inverse of the automorphic factor f_n, which is equal to the number of ways these operators can be joined together to generate the same excitation. Unlike the VSCC method, the COS-CC method incorporates the effect of three-body correlation operators without explicitly including them in the cluster Ansatz. Comparison with the FCI results reveals that the COS-CC method performs consistently much better than the VSCC theory. Also, the applications clearly show the efficiency of our COS-CC Ansatz in providing an adequate description of the orbital relaxation effects.

ACKNOWLEDGEMENT

DD acknowledges the Council of Scientific and Industrial Research (CSIR), New Delhi, India for his research fellowship. DM thanks the Department of Science and Technology for awarding him the J. C. Bose Fellowship. DM also thanks Jawaharlal Nehru Centre for Advanced Scientific Research, Bangalore, INDIA for conferring on him the Honorary Professorship.

REFERENCES

1. J. Cizek, J. Chem. Phys. 45 (1966) 4256; J. Cizek, Adv. Chem. Phys. 14 (1969) 35.
2. R. J. Bartlett, in: D. R. Yarkony (Ed.), Modern electronic structure theory, Advanced Series in Physical Chemistry, vol. 2, World Scientific Publishing, Singapore, 1995.
3. T. D. Crawford, H. F. Schaefer III, in: K. B. Lipkowitz, D. B. Boyd (Eds.), Reviews in Computational Chemistry, vol. 14, Wiley-VCH, John Wiley and Sons. Inc, New York, 2000.
4. J. P. Malrieu, Ph. Durand, J. P. Daudey, J. Phys. A 18 (1985) 809; Ph. Durand, J. P. Malrieu, Adv. Chem. Phys. 67 (1987) 321.
5. For an extensive survey see D. Mukherjee, S. Pal, Adv. Quantum Chem. 20 (1989) 291 and references therein.
6. I. Lindgren, Int. J. Quantum Chem. S12 (1978) 33; I. Lindgren, D. Mukherjee, Phys. Rep. 151 (1987) 93.
7. A. Haque, U Kaldor, Chem. Phys. Lett. 120 (1989) 261.
8. B. Jeziorski, H. J. Monkhorst, Phys. Rev. A 24 (1981) 1668; L. Stolarczyk, H. J. Monkhorst, Phys. Rev. A 32 (1985) 743; B. Jeziorski, J. Paldus, J. Chem. Phys. 88 (1988) 5673.
9. M. Rittby, R. J. Bartlett, J. Phys. Chem. 92 (1988) 3033.
10. G. E. Scuseria, Chem. Phys. Lett. 176 (1991) 27.
11. T. J. Lee, D. Jayatilaka, Chem. Phys. Lett. 201 (1993) 1; D. Jayatilaka, T. J. Lee, J. Chem. Phys. 98 (1993) 9734.
12. H. Nakatsuji, K. Hirao, J. Chem. Phys. 68 (1978) 2053 ; H. Nakatsuji, Chem. Phys. Lett. 59 (1978) 362; 67 (1979) 329.
13. C. L. Janssen, H. F. Schaefer III, Theor. Chim. Acta 79 (1991) 1.
14. X. Li, J. Paldus, J. Chem. Phys. 101 (1994) 8812; X. Li, J. Paldus, in: R. J. Bartlett (Ed.), Modern Ideas in Coupled Cluster Methods, vol. 183, World Scientific, Singapore, 1997.
15. P. G. Szalay, J. Gauss, J. Chem. Phys. 107 (1997) 9028.
16. M. Heckert, O. Heun, J. Gauss, P. G. Szalay, J. Gauss J. Chem. Phys. 124 (2006) 124105.
17. D. Jana, B. Bandyopadhyay, D. Mukherjee, Theo. Chem. Acc. 102 (1999) 317; D. Jana, U. S. Mahapatra, D. Mukherjee, Int. J. Mol. Sci. 3 (2002) 550.
18. D. Jana, D. Mukherjee, J. Chem. Phys. 122 (2005) 234101.
19. D. Jana, D. Datta, D. Mukherjee, Chem. Phys. 329 (2006) 290.
20. D. Jana, U. S. Mahapatra, D. Mukherjee, Chem. Phys. Lett. 353 (2002) 100.
21. D. Datta, D. Mukherjee (in preparation).
22. D. Mukherjee, A. Zaitsevskii, Chem. Phys. Lett. 233 (1995) 605.
23. M. Kállay, J. Gauss, P. G. Szalay, J. Chem. Phys. 119 (2003) 2991.

Extrapolating Potential Energy Surfaces by Scaling Electron Correlation: Isomerization of Bicyclobutane to Butadiene

Jesse J. Lutz* and Piotr Piecuch*,†

*Department of Chemistry, Michigan State University, East Lansing, MI 48824, USA
†Department of Physics and Astronomy, Michigan State University, East Lansing, MI 48824, USA

Abstract. The recently proposed potential energy surface (PES) extrapolation scheme, which predicts smooth molecular PESs corresponding to larger basis sets from the relatively inexpensive calculations using smaller basis sets by scaling electron correlation energies [A.J.C. Varandas and P. Piecuch, Chem. Phys. Lett. **430**, 448 (2006)], is applied to the PESs associated with the conrotatory and disrotatory isomerization pathways of bicyclo[1.1.0]butane to buta-1,3-diene. The relevant electronic structure calculations are performed using the completely renormalized coupled-cluster method with singly and doubly excited clusters, and a non-iterative treatment of connected triply excited clusters, termed CR-CC(2,3). A comparison with the explicit CR-CC(2,3) calculations using the large correlation-consistent basis set of the cc-pVQZ quality shows that the cc-pVQZ PESs obtained by the extrapolation from the smaller basis set calculations employing the cc-pVDZ and cc-pVTZ basis sets are practically identical, to within fractions of a millihartree, to the true cc-pVQZ PESs. It is also demonstrated that one can use a similar extrapolation procedure to accurately predict the complete basis set (CBS) limits of the calculated PESs from the results of smaller basis set calculations at a fraction of the effort required by the conventional point-wise CBS extrapolations.

Keywords: Potential energy surfaces, basis sets, electron correlation effects, coupled-cluster methods, complete basis set limit
PACS: 31.10.+z,31.15.bw,31.15.vq,31.50.Bc,82.20.Kh,82.30.Qt

INTRODUCTION

The conventional procedure for generating molecular potential energy surfaces (PESs) or chemical reaction pathways using high accuracy *ab initio* methods of quantum chemistry, in which one is required to solve the electronic Schrödinger equation point-by-point, is very tedious, particularly for larger polyatomic molecules. In many cases, obtaining even a one-dimensional PES cut corresponding, for example, to a minimum reaction pathway, becomes computationally impractical using anything but small basis sets. This is due to the fact that computer costs (meaning, the CPU operation count) of the accurate *ab initio* electronic structure methods that are appropriate for calculations of molecular PESs scale at least as n^4 with the number n of unoccupied orbitals in a basis set. The cost of generating PESs also grows as m^{τ}, where m is a typical number of points (nuclear geometries) required to represent a one-dimensional PES cut (typically, m is on the order 10) and τ is the PES dimension (for the one-dimensional PES cuts along the relevant intrinsic reaction coordinates, $\tau = 1$; for the global, multi-dimensional PESs, $\tau = 3N - 6$, where N is the number of nuclei or atoms in a molecular system of interest). The total cost of generating a PES for a given molecular system is then at least $m^{\tau}n^4$. This being the case, the traditional point-by-point calculations of accurate PESs are usually not practical, particularly when polyatomic species are involved.

To overcome this difficulty, an *ab initio* scheme has been proposed that predicts the PES corresponding to a larger basis set from the results of smaller basis set calculations by scaling electron correlation energies [1]. In this approach, a universal correlation energy scaling factor is determined at a single nuclear geometry, called a reference geometry, over a series of systematically constructed basis sets. This scaling factor is then applied to the electron correlation energies of the remaining geometries defining the PES of interest obtained with smaller basis sets in order to predict the PES corresponding to the larger target basis set, offering huge savings in the computer effort.

So far, the PES extrapolation scheme developed in [1] has been tested on small molecular examples involving two or three atoms. The major goal of the present study is to assess the accuracy and consistency of the extrapolation scheme developed in [1] in predicting molecular PESs corresponding to larger basis sets for more complex polyatomic systems as the inherent flexibilities of the scheme are exercised.

Another goal of this work is to develop a cost-effective computational procedure which would allow one to obtain a smooth molecular PES corresponding to the complete basis set (CBS) limit, i.e., without performing the laborious CBS extrapolations at each point on the PES of interest. The existing CBS extrapolation methods, such as those described in [2–5], employing the correlation consistent (cc) basis sets [6, 7], abbreviated as cc-pVXZ where X is the

CP995, *Nuclei and Mesoscopic Physics - WNMP 2007*, edited by P. Danielewicz, P. Piecuch, and V. Zelevinsky

cardinal number of a basis set, require that one performs a sequence of electronic structure calculations for the cc-pVXZ basis sets of varying size, such as, for example, cc-pVDZ ($X=2$) and cc-pVTZ ($X=3$) or, better, cc-pVTZ and cc-pVQZ ($X=4$), or their augmented analogs, at each point on the PES. One of the existing empirical laws defining the dependence of the electron correlation energy ΔE on X, such as [2]

$$\Delta E_X(\mathbf{R}) = \Delta E_{CBS}(\mathbf{R}) + A(\mathbf{R})X^{-3}, \tag{1}$$

where $\Delta E_X(\mathbf{R})$ is the correlation energy obtained with the cc-pVXZ basis set at the nuclear geometry $\mathbf{R}$ and $\Delta E_{CBS}(\mathbf{R})$ and $A(\mathbf{R})$ are the parameters determined from fitting $\Delta E_X(\mathbf{R})$ to the calculated correlation energies, is then used to determine the CBS value of the correlation energy $\Delta E_{CBS}(\mathbf{R})$. Since the dependence of the correlation energy $\Delta E_X(\mathbf{R})$ on the cardinal number X is geometry-dependent (as indicated by the presence of ($\mathbf{R}$), representing the relevant nuclear coordinates, in Eq. (1)), the CBS extrapolation of a molecular PES must be repeatedly carried out for each nuclear geometry, resulting in a PES which is no longer a smooth function of the nuclear coordinates $\mathbf{R}$ that define it due to a numerical noise introduced by point-wise extrapolations. Moreover, the need to carry out the sequence of electronic structure calculations using basis sets of varying size to perform the CBS extrapolations at each point on the PES makes the point-wise CBS extrapolation procedure too expensive for the majority of interesting chemical applications involving polyatomic systems, where the cost of a single-point calculation is no longer trivial. The procedure developed in this work eliminates or considerably reduces these problems by determining the ΔE_{CBS} value at only one (reference) geometry, using one of the CBS extrapolation laws, such as that given by Eq. (1), and by scaling the correlation energies at the remaining geometries defining the PES of interest obtained with smaller basis sets to predict the ΔE_{CBS} values at the remaining points.

The pericyclic rearrangement of bicyclo[1.1.0]butane (abbreviated as **bicbut**) to *trans*-buta-1,3-diene (abbreviated as **t-but**) was chosen as the prototypical polyatomic molecular system to test the accuracy and effectiveness of the PES extrapolation schemes examined in this work. We chose this system due to the intrinsic complexity of the associated isomerization pathways, which involve polyatomic structures with a rapidly varying degree of biradical character that require an accurate and balanced description of the dynamical and non-dynamical correlation effects, and due to the size of the system, which is not too large, enabling the explicit highly correlated calculations with larger basis sets of the cc-pVQZ quality that are needed to benchmark the results of the PES extrapolation scheme of [1] tested in this work. Let us recall that early experimental studies suggest that the **bicbut→t-but** isomerization proceeds by concerted conrotatory movement of the methylene groups [8, 9], as predicted by Woodward-Hoffman symmetry rules. Computational studies confirm this, predicting, in addition, that near the end the reaction pathway passes through the *gauche*-buta-1,3-diene configuration (abbreviated as **g-but**) before reaching the final **t-but** configuration [10, 11]. Theoretical studies have also considered the concerted disrotatory [10–12] and non-concerted [13] pathways, finding the concerted disrotatory transition state (TS) to be ~ 20 kcal/mol higher in energy than the conrotatory TS and the non-concerted pathway to be much too high in energy to be even considered as a plausible mechanism. The conrotatory TS was found to have a 24 % biradical character while the disrotatory TS was found to have a 90 % biradical character, according to the high level electronic structure calculations reported in [11]. The current consensus on the mechanism for the isomerization of bicyclo[1.1.0]butane into buta-1,3-diene is that both concerted pathways begin at **bicbut** and, after passing through the corresponding conrotatory or disrotatory TS (**con_TS** and **dis_TS**, respectively), they converge at the local minima defining the intermediate **g-but** configuration. The **g-but** intermediate isomerizes via a low-energy rotational barrier, defined by the TS structure labeled as **gt_TS**, before the final product, **t-but**, is reached. The conrotatory and disrotatory concerted pathways describing the isomerization of bicyclo[1.1.0]butane into buta-1,3-diene, along with their available experimental activation and reaction enthalpies [14] and the theoretical enthalpy values obtained in this work, are illustrated in Figure 1.

In theoretical studies of chemical reaction pathways, particularly those involving a mixture of closed-shell and biradical structures, one must be careful about the method used in electronic structure calculations. The conventional single-reference methods based on coupled-cluster (CC) theory [15–19], including the widely used CC approach with singles, doubles, and a quasi-perturbative non-iterative treatment of triples, abbreviated as CCSD(T) [20], are very accurate in describing closed-shell structures, for which the electron correlation effects are dominated by dynamical correlations, but fail in the biradical and bond breaking regions of molecular PESs which require an accurate and balanced treatment of both dynamical and non-dynamical correlation effects [21–24]. On the other hand, the least expensive and most frequently used multi-reference methods, such as the complete-active-space self-consistent-field (CASSCF) approach and methods that correct CASSCF results for the leading dynamical correlation effects through the second-order many-body perturbation theory (MBPT) [25], are reasonably accurate in the biradical and bond breaking regions, but relatively inaccurate in the closed-shell regions, leading to relatively poor reaction path energetics

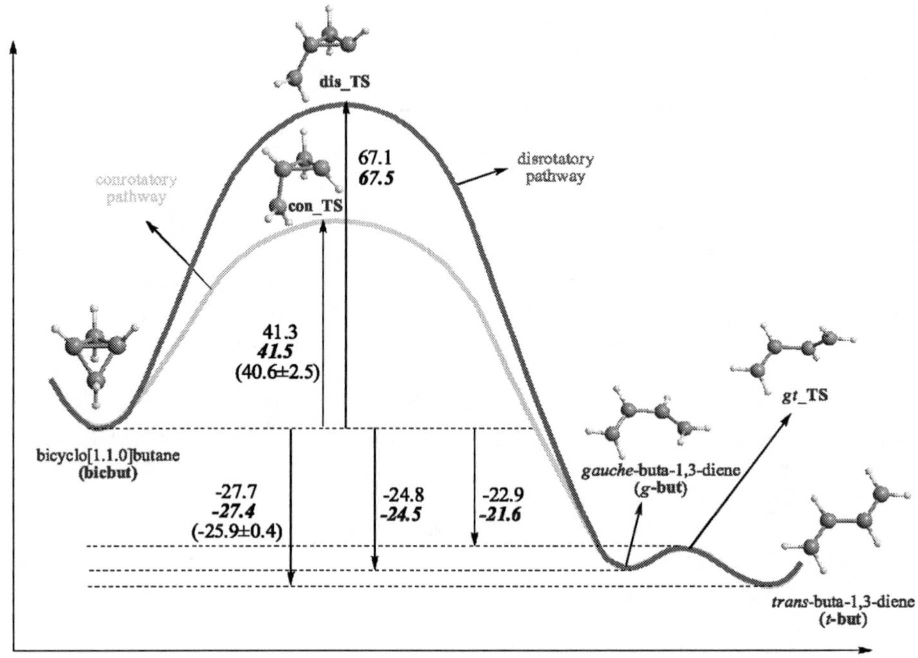

FIGURE 1. The conrotatory and disrotatory pathways describing the isomerization of bicyclo[1.1.0]butane to *trans*-buta-1,3-diene, along with the enthalpy values relative to the reactant at all stationary points obtained in the explicit CR-CC(2,3)/cc-pVQZ/CASSCF(10,10)/cc-pVDZ calculations (numbers in roman) and the enthalpy values obtained with the PES extrapolation procedure using the CR-CC(2,3)/cc-pVDZ, CR-CC(2,3)/cc-pVTZ, and RHF/cc-pVQZ energies at all stationary points resulting from the CASSCF(10,10)/cc-pVDZ optimizations, and the CR-CC(2,3)/cc-pVQZ correlation energy at the reactant geometry (numbers in bold italics). The available experimental enthalpy values are in parentheses. All enthalpies are in kcal/mol.

when both closed-shell and biradical structures are present along the reaction coordinate. This is exactly what happens when various methods are applied to the conrotatory and disrotatory pathways describing the isomerization of bicyclo[1.1.0]butane to *trans*-buta-1,3-diene. As shown in [11], the CCSD(T) approach fails by placing the disrotatory pathway defined by the strongly biradical TS, which we call **dis_TS**, about 20 kcal/mol below the conrotatory pathway, contradicting experiment which states that the conrotatory pathway represents a true mechanism. At the same time, the CASSCF and second-order multi-reference MBPT approaches, which correctly place the conrotatory pathway below the disrotatory one, provide the relatively poor energetics when compared to the available experimental and more accurate electronic structure data [11], although CASSCF produces reasonable geometries of the corresponding stationary points.

As shown in [11], the electronic structure method that offers the well-balanced and highly accurate description of both isomerization pathways is the recently developed rigorously size extensive completely renormalized (CR) CC approach with singly and doubly excited clusters, and a non-iterative treatment of connected triply excited clusters, termed CR-CC(2,3) [26, 27], which is known to provide a highly accurate description of chemical reaction profiles involving closed-shell and biradical TSs and intermediates, and single bond stretching or breaking [11, 26–33]. In analogy to all CR-CC approaches and other techniques based on the asymmetric energy expressions defining the method of moments of CC equations (cf., e.g., [21–23, 33] for selected reviews), which is the underlying formalism that leads to CR-CC schemes, the basic idea of the CR-CC(2,3) approach is to correct the results of the CCSD (CC singles and doubles) [34–36] calculations for the effects of higher-order correlations via non-iterative corrections added to the CCSD energy. In the specific case of the CR-CC(2,3) approach, the CCSD energy is corrected for the effects of connected triply excited clusters. The CR-CC(2,3) approach eliminates the failures of the traditional single-reference CC methods, such as CCSD(T), in the bond breaking/biradical regions of molecular PESs, while preserving the ease-of-use ("black-box" character) and the relatively low computer costs of the CCSD(T) approximation. As already mentioned, the CR-CC(2,3) approach provides an accurate and balanced description of both the conrotatory and disrotatory pathways describing the isomerization of bicyclo[1.1.0]butane to *trans*-buta-1,3-diene, with a computational effort similar to the conventional CCSD(T) calculations, making it the method of choice for the present study. In this work, we use the CR-CC(2,3) approach of [26, 27] to test the effectiveness of the PES extrapolation procedure developed in [1] and its extension proposed in this work, which enables one to predict smooth PESs corresponding to the CBS limit from the results of smaller basis set calculations, by applying it to the PESs associated with the conrotatory and disrotatory isomerization pathways of bicyclo[1.1.0]butane to buta-1,3-diene.

THEORY AND COMPUTATIONAL DETAILS

In the following description, we focus on the PES extrapolation procedures employing the ideas laid down in [1]. For the details of the CR-CC(2,3) approach, which is used to perform the electronic structure calculations reported in this study, and the equations defining the CR-CC(2,3) energy, we refer the reader to [26, 27]. Here, we only mention that the CR-CC(2,3) energy, $E^{\text{CR-CC(2,3)}}$, is a sum of the independent-particle-model reference energy, in our case, the energy obtained in the restricted Hartree-Fock (RHF) calculations, E^{RHF}, and the correlation energy, $\Delta E^{\text{CR-CC(2,3)}}$, which itself splits into the CCSD part, ΔE^{CCSD}, and the triples correction part, $\delta E^{(2,3)}$, so that

$$E^{\text{CR-CC(2,3)}} = E^{\text{RHF}} + \Delta E^{\text{CR-CC(2,3)}}, \tag{2}$$

where

$$\Delta E^{\text{CR-CC(2,3)}} = \Delta E^{\text{CCSD}} + \delta E^{(2,3)}. \tag{3}$$

We can also write,

$$E^{\text{CR-CC(2,3)}} = E^{\text{CCSD}} + \delta E^{(2,3)}, \tag{4}$$

where

$$E^{\text{CCSD}} = E^{\text{RHF}} + \Delta E^{\text{CCSD}} \tag{5}$$

is the total CCSD energy. Both ways of decomposing the total CR-CC(2,3) energy, $E^{\text{CR-CC(2,3)}}$, given by Eqs. (2) and (4), which use, respectively, the RHF and CCSD energies as reference energies for defining the PES extrapolations, are tested in this work.

The PES Extrapolation Approach Employing Correlation Energy Scaling

In general, the PES extrapolation scheme proposed in [1] focuses on scaling the difference ΔE between the total electronic energy, E, and the energy of some reference wave function, E^{ref}, which represents the correlation energy or part of it. In the specific case of the PES calculations performed in this study, where the total electronic energies are obtained using the CR-CC(2,3) approach, we consider two alternative definitions of the reference energy E^{ref}, namely, $E^{\text{ref}} = E^{\text{RHF}}$ and $E^{\text{ref}} = E^{\text{CCSD}}$. In the former case, where we decompose the CR-CC(2,3) energy using Eq. (2), ΔE represents the total correlation energy of the CR-CC(2,3) approach, $\Delta E^{\text{CR-CC(2,3)}}$, defined by Eq. (3). In the latter case, where we decompose the CR-CC(2,3) energy using Eq. (4), ΔE is identified with the part of the CR-CC(2,3) correlation energy that defines the triples correction $\delta E^{(2,3)}$.

Suppose the correlated PES calculations are performed using basis sets indexed by numbers $m-1$ and m. In the case of the CR-CC(2,3) calculations reported in this work, where we used the cc-pVXZ basis sets of Dunning, index m is identified with the cardinal number X of a given cc-pVXZ basis ($m=2$ for cc-pVDZ, $m=3$ for cc-pVTZ, and $m=4$ for cc-pVQZ). Let us designate the resulting PESs obtained for basis sets $m-1$ and m as $E_{m-1}(\mathbf{R})$ and $E_m(\mathbf{R})$, respectively. The extrapolated PES for the larger $(m+1)$-th basis set is obtained as [1]

$$E_{m+1}(\mathbf{R}) = E^{\text{ref}}_{m+1}(\mathbf{R}) + \chi_{m+1,m}(\mathbf{R})\Delta E_m(\mathbf{R}). \tag{6}$$

Here, $E^{\text{ref}}_{m+1}(\mathbf{R})$ is the energy of the reference wave function calculated with the $(m+1)$-th basis set, $\Delta E_m(\mathbf{R}) = E_m(\mathbf{R}) - E^{\text{ref}}_m(\mathbf{R})$, $\mathbf{R}$ denotes the τ-dimensional vector of the nuclear space coordinates defining the PES, and the scaling factor $\chi_{m+1,m}(\mathbf{R})$ is defined as [1]

$$\chi_{m+1,m}(\mathbf{R}) = 1 + \left[\frac{S(\mathbf{R})_{m,m-1} - 1}{S(\mathbf{R}_e)_{m,m-1} - 1}\right][S(\mathbf{R}_e)_{m+1,m} - 1], \tag{7}$$

with

$$S(\mathbf{R})_{m,m-1} = \Delta E_m(\mathbf{R})/\Delta E_{m-1}(\mathbf{R}). \tag{8}$$

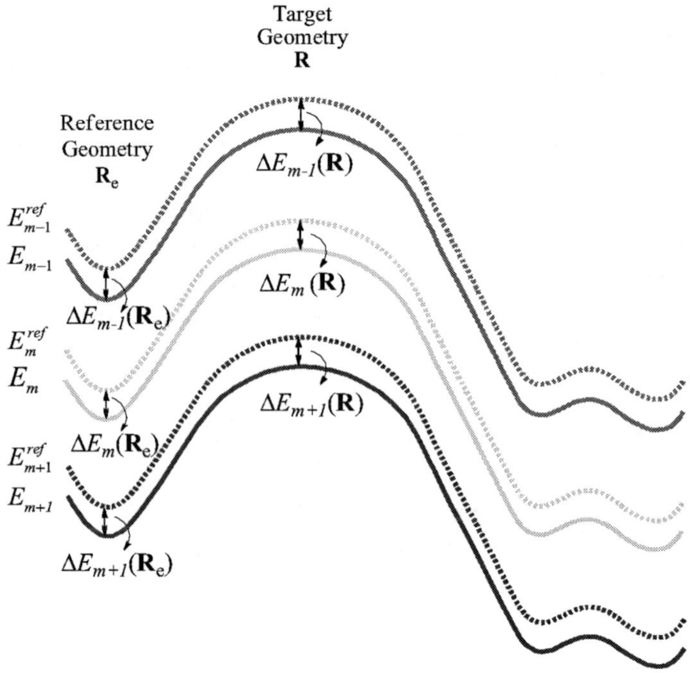

FIGURE 2. The differences between the total electronic energies in a given basis set series, $m-1$, m, and $m+1$, designated by E_{m-1}, E_m and E_{m+1}, and the corresponding energies of some reference wave function in the same series of basis sets, E^{ref}_{m-1}, E^{ref}_{m}, and E^{ref}_{m+1}, defining the correlation energy corrections, ΔE_{m-1}, ΔE_m, and ΔE_{m+1}, respectively, along the reaction coordinate. In the PES extrapolation procedure introduced in [1] and examined in the present study, the total energy $E_{m+1}(\mathbf{R})$ at the geometry $\mathbf{R}$ along the reaction pathway corresponding to the target basis set $m+1$ is obtained by scaling the correlation energy correction $\Delta E_m(\mathbf{R})$ at the geometry $\mathbf{R}$ using the information about the correlation energies $\Delta E_{m-1}(\mathbf{R}_e)$, $\Delta E_m(\mathbf{R}_e)$, and $\Delta E_{m+1}(\mathbf{R}_e)$ determined at the reference geometry $\mathbf{R}_e$, and adding the predicted $\Delta E_{m+1}(\mathbf{R})$ value to the reference energy $E^{\text{ref}}_{m+1}(\mathbf{R})$, as in Eqs. (6)–(8).

The reference geometry $\mathbf{R}_e$ can be the equilibrium geometry, the geometry corresponding to the reactants or products, or any other point on the PES. Our experience to date [1], confirmed in the calculations of the conrotatory and disrotatory isomerization pathways of bicyclo[1.1.0]butane to buta-1,3-diene reported in this work, shows that the extrapolated PES $E_{m+1}(\mathbf{R})$ is virtually independent of the choice of the reference geometry. Thus, the PES $E_{m+1}(\mathbf{R})$ is extrapolated from the PESs $E_{m-1}(\mathbf{R})$ and $E_m(\mathbf{R})$, the reference energies $E^{\text{ref}}_{m-1}(\mathbf{R})$, $E^{\text{ref}}_{m}(\mathbf{R})$, and $E^{\text{ref}}_{m+1}(\mathbf{R})$, and a single correlated energy, $E_{m+1}(\mathbf{R}_e)$, calculated at the reference geometry $\mathbf{R}_e$. Figure 2 gives a pictorial representation of the key features of the PES extrapolation procedure described here. An example of the application of this procedure is a specific situation explored in this study, where the PES corresponding to the cc-pVQZ basis set is extrapolated from the PESs obtained in the cc-pVDZ and cc-pVTZ calculations, the reference energies obtained in the cc-pVDZ, cc-pVTZ, and cc-pVQZ calculations, and a single correlated energy obtained with the cc-pVQZ basis set at the reference geometry $\mathbf{R}_e$. This would correspond to setting m in the above equations at 3.

Although in this work we focus on the cc-pVXZ basis sets of Dunning, when performing the PES extrapolations, which are very systematic in terms of the dependence of the correlation energies on the cardinal number X that can be approximated by expressions such as Eq. (1), we would like to emphasize that the PES extrapolation method of [1] is not tailored to fit any single class of electronic structure approaches nor is it limited to any family of basis sets. Thus, instead of the CR-CC(2,3) method used in this work one could choose another electronic structure approach that includes electron correlation effects in the process of determining the relevant electronic energies and instead of the cc-pVXZ basis sets, one could use another sequence of basis sets of increasing size to predict the PES. Examples illustrating the use of other electronic structure methods and other basis sets in the context of the PES extrapolations based on Eq. (6)–(8) can be found in [1].

As already mentioned, we also have a flexibility in defining what is the reference energy for determining the correlation energy corrections ΔE_{m-1}, ΔE_m, and ΔE_{m+1} that enter Eqs. (6)–(8). In the case of the PES calculations using the CR-CC(2,3) approach, one can decompose the CR-CC(2,3) energy using Eq. (2), so that ΔE_{m-1}, ΔE_m, and ΔE_{m+1} represent the total correlation energies of the CR-CC(2,3) approach corresponding to basis sets $m-1$, m, and $m+1$, respectively, and defined by Eq. (3). In this case, the PES extrapolation procedure defined by Eqs. (6)–(8) determines the target PES $E_{m+1}(\mathbf{R})$ by scaling the CR-CC(2,3) correlation energies. Another natural idea is to decompose the CR-CC(2,3) energy using Eq. (4), so that ΔE_{m-1}, ΔE_m, and ΔE_{m+1} represent the triples contributions to the CR-CC(2,3) correlation energies associated with basis sets $m-1$, m, and $m+1$, respectively, which are much smaller than the corresponding total correlation energies, and thus easier to extrapolate. In general, choosing a reference wave function with some inherently correlated nature, such as the CCSD wave function in the context of CR-CC(2,3) or CCSD(T) calculations, or the CASSCF wave function in the context of multi-reference configuration interaction, multi-reference MBPT, and multi-reference CC calculations, reduces the fraction of the correlation energy to be extrapolated, decreasing the magnitude of the potential extrapolation error for the resulting energy $E_{m+1}(\mathbf{R})$. On the other hand, this advantage of using part of the correlation energy rather than the total correlation energy in defining the corrections ΔE_m to be extrapolated may be outweighed by the expense of performing correlated calculations at all of the points of interest on a particular PES (say, relatively expensive CCSD calculations with the basis set $m+1$ when one uses the triples contribution to the CR-CC(2,3) correlation energy, $\delta E^{(2,3)}$, to define the ΔE_m values). One of the objectives of this work is to show that the final extrapolated PESs $E_{m+1}(\mathbf{R})$ are not very sensitive to the choice of the reference energies in defining the corrections ΔE_{m-1}, ΔE_m, and ΔE_{m+1}.

Extrapolating Potential Energy Surfaces to the Complete Basis Set Limit

The PES extrapolation scheme outlined above is not exclusive to groups of finite basis sets immediately sequential in size. The target basis set $m+1$ in Eq. (6) could also be an infinite basis set, as long as we know how to estimate the CBS limits of the reference energies $E^{\text{ref}}(\mathbf{R})$ at each point $\mathbf{R}$ on the PES of interest, which would define the necessary $E^{\text{ref}}_{m+1}(\mathbf{R})$ values, and as long as we can determine the CBS limit of the correlation energy $\Delta E(\mathbf{R}_e)$ at the reference geometry $\mathbf{R}_e$, which we could use to define the $\Delta E_{m+1}(\mathbf{R}_e)$ value for the purpose of calculating the $\chi_{m+1,m}(\mathbf{R})$ scaling factor according to Eqs. (7) and (8).

For example, we can use a sequence of basis sets of the cc-pVXZ quality, with higher values of X, to determine the CBS limits of the reference energies $E^{\text{ref}}(\mathbf{R})$. Since the reference energies $E^{\text{ref}}(\mathbf{R})$ could be determined with an inexpensive mean-field or SCF theory, such as the Hartree-Fock approach or its RHF variant, and since it is well established that the convergence of the Hartree-Fock energies with X is very fast (exponential), one can determine the smooth reference PES $E^{\text{ref}}_{m+1}(\mathbf{R})$ that one needs in Eq. (6) and that represents in this case the CBS limit of the reference surface by using a large basis set of the cc-pV5Z or cc-pV6Z quality which effectively provides the CBS values of $E^{\text{ref}}(\mathbf{R})$ [37] (see Table 1 for an example). As shown in Table 1, the differences between the RHF single-point reference energies calculated at the cc-pV5Z and cc-pV6Z levels do not exceed 0.5 millihartree (~ 0.3 kcal/mol) at all stationary points defining the conrotatory and disrotatory pathways describing the isomerization of bicyclo[1.1.0]butane to *trans*-buta-1,3-diene. Hence, it is safe to conclude that the cc-pV6Z RHF energies approximate the reference energies corresponding to the CBS-limit to within a very small error, acceptable for the majory of applications. One can also determine the approximate CBS limit of the correlation energy $\Delta E(\mathbf{R}_e)$ at the reference geometry $\mathbf{R}_e$ using, for example, the correlation energies obtained with smaller basis sets and one of the CBS extrapolation laws, such as Eq. (1), and use the resulting CBS value of $\Delta E(\mathbf{R}_e)$ as $\Delta E_{m+1}(\mathbf{R}_e)$ to determine the $S(\mathbf{R}_e)_{m+1,m}$ ratio and, ultimately, the scaling factor $\chi_{m+1,m}(\mathbf{R})$. According to Eqs. (6)–(8), the CBS reference energies $E^{\text{ref}}_{m+1}(\mathbf{R})$ at all points $\mathbf{R}$ of interest, the correlation energy scaling factor $\chi_{m+1,m}(\mathbf{R})$ from the finite basis set m to the infinite basis set $m+1$, and the correlation energies $\Delta E_m(\mathbf{R})$ obtained with a finite basis set m at all points $\mathbf{R}$ can be used to predict the CBS limit $E_{m+1}(\mathbf{R})$ of the PES of interest.

A good example of this procedure is the calculation of the CBS limits of the PESs defining the conrotatory and disrotatory pathways describing the isomerization of bicyclo[1.1.0]butane to *trans*-buta-1,3-diene discussed in this paper (see the next section). In this case, corresponding to choosing the cc-pVDZ basis as basis set $m-1$, the cc-pVTZ basis as basis set m, and an infinite basis set as basis set $m+1$ in the PES extrapolation procedure based on Eqs. (6)–(8), we use the results of the CR-CC(2,3) calculations obtained with the cc-pVTZ and cc-pVQZ basis sets at the reference geometry $\mathbf{R}_e$ to determine the CBS limit of the CR-CC(2,3) correlation energy at $\mathbf{R}_e$ using Eq. (1). The resulting CBS value of the CR-CC(2,3) correlation energy at $\mathbf{R}_e$ is used to define $\Delta E_{m+1}(\mathbf{R}_e)$. We then use the $\Delta E_{m+1}(\mathbf{R}_e)$ correlation energy at $\mathbf{R}_e$ determined in this way and the CR-CC(2,3)/cc-pVDZ and CR-CC(2,3)/cc-pVTZ correlation energies, $\Delta E_{m-1}(\mathbf{R})$ and $\Delta E_m(\mathbf{R})$, respectively, at all geometries $\mathbf{R}$ of interest to determine the scaling factor $\chi_{m+1,m}(\mathbf{R})$ from the cc-pVTZ basis set to the CBS limit according to Eqs. (7) and (8). Once the scaling factor $\chi_{m+1,m}(\mathbf{R})$ is known, we use it to determine the CBS limit of the CR-CC(2,3) PES, $E_{m+1}(\mathbf{R})$, by multiplying the CR-CC(2,3) correlation energies $\Delta E_m(\mathbf{R})$ obtained with the cc-pVTZ basis set by $\chi_{m+1,m}(\mathbf{R})$ and by adding the resulting

TABLE 1. RHF energies, in hartree, obtained with various cc-pVXZ basis sets for the stationary points defining the conrotatory and disrotatory pathways characterizing the **bicbut**→**t-but** isomerization.

	RHF Energies					
	bicbut	con_TS	dis_TS	gbut	gt_TS	t-but
cc-pVDZ	-154.880384	-154.786391	-154.710770	-154.926057	-154.921389	-154.931310
cc-pVTZ	-154.921100	-154.828555	-154.754302	-154.967865	-154.963458	-154.973471
cc-pVQZ	-154.930815	-154.838710	-154.765231	-154.977796	-154.973430	-154.983441
cc-pV5Z	-154.933078	-154.841237	-154.768189	-154.980238	-154.975868	-154.985888
cc-pV6Z	-154.933393	-154.841597	-154.768653	-154.980562	-154.976188	-154.986214

energies to the reference energies $E^{\rm ref}_{m+1}(\mathbf{R})$ obtained in the RHF/cc-pV6Z calculations which, as already explained, are practically equivalent to the reference energies in the CBS limit. By avoiding the point-wise CBS extrapolations of the correlation energies from the cc-pVTZ and cc-pVQZ basis sets at all geometries **R** (we do it only at the reference geometry) and by using the cc-pV6Z values of the RHF energies, we save a lot of CPU cycles, while producing the smooth PESs corresponding to the CBS limits, since the reference PES obtained with a fixed finite basis set and the scaling factor $\chi_{m+1,m}(\mathbf{R})$ are smooth functions of **R**. Point-wise CBS extrapolations of the correlation energies and/or point-wise extrapolations of the RHF reference energies would lead to a numerical noise in the extrapolated PESs. As shown in the next section, the accuracy of the above procedure is as good as the accuracy of the point-wise CBS extrapolation procedure, but our new procedure discussed here significantly reduces a computer effort and eliminates a numerical noise resulting from point-wise CBS extrapolations.

The Remaining Computational Details

The geometrical parameters of the stationary points along the conrotatory and disrotatory isomerization pathways of bicyclo[1.1.0]butane to buta-1,3-diene shown in Figure 1, obtained with the 10-electron, 10-orbital CASSCF (CASSCF(10,10)) method and the cc-pVDZ basis set, were taken from [11]. The required CCSD and CR-CC(2,3) calculations using the cc-pVDZ, cc-pVTZ, and cc-pVQZ basis sets were performed with the CC computer programs developed by our group [26, 27, 31, 38] and incorporated in the GAMESS package [39]. GAMESS was also used in the corresponding RHF calculations. The RHF calculations with the cc-pV5Z and cc-pV6Z basis sets, needed to determine the approximate CBS limits of the reference RHF PESs, were performed with the MOLPRO package [40]. When converting the results of electronic structure calculations into the enthalpy values shown in Figure 1, we used the zero-point vibrational energies at all stationary points shown in Figure 1 obtained with the CASSCF(10,10)/cc-pVDZ approach, reported in [11].

RESULTS AND DISCUSSION

Extrapolation of the CR-CC(2,3)/cc-pVQZ Reaction Profiles

To assess the performance of the PES extrapolation scheme discussed in the previous section in studies of chemical reaction pathways involving polyatomic molecules, we applied it to the stationary points defining the controtatory and disrotatory paths characterizing the isomerization of bicyclo[1.1.0]butane to buta-1,3-diene shown in Figure 1. The primary objective was to determine if the PES extrapolation scheme developed in [1] is capable of recovering the results of the laborious point-wise CR-CC(2,3)/cc-pVQZ calculations from the PESs obtained in the CR-CC(2,3)/cc-pVDZ and CR-CC(2,3)/cc-pVTZ calculations, the reference energies obtained in the cc-pVDZ, cc-pVTZ, and cc-pVQZ calculations, and a single CR-CC(2,3) energy obtained with the cc-pVQZ basis set at the reference geometry $\mathbf{R}_e$. As mentioned in the previous section, two choices of the reference energy for defining the ΔE values that enter Eqs. (6)–(8) were considered, namely, the RHF energy and the CCSD energy. The results obtained with both choices of the reference energy are collected in Table 2. We also considered different choices of the reference geometry $\mathbf{R}_e$, corresponding to all six stationary points that define both isomerization pathways, which we do not analyze in Table 2, since the resulting maximum differences between the calculated and extrapolated CR-CC(2,3)/cc-pVQZ energies characterizing both pathways are virtually independent of the choice of the reference geometry $\mathbf{R}_e$. For this reason, in Table 2 we focus on one specific choice of $\mathbf{R}_e$, namely, the geometry of the **bicbut** reactant (see Figure 1).

As one can see, independent of the choice of the reference energies (RHF or CCSD), there is virtually no difference between the calculated and extrapolated CR-CC(2,3)/cc-pVQZ energies. The differences between the calculated

TABLE 2. The calculated CR-CC(2,3)/cc-pVDZ and CR-CC(2,3)/cc-pVTZ energies and the calculated and extrapolated CR-CC(2,3)/cc-pVQZ energies at the stationary points defining the conrotatory and disrotatory pathways characterizing the **bicbut→t-but** isomerization. The **bicbut** reactant defines the reference geometry for the PES extrapolations.

	RHF Reference					
	E_2[a]	E_3[a]	E_4^{calc}[a]	E_4^{extr}[b]	ΔE_{43}[c]	ε_4[d]
bicbut	-155.493284	-155.651348	-155.695497	-155.695497	-44.149	0.000
con_TS	-155.424792	-155.581666	-155.625392	-155.625147	-43.727	0.245
dis_TS	-155.388352	-155.540034	-155.582481	-155.581850	-42.447	0.631
gbut	-155.533086	-155.689327	-155.733236	-155.732754	-43.909	0.482
gt_TS	-155.528145	-155.684308	-155.728073	-155.727663	-43.764	0.410
t-but	-155.537788	-155.694080	-155.738043	-155.737480	-43.963	0.563
	CCSD Reference					
	E_2[a]	E_3[a]	E_4^{calc}[a]	E_4^{extr}[b]	ΔE_{43}[c]	ε_4[d]
bicbut	-155.493284	-155.651348	-155.695497	-155.695497	-44.149	0.000
con_TS	-155.424792	-155.581666	-155.625392	-155.625212	-43.727	0.180
dis_TS	-155.388352	-155.540034	-155.582481	-155.582364	-42.447	0.117
gbut	-155.533086	-155.689327	-155.733236	-155.732971	-43.909	0.265
gt_TS	-155.528145	-155.684308	-155.728073	-155.727894	-43.764	0.179
t-but	-155.537788	-155.694080	-155.738043	-155.737766	-43.963	0.277

[a] Total energies, in hartree, calculated at the CR-CC(2,3)/cc-pVDZ (E_2), CR-CC(2,3)/cc-pVTZ (E_3), and CR-CC(2,3)/cc-pVQZ (E_4) levels.
[b] Total CR-CC(2,3)/cc-pVQZ energies, in hartree, resulting from the PES extrapolation procedure discussed in the text.
[c] Differences, in millihartree, between the actual CR-CC(2,3)/cc-pVQZ and CR-CC(2,3)/cc-pVTZ energies.
[d] Deviations, in millihartree, between the calculated and extrapolated CR-CC(2,3)/cc-pVQZ energies.

and extrapolated CR-CC(2,3)/cc-pVQZ energies characterizing both isomerization pathways do not exceed 0.631 millihartree, when $E^{\text{ref}} = E^{\text{RHF}}$, and 0.277 millihartree, when $E^{\text{ref}} = E^{\text{CCSD}}$. This should be compared to the changes in the total electronic energies when going from the cc-pVTZ to cc-pVQZ basis sets, which are on the order of 42–44 millihartree. Thus, the PES extrapolation procedure proposed in [1] reproduces changes in the total energies when going from the cc-pVTZ to cc-pVQZ basis sets to within 1 %, when the RHF energies are used as a reference energies, and 0.6 %, when the CCSD energies ate employed instead to define the reference energies. As one might expect, the use of the CCSD wave function in determining the reference energies reduces the observed differences between the calculated and extrapolated CR-CC(2,3)/cc-pVQZ energies. As pointed out in the previous section, the use of the correlated reference energies, such as those obtained in the CCSD calculations, reduces the fraction of the correlation energy to be extrapolated, decreasing the magnitude of the extrapolation errors. On the other hand, it is quite remarkable that the use of uncorrelated RHF reference energies in determining the ΔE values that enter Eqs. (6)–(8) leads to the extrapolated CR-CC(2,3)/cc-pVQZ PESs which are identical to the calculated CR-CC(2,3)/cc-pVQZ PESs to within ~ 0.6 millihartree. These small differences between the calculated and extrapolated CR-CC(2,3)/cc-pVQZ electronic energies are also reflected in the insignificant differences between the calculated and extrapolated CR-CC(2,3)/cc-pVQZ enthalpy values characterizing the stationary points along the controtatory and disrotatory pathways that define the isomerization of bicyclo[1.1.0]butane to buta-1,3-diene, shown in Figure 1, which are on the order of 1 kcal/mol or less.

To illustrate the savings in the computer effort offered by the PES extrapolation scheme of [1], we compare the CPU time required to construct the CR-CC(2,3)/cc-pVQZ PESs examined in this study in the conventional, point-by-point CR-CC(2,3)/cc-pVQZ calculations and the CPU time needed to obtain the equivalent data from the PESs obtained in the CR-CC(2,3)/cc-pVDZ and CR-CC(2,3)/cc-pVTZ calculations, the RHF reference energies obtained in the cc-pVDZ, cc-pVTZ, and cc-pVQZ calculations, and a single CR-CC(2,3) energy obtained with the cc-pVQZ basis set at the reference geometry. The total CPU time required to calculate the CR-CC(2,3)/cc-pVQZ energies corresponding to all six stationary points defining the controtatory and disrotatory paths characterizing the isomerization of bicyclo[1.1.0]butane to buta-1,3-diene was 39,323 minutes (≈ 27.3 days). The CPU time needed to construct the equivalent PES using the extrapolation procedure of [1], as discussed above, was 7,714 minutes, a factor of five times less than the conventional approach. This enormous saving in the computer effort is accompanied by an

TABLE 3. The calculated CR-CC(2,3)/cc-pVDZ, CR-CC(2,3)/cc-pVTZ, and CR-CC(2,3)/cc-pVQZ energies and the CBS values of the CR-CC(2,3) energies obtained using the point-wise extrapolations exploiting Eq. (1) and the CBS extrapolation procedure combining Eqs. (6)–(8) with Eq. (1) discussed in the text at the stationary points defining the conrotatory and disrotatory pathways characterizing the **bicbut→t-but** isomerization.

	E_2 [a]	E_3 [a]	E_4 [a]	E_{CBS}^{calc} [b]	E_{CBS}^{extr} [c]	$\Delta E_{CBS,3}$ [d]	ε_{CBS} [e]
bicbut	-155.493284	-155.651348	-155.695497	-155.723202	-155.723202	-71.854	0.000
con_TS	-155.424792	-155.581666	-155.625392	-155.652777	-155.652353	-71.112	0.424
dis_TS	-155.388352	-155.540034	-155.582481	-155.608902	-155.607810	-68.868	1.092
gbut	-155.533086	-155.689327	-155.733236	-155.760795	-155.759962	-71.468	0.833
gt_TS	-155.528145	-155.684308	-155.728073	-155.755491	-155.754781	-71.182	0.710
t-but	-155.537788	-155.694080	-155.738043	-155.765622	-155.764647	-71.541	0.975

[a] Total energies, in hartree, calculated at the CR-CC(2,3)/cc-pVDZ (E_2), CR-CC(2,3)/cc-pVTZ (E_3) and CR-CC(2,3)/cc-pVQZ (E_4) levels.
[b] Total energies, in hartree, obtained by adding the RHF/cc-pV6Z energies to the CBS correlation energies resulting from the point-wise CBS extrapolations employing the CR-CC(2,3)/cc-pVTZ and CR-CC(2,3)/cc-pVQZ calculations and Eq. (1).
[c] Total energies, in hartree, obtained by adding the RHF/cc-pV6Z energies to the CBS values of the correlation energy resulting from the extrapolation procedure combining Eqs. (6)–(8) with Eq. (1), as discussed in the text.
[d] Differences, in millihartree, between the CR-CC(2,3)/CBS energies resulting from the point-wise CBS extrapolations employing Eq. (1) and the corresponding CR-CC(2,3)/cc-pVTZ energies.
[e] Differences, in millihartree, between the CR-CC(2,3)/CBS energies resulting from the point-wise CBS extrapolations employing Eq. (1) and the CBS PES extrapolation combining Eqs. (6)–(8) with Eq. (1).

insignificant loss of accuracy which, as explained above, does not exceed 1 % of the changes of the energy when going from the cc-pVTZ to cc-pVQZ basis sets. Clearly, we would observe even larger savings if the number of stationary points along reaction pathways of interest was larger than in the case of the controtatory and disrotatory paths defining the isomerization of bicyclo[1.1.0]butane to buta-1,3-diene examined in this work.

Extrapolation of the CR-CC(2,3) Reaction Profiles Corresponding to the CBS Limit

Extension of the PES extrapolation scheme of [1] to the CBS-limit described in the previous section, in which we use the CR-CC(2,3)/cc-pVTZ and CR-CC(2,3)/cc-pVQZ single-point calculations at the reference geometry to determine the CBS limit of the CR-CC(2,3) correlation energy at $\mathbf{R}_e$ using Eq. (1) and exploit the resulting CBS value of the CR-CC(2,3) correlation energy at $\mathbf{R}_e$, along with the CR-CC(2,3)/cc-pVDZ and CR-CC(2,3)/cc-pVTZ correlation energies at all geometries $\mathbf{R}$, to determine the scaling factor $\chi_{m+1,m}(\mathbf{R})$ from the cc-pVTZ basis set to the CBS limit that enters Eq. (6), and the results of the RHF/cc-pV6Z calculations that provide the approximate CBS RHF energy values, works excellent as well. Indeed, as shown in Table 3, the differences between the energies resulting from the CBS extrapolation scheme that combines Eqs. (6)–(8) with Eq. (1), proposed in this study, and the energies resulting from the conventional point-wise CBS extrapolations using the CR-CC(2,3)/cc-pVTZ and CR-CC(2,3)/cc-pVQZ data at each stationary point defining the controtatory and disrotatory pathways shown in Figure 1 do not exceed 1.092 millihartree, i.e., they are on the same order as the intrinsic error of the CBS extrapolations based on Eq. (1) [5]. We must remember though that the point-wise CBS extrapolations of the correlation energies using Eq. (1) would result in a numerical noise in the extrapolated PESs, whereas the CBS PES extrapolation procedure that combines Eqs. (6)–(8) with Eq. (1) proposed in this work leads to smooth PESs. Moreover, the CBS PES extrapolation procedure combining Eqs. (6)–(8) with Eq. (1) requires only one CR-CC(2,3)/cc-pVQZ calculation (at the reference geometry), whereas the conventional point-wise CBS extrapolations require the CR-CC(2,3)/cc-pVQZ calculations at each point on the PES of interest, enormously increasing the computer costs, as illustrated in the previous subsection.

Finally, it is worth emphasizing the superb accuracies of the CR-CC(2,3)/CBS calculations when compared to the available experimental data. The activation enthalpy resulting from the CBS extrapolation scheme discussed in this work, which combines the PES extrapolation scheme developed in [1] with the CBS extrapolation law defined by Eq. (1), corresponding to the conrotatory TS **con_TS** is 41.8 kcal/mol, in excellent agreement with the experimental value of 40.6 ± 2.5 kcal/mol (see Figure 1). As shown in Figure 1, the CR-CC(2,3)/cc-pVQZ data obtained with the PES extrapolation scheme developed in [1] lead a very similar result (41.5 kcal/mol), indicating that for all practical purposes the results of the cc-pVQZ calculations can be viewed as virtually identical to the CBS limit. The enthalpy of isomerization of bicyclo[1.1.0]butane into buta-1,3-diene, the experimental value of which is -25.9 ± 0.4 kcal/mol, resulting from the CBS PES extrapolation procedure combining Eqs. (6)–(8) with Eq. (1) is -27.1 kcal/mol, when *trans*-buta-1,3-diene (**t-but**) is used as a product, and -24.1 kcal/mol when the nearly isoenergetic *gauche*-buta-1,3-diene (**g-but**) rotamer is used instead (for the corresponding CR-CC(2,3)/cc-pVQZ values, see Figure 1).

CONCLUSIONS

The PES extrapolation scheme developed in [1], which predicts smooth molecular PESs corresponding to larger basis sets from the relatively inexpensive calculations performed using smaller basis sets by scaling electron correlation energies, was overviewed and applied to the reaction profiles corresponding to the conrotatory and disrotatory isomerization pathways of bicyclo[1.1.0]butane to buta-1,3-diene. The relevant electronic structure calculations were performed using the CR-CC(2,3) approach. We demonstrated that the PESs resulting from the explicit CR-CC(2,3) calculations using the cc-pVQZ basis set and the analogous PESs obtained with the PES extrapolation scheme of [1], which offers considerable savings in the computer effort, are virtually identical. We also demonstrated that the use of a similar extrapolation procedure enables one to accurately predict smooth PESs corresponding the CBS limit at the small fraction of an effort required by the conventional point-wise CBS extrapolation schemes which lead to non-smooth PESs characterized by numerical noise.

REFERENCES

1. A. J. C. Varandas and P. Piecuch, Chem. Phys. Lett. **430**, 448 (2006).
2. T. Helgaker, W. Klopper, H. Koch, and J. Noga, J. Chem. Phys. **106**, 9639 (1997).
3. A. Halkier, T. Helgaker, P. Jørgensen, W. Klopper, H. Koch, J. Olsen and A. K. Wilson, Chem. Phys. Lett. **286**, 243 (1998).
4. D. G. Truhlar, Chem. Phys. Lett. **294**, 45 (1998).
5. A. J. C. Varandas, J. Chem. Phys. **113**, 8880 (2000).
6. T. H. Dunning, Jr., J. Chem. Phys. **90**, 1007 (1989).
7. R. A. Kendall, T. H. Dunning Jr., and R. J. Harrison, J. Chem. Phys. **96**, 6796 (1992).
8. K. B. Wiberg and J. M. Lavanish, J. Am. Chem. Soc. **88**, 5272 (1966).
9. G. L. Closs and P. E. Pfeffer, J. Am. Chem. Soc. **90**, 2452 (1968).
10. K. A. Nguyen and M.S. Gordon, J. Am. Chem. Soc. **117**, 3835 (1995).
11. A. Kinal and P. Piecuch, J. Phys. Chem. A **111**, 734 (2007).
12. P. B. Shevlin and M.L. Mckee, J. Am. Chem. Soc. **110**, 1666 (1988).
13. M. J. S. Dewar and S. Kirschner, J. Am. Chem. Soc. **97**, 2931 (1975).
14. R. Srinivasan, A. Levi and I. Haller, J. Phys. Chem. **58**, 1775 (1965).
15. F. Coester, Nucl. Phys. **7**, 421 (1958).
16. F. Coester, H. Kümmel, Nucl. Phys. **17**, 477 (1960).
17. J. Čížek, J. Chem. Phys. **45**, 4256 (1966).
18. J. Čížek, Adv. Chem. Phys. **14**, 35 (1969).
19. J. Čížek and J. Paldus, Int. J. Quantum Chem. **5**, 359 (1971).
20. K. Raghavachari, G.W. Trucks, G.W. Pople, and M. Head-Gordon, Chem. Phys. Lett. **157**, 479 (1989).
21. P. Piecuch, K. Kowalski, in *Computational Chemistry: Reviews of Current Trends*, edited by J. Leszczyński (World Scientific, Singapore, 2000), Vol. 5, pp. 1-104.
22. P. Piecuch, K. Kowalski, I.S.O. Pimienta, and M.J. McGuire, Int. Rev. Phys. Chem. **21**, 527 (2002).
23. P. Piecuch, K. Kowalski, I.S.O. Pimienta, P.-D. Fan, M. Lodriguito, M.J. McGuire, S.A. Kucharski, T. Kuś, and M. Musiał, Theor. Chem. Acc. **112**, 349 (2004).
24. J. Paldus and X. Li, Adv. Chem. Phys. **110**, 1 (1999).
25. M. W. Schmidt and M. S. Gordon, Annu. Rev. Phys. Chem. **49**, 233 (1998).
26. P. Piecuch and M. Włoch, J. Chem. Phys. **123**, 224105 (2005).
27. P. Piecuch, M. Włoch, J.R. Gour and A. Kinal, Chem. Phys. Lett. **418**, 467 (2006).
28. M. Włoch, M.D. Lodriguito, P. Piecuch and J.R. Gour, Mol. Phys. **104**, 2149 (2006).
29. C.J. Cramer, M. Włoch, P. Piecuch, C. Puzzarini, and L. Gagliardi, J. Phys. Chem. A **110**, 1991 (2006).
30. C.J. Cramer, A. Kinal, M. Włoch, P. Piecuch, and L. Gagliardi, J. Phys. Chem. A **110**, 11557 (2006).
31. M. Włoch, J.R. Gour, and P. Piecuch, J. Phys. Chem. A **111**, 11359 (2007).
32. Y. Ge, M.S. Gordon, and P. Piecuch, J. Chem. Phys. **127**, 174106 (2007).
33. P. Piecuch, M. Włoch, and A.J.C. Varandas, in *Progress in Theoretical Chemistry and Physics*, Vol. 16, edited by S. Lahmar, J. Maruani, S. Wilson, and G. Delgado-Barrio (Springer, Berlin, 2007), pp. 63-121.
34. G.D. Purvis, III and R.J. Bartlett, J. Chem. Phys. **76**, 1910 (1982).
35. G.E. Scuseria, A.C. Scheiner, T.J. Lee, J.E. Rice, and H.F. Schaefer, III, J. Chem. Phys. **86**, 2881 (1987).
36. P. Piecuch and J. Paldus, Int. J. Quantum Chem. **36**, 429 (1989).
37. A. Halkier, T. Helgaker, P. Jørgensen, W. Klopper, and J. Olsen, Chem. Phys. Lett. **302**, 437 (1999).
38. P. Piecuch, S.A. Kucharski, K. Kowalski, and M. Musiał, Comp. Phys. Commun. **149**, 71 (2002).
39. M. W. Schmidt et al., J. Comput. Chem. **14**, 1347 (1993).
40. H.-J. Werner et al., MOLPRO, version 2002.6, a package of ab initio programs, (2003), see http://www.molpro.net.

Some Recent Developments of Quantum Monte Carlo to Molecular Systems.

William A. Lester, Jr.[*,†] and Dominik Domin[*]

[*]*Kenneth S. Pitzer Center for Theoretical Chemistry, Department of Chemistry, University of California, Berkeley, CA 94720-1460, USA.*
[†]*Chemical Sciences Division, Lawrence Berkeley National Laboratory, Berkeley, CA 94720, USA*

Abstract. A brief review of recent advances in quantum Monte Carlo for the electronic structure of molecules at the University of California, Berkeley, is given based on an invited talk presented at the Workshop on Nuclear and Mesoscopic Physics.

Keywords: electronic structure; quantum Monte Carlo; linear diffusion Monte Carlo; quantum chemistry; computational chemistry; photo-protection
PACS: 02.70.Ss;87.16.af;02.70.Uu

INTRODUCTION

A continuing major challenge in molecular electronic structure theory is the accurate description of electron correlation - the instantaneous mutual interaction of electrons of a system with each other. The Schrödinger equation, a 3N partial differential equation does, in principle, characterize such systems. Various *ab initio* methods provide alternative ways for generating approximate solutions to this equation [1] that can be only solved exactly for the simplest atom. For many-electron systems, however, one finds that highly accurate approaches, such as coupled-cluster method [2, 3, 4], suffer from the growth in computational effort with system size that renders unfeasible calculations on very large systems in the absence of high symmetry.

For many years, the Hartree-Fock (HF) method and more recently density functional theory (DFT) have been used for a broad range of molecular applications. The HF method is a mean-field approximation that by definition cannot account for electron correlation. Although in principle exact, DFT in its various implementations introduces uncertainty associated with the description chosen to describe the exchange-correlation functional. The latter issue continues to be an open area of research.

The methods known as quantum Monte Carlo (QMC) solve the Schrödinger equation stochastically [5, 6, 7, 8, 9, 10, 11, 12, 13, 14, 15, 16]. These approaches have been successful in calculating ground-state and low-lying excited-state energies, as well as other observables, for atoms, molecules and solids to high accuracy [5, 6, 7, 8, 9, 10, 12, 13, 14, 15, 16]. In these approaches, electron correlation is described explicitly by functions that depend directly on interparticle distances. Two of the most popular QMC algorithms for electronic structure calculations are variational Monte Carlo (VMC)[5, 11] and diffusion Monte Carlo (DMC) [5, 6, 7, 8, 9, 10, 12, 13, 14, 15, 16, 17]. The VMC method computes the expectation value of a given trial function Ψ_T via Monte Carlo integration [18]. The DMC method is presently the more commonly used version of QMC and asymptotically yields time-independent solutions of the Schrödinger equation from an imaginary time stochastic formalism [5, 6].

DMC METHOD AND APPLICATIONS

The DMC method is derived from an exact Green's function approach for solving the Schrödinger equation. For molecules, one seeks fermion solutions that reflect the antisymmetry of the electron under exchange. The approach for obtaining these solutions that characterizes the DMC method is to introduce the fixed-node approximation. This step involves imposing the nodes of the approximate wave function Ψ_T onto the unknown exact wave function Ψ. The Schrödinger equation is solved exactly within the volumes defined by the many-electron nodes of Ψ_T [19]. The error generated by this approximation is generally quite small; DMC consistently recovers between 95 – 100% of the

CP995, *Nuclei and Mesoscopic Physics - WNMP 2007*, edited by P. Danielewicz, P. Piecuch, and V. Zelevinsky

correlation energy. The fixed-node approximation has become a fundamental part of the DMC method and we shall refer to fixed-node DMC hereafter as DMC.

QMC methods for molecules date back some thirty years to the pioneering studies of Anderson [20, 21]. Since this initial research, algorithmic improvements and dramatically increased computational power have made possible the application of QMC to much larger systems.

DMC FOR LARGE SYSTEMS

With recent advances in algorithms, the development of better codes, and access to greater computer power, the stage was set for the study of larger systems with DMC. The first large-system calculation in this laboratory was of porphyrin for which the vertical transition between the ground-state singlet and the second excited-state singlet, and the adiabatic transition between the ground state and the lowest triplet state were determined [22]. The results obtained for this 162-electron system are in excellent agreement with experiment, lying within 0.1eV in each case.

More recently, efforts in this laboratory have moved to the study of molecules involved in photosystem II, in particular, the molecules bacteriochlorophyll (BChl) and spheroidene (Spo), which are important in light harvesting and photo-protection. These molecules contain a total of 304 and 314 electrons, respectively, making them the largest systems to be studied using all-electron DMC.

Several scientific questions remain unanswered about the photoprotection mechanism including the ground-state to lowest-triplet-state excitation energies of both molecules as well as the energy transfer rate between the systems. At present, DFT is not sufficiently accurate to answer to these questions. A more rigorous electronic structure method that accurately treats electron correlation is needed, but, as noted, basis-set correlated methods scale too steeply with system size to be applicable. DMC presently offers the only feasible option for the accurate calculation of these quantities.

For the past five years, a new QMC code, Zori [23, 24] has been under development in this laboratory to make possible the calculation of large molecular systems. The goal with this code is to provide both, a research platform for developing and testing new QMC methodologies, as well as a robust computational environment for the user. The development model is based on the GNU General Public License (GNU/GLP). Zori includes improved random walk algorithms and has been developed to have improved I/O performance.

QMC calculations for large systems demand large computational resources. Zori has been developed to take maximum advantage of highly parallel computers. Although QMC algorithms are intrinsically parallelizable, a load balancing strategy is required to maintain population control in the DMC algorithm. Zori utilizes a dynamic approach for this purpose, performing the balance over the course of program execution that also enables the user to select flexibly the balancing strategy to be followed. Zori has been ported to several platforms including AIX and a variety of Linux distributions. it has been successfully tested for parallel execution of up to 6,000 concurrent tasks on an IBM SP system.

For increasingly larger molecules, the number of operations needed to evaluate the wave function and the energy at each point of the simulation increases. Because the interaction between electrons diminishes exponentially with distance [25], several approaches have been introduced to take advantage of this property leading to reduction of computational effort to a linear function of system size. In Zori, a number of steps have been taken in order to obtain nearly linear scaling performance. [26, 27] In particular, wave function evaluation in Zori follows the procedure of Aspuru-Guzik et al. [28] in which a sparse representation of the Slater matrix is used that leads to evaluation of only the non-zero elements of the wave function. This approach, combined with the use of localized orbitals in a Slater-type-orbital basis set, significantly extends the size molecule that can be treated with the QMC method. Tests confirm that this approach leads to linear scaling of Slater determinant evaluation with system size [28]. With this approach, the calculation of excitation energies scales as N^2.

To treat systems of increasing chemical interest, there is the need in QMC to address molecules in media and in environments for which explicit treatment at the molecular level leads to too large a problem for present methods. With this in mind, we have initiated studies of several QM/MM (quantum mechanics/molecular mechanics) schemes [29]. This research is ongoing and will be reported at a later time.

ACKNOWLEDGMENTS

W. A. L. was supported by the Director, Office of Science, Office of Basic Energy Sciences, Chemical Sciences Division of the U. S. Department of Energy under Contract No. DE-AC03-76SF0098. D. D. was supported by the National Science Foundation through the CREST Program.

REFERENCES

1. E. Cancès, M. Defranceschi, W. Kutzelnigg, C. Le Bris, and Y. Maday, "The models of quantum chemistry and their discretizations," in *Computational Chemistry*, edited by C. Le Bris, Elsevier, Amsterdam, The Netherlands, 2003, vol. X of *Handbook of Numerical Analysis*, chap. 1, first. edn.
2. G. D. Purvis, III, and R. J. Bartlett, *J. Chem. Phys.* **76**, 1910–1918 (1982).
3. M. Head-Gordon, *J. Phys. Chem.* **100**, 13213 (1996).
4. T. D. Crawford, and H. F. Schaefer III, *Rev. Comp. Chem.* **14**, 33–136 (2000).
5. B. L. Hammond, W. A. Lester, Jr., and P. J. Reynolds, *Monte Carlo Methods in Ab Initio Quantum Chemistry*, World Scientific, Singapore, 1994.
6. J. B. Anderson, *Int. Rev. Phys. Chem.* **14**, 85–112 (1995).
7. D. M. Ceperley, and L. Mitas, "Quantum Monte Carlo Methods in Chemistry," in *New Methods in Computational Quantum Mechanics*, edited by I. Prigogine, and S. A. Rice, John Wiley and Sons, 1996, vol. XCIII of *Adv. Chem. Phys.*
8. S. A. Alexander, and R. L. Coldwell, "Atomic Calculations Using Variational Monte Carlo," in *Recent Advances in Quantum Monte Monte Carlo Methods*, edited by W. A. Lester, Jr., World Scientific, Singapore, 1997, p. 39.
9. P. H. Acioli, *J. of Mol. Struct. (Theochem)* **394**, 75 (1997).
10. L. Mitas, "Diffusion Monte Carlo," in *Quantum Monte Carlo Methods in Physics and Chemistry*, edited by M. P. Nightingale, and C. J. Umrigar, Kluwer Academic Publishers, 1998, vol. 525, p. 247.
11. D. Bressanini, and P. J. Reynolds, *Adv. Chem. Phys.* **105**, 37 (1998).
12. J. B. Anderson, "Quantum Monte Carlo: Atoms, Molecules, Clusters, Liquids and Solids," in *Reviews in Computational Chemistry*, edited by K. B. Lipkowitz, and D. B. Boyd, John Wiley and Sons, New York, 1999, vol. 13, p. 133.
13. A. Lüchow, and J. B. Anderson, *Ann. Rev. Phys. Chem.* **51**, 501 (2000).
14. M. Foulkes, L. Mitas, R. Needs, and G. Rajagopal, *Rev. Mod. Phys.* **73**, 33 (2001).
15. C. Filippi, and C. J. Umrigar, "Interatomic Forces and Correlated Sampling in Quantum Monte Carlo," in *Recent Advances in Quantum Monte Carlo Methods, Part II*, edited by S. M. Rothstein, W. A. Lester, Jr., and S. Tanaka, World Scientific, Singapore, 2002.
16. S. E. Koonin, and D. C. Meredith, *Computational physics, FORTRAN version*, Addison Wesley, 1995, third edn.
17. L. Mitas, *Computer Physics Communications* **97**, 107 (1996).
18. N. Metropolis, A. W. Rosenbluth, M. N. Rosenbluth, N. M. Teller, and E. Teller, *J. Chem. Phys.* **21**, 1087–1091 (1953).
19. P. J. Reynolds, D. M. Ceperley, B. Alder, and W. A. Lester, Jr., *J. Chem. Phys.* **77**, 5593–603 (1982).
20. J. B. Anderson, *J. Chem. Phys.* **65**, 4121–4127 (1976).
21. J. B. Anderson, *J. Chem. Phys.* **63**, 1499–1503 (1975).
22. A. Aspuru-Guzik, O. El Akramine, J. C. Grossman, and W. A. Lester, Jr., *J. Chem. Phys.* **120**, 3049–3050 (2004), URL `http://link.aip.org/link/?JCP/120/3049/1`.
23. A. Aspuru-Guzik, R. Salomon-Ferrer, B. Austin, and W. A. Lester, Jr., *Zori: a quantum Monte Carlo program.* (2004), URL `http://zori-code.com`, internet site: (`http://zori-code.com`).
24. A. Aspuru-Guzik, R. Salomon-Ferrer, B. Austin, R. Perusquia-Flores, M. A. Griffin, R. Oliva, D. Skinner, D. Domin, and W. A. Lester, Jr., *J. Comp. Chem.* **26**, 856–862 (2005).
25. P. Maslen, C. Ochsenfeld, C. White, M. Lee, and M. Head-Gordon, *J. Phys. Chem. A* **102**, 2215–2222 (1998).
26. A. Luchow, and S. Manten, *J. Chem. Phys.* **119**, 1307–1312 (2003).
27. A. Williamson, R. Hood, and J. Grossman, *Phys. Rev. Lett.* **8724** (2001).
28. A. Aspuru-Guzik, R. Salomon-Ferrer, B. Austin, and W. A. Lester, Jr., *J. Comp. Chem.* **26**, 708–715 (2005).
29. Hyung-Min Cho, and William A. Lester Jr., *research in progress*.

Continuum shell model: From Ericson to conductance fluctuations

G.L. Celardo*, F.M. Izrailev†,**, S. Sorathia†, V.G. Zelevinsky** and G.P. Berman‡

*Physics Department, Tulane University, New Orleans, LA 70118, USA
†Instituto de Física, Universidad Autónoma de Puebla, Apartado Postal J-48, Puebla, Pue., 72570, México
**NSCL and Department of Physics and Astronomy, Michigan State University, East Lansing, Michigan 48824-1321, USA
‡Theoretical Division and CNLS, Los Alamos National Laboratory, Los Alamos, New Mexico 87545, USA

Abstract. We discuss an approach for studying the properties of mesoscopic systems, where discrete and continuum parts of the spectrum are equally important. The approach can be applied (i) to stable heavy nuclei and complex atoms near the continuum threshold, (ii) to nuclei far from the region of nuclear stability, both of the regions being of great current interest, and (iii) to mesoscopic devices with interacting electrons. The goal is to develop a new consistent version of the continuum shell model that simultaneously takes into account strong interaction between fermions and coupling to the continuum. Main attention is paid to the formation of compound resonances, their statistical properties, and correlations of the cross sections. We study the Ericson fluctuations of overlapping resonances and show that the continuum shell model nicely describes universal properties of the conductance fluctuations.

Keywords: shell model, two-body random interaction, Ericson fluctuations, conductance fluctuations
PACS: 05.50.+q, 75.10.Hk, 75.10.Pq

Introduction

The focus of interest in low-energy nuclear physics has recently concentrated on nuclei far from the region of stability. The success of the nuclear shell model with strong interaction between nucleons urges one to look for ways of incorporating this rich experience into a more general context that would properly include the continuum. The continuum shell model is expected to combine the physics of intrinsic structure and reactions in contrast to the traditional spectroscopy and reaction theory which were historically developed as different branches of nuclear physics. In this approach, the properties of open channels, threshold behavior, symmetries, and spectroscopic factors enter the overall theory, so that a fully consistent picture uniting structure and reaction mechanisms is created. With great successes of mesoscopic physics (complex atoms and molecules, micro- and nano-devices of condensed matter, atomic and molecular traps, and prototypes of quantum computers), the nuclear problem becomes a typical part of general science of open mesoscopic systems.

In this paper we discuss a new approach [1] for studying collective phenomena in nuclei and quantum dots which emerge as a result of strong interactions between fermions coupled to the continuum. In this problem the standard shell model techniques, adjusted for the bound states, cannot work and continuum effects must be properly accounted for. At a certain strength of the continuum coupling, the system of strongly interacting fermions undergoes a collective restructuring. This is an area where the conventional division of nuclear physics into "structure" and "reactions" becomes inappropriate and two views of the process, from the inside (structure and properties of bound states) and from the outside (cross sections of reactions), have to be unified.

Conventional theory of statistical reactions does not answer the question of interplay between reactions and internal structure determined by the character of interactions between the constituents. Here more detailed considerations are required based on the generalization of the shell model of nuclear reactions [2]. Such an extension requires statistical assumptions concerning intrinsic dynamics and the coupling to the continuum [3, 4]. To account for specificity of the system, one has to go beyond standard random matrix approaches [5, 6, 7] utilizing the Gaussian Orthogonal Ensemble (GOE). The consistent description based on the continuum shell model [2], as well as more phenomenological approaches [8, 9], indicates the presence of a sharp restructuring of the system when the widths of resonances become comparable to their energy spacings. This phenomenon carries features of a quantum phase transition with the strength of continuum coupling playing the role of a control parameter.

CP995, *Nuclei and Mesoscopic Physics - WNMP 2007,* edited by P. Danielewicz, P. Piecuch, and V. Zelevinsky

As was numerically observed in the shell model framework [10], the distribution of resonance widths rapidly changes in the transitional region in such a way that a number of very broad resonances equal to a number of open decay channels absorb the lion's share of the total width of all overlapped resonances, while the remaining states become very narrow. The corresponding theory was suggested in [11, 12, 13], where the mechanism of this restructuring was understood to be associated with the nature of the effective non-Hermitian Hamiltonian [2] that describes the intrinsic dynamics after eliminating the channel variables. The segregation of short-lived broad resonances from long-lived trapped states was shown to be similar to the superradiance [14] in quantum optics induced by the coupling of atomic radiators through the common radiation field, an analog of coherent coupling of many overlapped intrinsic states through continuum decay channels. Later a general character of the phenomenon was demonstrated for systems with the GOE intrinsic dynamics and many open channels [15, 16]. Modern versions of the shell model in continuum [17, 18] are based on the effective Hamiltonian and naturally reveal the superradiance phenomenon as an important element. The transition to this regime should be taken into account in all cases when a physical system is strongly coupled to the continuum, see for example [19] and references therein.

In the present work, using the same framework as in [1], we study the interplay between the intrinsic dynamics and statistical properties of cross sections comparing the results with those of conventional approaches, namely Hauser-Feshbach average cross sections and Ericson fluctuations and correlations. We also analyze how the universal conductance fluctuations appear in our model, how they depend on the intrinsic chaos and what is the role of correlations between cross sections of various processes.

The model

We consider a generic system of n interacting fermions distributed over m single-particle levels of energies ε_s ("mean-field orbitals"). The fermions are coupled to continuum through the finite number M of open decay channels. In the many-body representation, the Hamiltonian matrix of size $N = m!/[n!(m-n)!]$ entirely determines the intrinsic dynamics of the closed system, and our main interest is to reveal how the properties of the reactions depend on the degree of intrinsic chaos that is due to the inter-particle interaction. According to the well developed formalism [2, 12, 19], the behavior of an open system can be described in terms of the effective non-Hermitian Hamiltonian consisting of two terms,

$$\mathscr{H} = H - \frac{i}{2}W; \quad W_{ij} = \sum_{c=1}^{M} A_i^c A_j^c. \tag{1}$$

Here H is a Hermitian matrix that can be also influenced by the presence of the continuum [17, 18, 20], and W stands for the coupling between intrinsic many-body states labeled as $i, j, ...$, and open decay channels $a, b, c....$ The factorized structure of W is dictated by the requirement of the unitarity of the scattering matrix. In what follows we assume that the transition amplitudes A_i^c between intrinsic states $|i\rangle$ and channels c are real quantities, therefore, we restrict ourself by the time-invariant systems. As a result, both H and W are real symmetric matrices.

The Hermitian part, $H = H_0 + V$, of the full Hamiltonian (1) is modeled by the so-called two-body random interaction (TBRI) (see, for example, [21] and references therein),

$$H = H_0 + V; \qquad H_0 = \sum_{s=1}^{m} \varepsilon_s a_s^\dagger a_s; \qquad V = \frac{1}{2}\sum V_{s_1 s_2 s_3 s_4} a_{s_1}^\dagger a_{s_2}^\dagger a_{s_3} a_{s_4}. \tag{2}$$

Here the mean-field part H_0 describes non-interacting particles (or quasi-particles), and the interaction between particles is given by V. In the many-particle basis $|k\rangle$ the matrix H is constructed by the Slater determinant, $|k\rangle = a_{s_1}^\dagger \dots a_{s_n}^\dagger |0\rangle$, where $a_s^\dagger$ and a_s are the creation and annihilation operators. The interaction between the particles is considered to be of two-body nature, therefore, each many-body matrix element $V_{lk} = \langle l|V|k\rangle$ is a sum of a number of two-body matrix elements $V_{s_1 s_2 s_3 s_4}$ involving at most four single-particle states $|s\rangle$. In our simulations we take $n = 6$, $m = 12$ that provides a sufficiently large dimension $N = 924$.

The *single-particle* energies, ε_s, are assumed to have a Poissonian distribution of spacings, with the mean level density $1/d_0$. The interaction V is characterized by the variance of the normally distributed two-body random matrix elements, $\langle V_{s_1,s_2;s_3,s_4}^2\rangle = v_0^2$. Without the interaction, $v_0 = 0$, the many-body states have also the Poissonian spacing distribution $P(s)$. In the opposite limit of an extremely strong interaction, $\lambda \equiv v_0/d_0 \to \infty$ (or, equivalently, for $d_0 = 0$), the function $P(s)$ is close to the Wigner-Dyson (WD) distribution that is typical for completely chaotic systems [5].

The critical value of the interaction for the onset of strong chaos was obtained in Ref. [21], it is estimated as

$$\lambda_{\rm cr} = \frac{v_{\rm cr}}{d_0} \approx \frac{2(m-n)}{N_s}, \tag{3}$$

where $N_s = n(m-n) + n(n-1)(m-n)(m-n-1)/4$ is the number of directly coupled many-body states in any row of the matrix H_{ij}. One can expect that for a very strong interaction, $\lambda \gg \lambda_{\rm cr}$, some of properties of the system may be close to those obtained with H taken from the Gaussian Orthogonal Ensemble (GOE). For this reason, we have also considered, for comparison, the GOE matrices in place of H.

In our study the amplitudes A_i^c are assumed to be random independent Gaussian variables with the zero mean and variance

$$\langle A_i^c A_j^{c'} \rangle = \delta_{ij}\delta^{cc'}\frac{\gamma^c}{N}. \tag{4}$$

This is compatible with the GOE or TBRI models where generic many-body states coupled to continuum have a very complicated structure, while the decay probes specific simple components of these states related to few open channels. Even for a weak intrinsic interaction, one needs to have in mind that in reality the states $|i\rangle$ have certain values of exact constants of motion, such as angular momentum and isospin in the nuclear case. Therefore, at sufficiently large dimension N, these states acquire *geometric chaoticity* [22] due to the almost random coupling of individual spins. Thus, the assumption of a random nature of the decay amplitudes is reasonable.

The parameters γ^c characterize the total coupling of all states to the channel c. The normalization used in Eq. (4) is convenient if the energy interval ND covered by decaying states is finite. Here D is the distance between the many-body states in the middle of the spectrum, $D = 1/\rho(0)$, where $\rho(E)$ is the many-body level density, and $E = 0$ corresponds to the center of the spectrum. Below we neglect a possible energy dependence of the amplitudes that is important near thresholds and is taken into account in realistic shell model calculations [18, 20]. The ratio γ^c/ND characterizes the degree of overlap of the resonances in the channel c. The control parameter determining the strength of the coupling to the continuum can be written as follows,

$$\kappa^c = \frac{\pi\gamma^c}{2ND}. \tag{5}$$

The transition from separated to strongly overlapped resonances corresponds to $\kappa^c \approx 1$. In order to keep the coupling to continuum fixed, in our numerical calculations we renormalize the absolute magnitude of the widths, γ^c, by varying the intrinsic interaction and, therefore, the level density ρ.

The effective Hamiltonian allows one to construct the scattering matrix,

$$S^{ab} = \delta^{ab} - i\mathscr{T}^{ab}, \tag{6}$$

where

$$\mathscr{T}^{ab}(E) = \sum_{i,j}^{N} A_i^a \left(\frac{1}{E-\mathscr{H}}\right)_{ij} A_j^b \tag{7}$$

are the scattering amplitudes determining the cross sections $\sigma^{ab}(E)$ of reactions,

$$\sigma^{ab}(E) = |\mathscr{T}^{ab}(E)|^2. \tag{8}$$

In our notations the cross sections are dimensionless since we omit the common factor π/k^2. In what follows we study both the elastic, $b = a$, and inelastic, $b \neq a$, cross sections. Note that we ignore the smooth potential phases that are irrelevant for our purposes.

Transition to superradiance

The complex eigenvalues $\mathscr{E}_r = \omega_r - (i/2)\Gamma_r$ of $\mathscr{H}$ coincide with the poles of the S-matrix and, for small γ^c, determine energies and widths of separated resonances. In the simulations we consider a real energy interval at the center of the spectrum of $\mathscr{H}$ with the constant many-body level density $\rho(0) = D^{-1}$. The transmission coefficient in the channel c,

$$T^c = 1 - |\langle S^{cc}\rangle|^2 = \frac{4\kappa^c}{(1+\kappa^c)^2}, \tag{9}$$

is maximal (equal to 1) at the critical point, $\kappa^c = 1$, and, for simplicity, we assume M equiprobable channels, $\kappa^c = \kappa$. For each value of κ we have used a large number of realizations of the Hamiltonian matrices, with further averaging over energy.

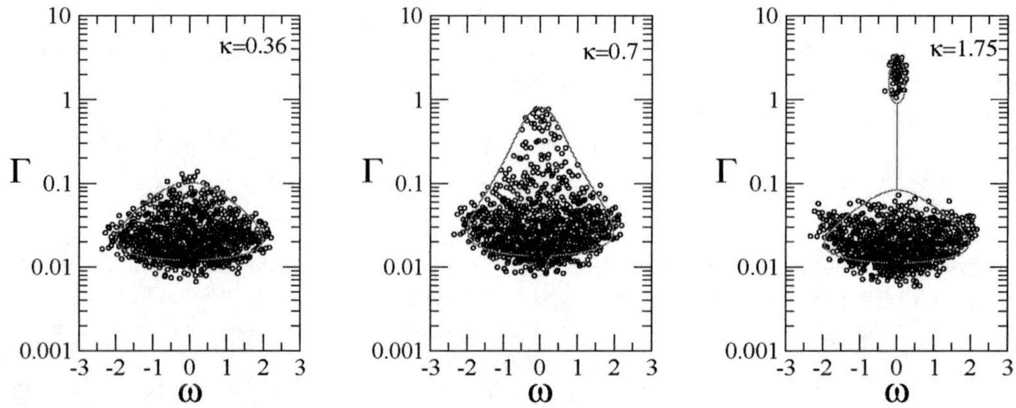

FIGURE 1. Distribution of widths, Γ, and positions, ω, of resonances (poles of the scattering matrix) as a function of κ. The results are shown for $n = 6$ particles, $m = 12$ orbitals, 50 open channels, and for a relatively strong interaction between particles, $v_0/d_0 = 1/10$, above the strong chaos threshold. The smooth curves are the boundaries due to the analytical results obtained in Ref.[23] for the GOE matrices in place of H.

As γ^c grows, the resonances start to overlap, leading to a segregation of the resonance widths [10, 11, 12, 23] occurring at $\kappa \approx 1$, see Fig. 1. The widths of M resonances are increasing at the expense of the remaining $N - M$ resonances. For a weak coupling, $\kappa \ll 1$, the widths are given by diagonal matrix elements, $\Gamma_i = \langle i|W|i\rangle = \sum_{c=1}^{M}(A_i^c)^2$, and the mean width is $\langle\Gamma\rangle = \gamma M/N$. In the limit of strong coupling, $\kappa \gg 1$, the widths of M broad resonances converge to the non-zero eigenvalues of the matrix W that has a rank M due to its factorized structure. As for the remaining "trapped" $(N - M)$ states, their widths decrease as $1/\gamma$. From our data one can conclude that the transition to the superradiance in dependence on the control parameter κ turns out to be quite sharp, and may be treated as a kind of the phase transition.

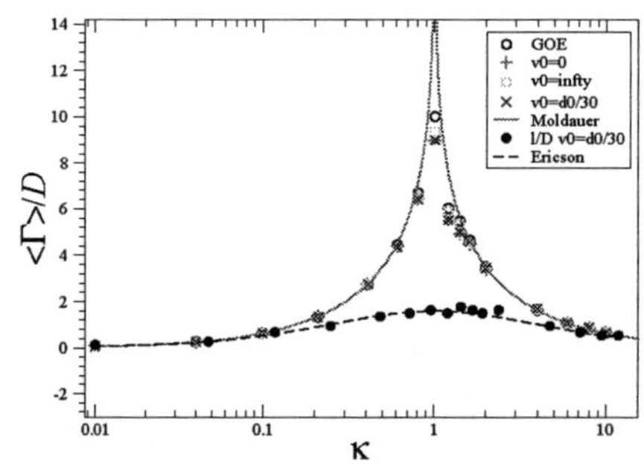

FIGURE 2. Normalized average width versus κ for $n = 6, m = 12$ and $M = 10$. The average was performed over the resonances around the center of the energy spectrum. Open circles refer to the GOE case, pluses to $\lambda = 0$, squares to $\lambda \to \infty$, and crosses to $\lambda = 1/30$. Solid curve shows the MS-expression (10). Full triangles stand for the normalized correlation length at $\lambda = 1/30$, see in the text. The dashed curve shows the Weisskopf relation (16).

In Fig. 2 we show how the *average width* (normalized to the mean level spacing D) depends on κ for different values of the interaction λ. The figure demonstrates a sharp change in the width distribution at the transition point $\kappa = 1$. The results are compared with the expression

$$\frac{\langle\Gamma\rangle}{D} = \frac{M}{\pi}\ln\left|\frac{1+\kappa}{1-\kappa}\right| \tag{10}$$

that is obtained from the Moldauer-Simonius (MS) relation [8, 24] for M equivalent channels. We have extrapolated the MS-relation to the region with $\kappa > 1$ for which only $N - M$ narrow resonances were taken into account. For one channel in the GOE case this result was obtained in Refs. [11, 12, 15]; here we see that the MS-expression is also valid for a large number of channels. Note that the results are independent of the degree of intrinsic chaoticity; this fact was explained in Ref.[1]. According to [16], the divergence of $\langle\Gamma\rangle$ at $\kappa = 1$ is due to the (non-integrable) power-law behavior for large Γ, see below; in the numerical simulation, see Fig. 2, there is no divergence because of the finite number of resonances, although the trend is clearly seen.

As for the width distribution $P(\Gamma)$, in Ref. [1] it was shown that for a weak coupling, $\kappa \ll 1$, the conventional χ^2_M-distribution is valid for any strength of the interaction. However, when κ grows, a clear dependence on the interaction strength emerges. As first noted in Ref. [25], as κ increases, $P(\Gamma)$ becomes broader than the χ^2_M distribution. For large κ, both for the GOE and for $\lambda \to \infty$, $P(\Gamma)$ is again given by the χ^2_M distribution, contrary to the cases of the finite interaction strength. Our results are in a good agreement with the analytical predictions [26, 16] obtained for the GOE case (neglecting the emergence of broad resonances in the regime of strong overlap at $\kappa > 1$). In particular, for large number of channels, the tails of $P(\Gamma)$ decrease as Γ^{-2}, thus leading to the divergence of $\langle\Gamma\rangle$ at $\kappa = 1$.

Ericson fluctuations

According to the conventional theory of Ericson fluctuations [27, 28, 29], the scattering amplitude can be written as the sum of the average and fluctuating parts, $\mathscr{T}^{ab}(E) = \langle\mathscr{T}^{ab}(E)\rangle + \mathscr{T}^{ab}_{\rm fl}(E)$, with $\langle\mathscr{T}^{ab}_{\rm fl}(E)\rangle = 0$. Correspondingly, the average cross section, $\sigma = |\mathscr{T}|^2$, can be also divided into two contributions, $\langle\sigma\rangle = \langle\sigma_{\rm dir}\rangle + \langle\sigma_{\rm fl}\rangle$. Here the direct reaction cross section, $\langle\sigma_{\rm dir}\rangle$, is determined by the average scattering amplitude only, while $\langle\sigma_{\rm fl}\rangle$ is the fluctuational cross section (also known as the compound nucleus cross section).

The main assumption of the standard theory is that in the regime of overlapping resonances, $\langle\Gamma\rangle > D$, the fluctuating amplitudes can be written as $\mathscr{T}^{ab}_{\rm fl}(E) = \xi + i\eta$ where both ξ and η are Gaussian random variables with zero mean. This is supported by the fact that for $\langle\Gamma\rangle \gg D$ both ξ and η are the sums of a large number of random variables. This assumption can be associated with the other one, namely, with statistical independence of poles (resonance energies) and residues (resonance amplitudes). As a result, the statistical methods can be developed that lead to various conclusions. In the frame of our model we have tested the most important of the Ericson predictions concerning the resonance widths and cross sections (see also Refs.[1]), paying special attention to the dependence on the intrinsic interaction strength, λ.

Fluctuations of resonance widths. In the theory of Ericson fluctuations it is usually assumed that the fluctuations of resonance widths are small for a large number of channels, $w(\Gamma)/\langle\Gamma\rangle^2 \ll 1$, where $w(\Gamma)$ stands for the variance of widths. In justification of this statement [27] it is argued that in the overlapping regime the width of a resonance can be presented as a sum of M partial widths. Assuming that they obey the Porter-Thomas distribution, the total width is expected to have a χ^2_M distribution, so that $w(\Gamma)/\langle\Gamma\rangle^2 = 2/M$ is small for $M \gg 1$.

However, recently we have found [1] that for large values of κ the distribution of the widths strongly differs from the χ^2_M distribution. The data show that, as κ increases, the normalized variance, $w(\Gamma)/\langle\Gamma\rangle^2$, also increases, remaining very large even for $M = 20 \gg 1$. Thus, contrary to the traditional belief, the relative variance of widths does not become small for large number of channels. The data confirm that for non-overlapping resonances (small $\kappa \ll 1$) there is no dependence on λ and $w(\Gamma)/\langle\Gamma\rangle^2$ decreases as $2/M$ for all the ensembles, as expected. However, as κ grows, the dependence on λ emerges: the weaker the intrinsic chaos (and, consequently, the more ordered is the intrinsic spectrum) the larger are the width fluctuations. For large $\kappa \gg 1$ and strong interaction between particles, $\lambda > \lambda_{cr}$, the width distribution is again given by the χ^2_M distribution, contrary to the cases of the finite interaction strength.

Cross sections. Speaking about cross sections and their fluctuations, it is convenient to use the so-called *elastic enhancement factor*, that for M equivalent channels takes the form,

$$F = \frac{\langle\sigma^{aa}_{\rm fl}\rangle}{\langle\sigma^{ab}_{\rm fl}\rangle}, \tag{11}$$

where $b \neq a$. Here $\langle\sigma^{ab}_{\rm fl}\rangle$ stands for the fluctuational part of cross sections (elastic for $a = b$ and inelastic for $a \neq b$). In the case of equal channels, we obtain

$$\langle\sigma^{ab}_{\rm fl}\rangle = \frac{1 - |\langle S^{aa}\rangle|^2}{F + M - 1} = \frac{T}{F + M - 1}, \tag{12}$$

therefore, $\langle\sigma^{aa}_{\rm fl}\rangle = F\langle\sigma^{ab}_{\rm fl}\rangle$, where T is the transmission coefficient, see Eq. (9). Since the transmission coefficient T does not depend on λ, the only dependence on λ in Eq. (12) is contained in the elastic enhancement factor F. The same seems to be correct even when the channels are non-equivalent, according to the results of Ref. [30]. It should be pointed out that F also depends on κ. Specifically, with an increase of κ from zero, the value of F decreases, being confined by the interval between 3 and 2, see discussion in Ref. [1].

As one can see from Eq.(12), with an increase of the number of channels the dependence on the interaction strength λ disappears for the fluctuational *inelastic* cross section. In contrast, the fluctuational *elastic* cross section manifests a clear dependence on λ, according to the data of Ref. [1]. This fact allows one to directly relate the value of the enhancement factor F to the strength λ of interaction between the particles. Specifically, the more regular is the intrinsic motion, the higher is the average cross section. For a large number of channels, the λ-dependence of the elastic cross section is in agreement with the estimate, $\langle\sigma_{\rm fl}^{aa}\rangle \to FT/M$.

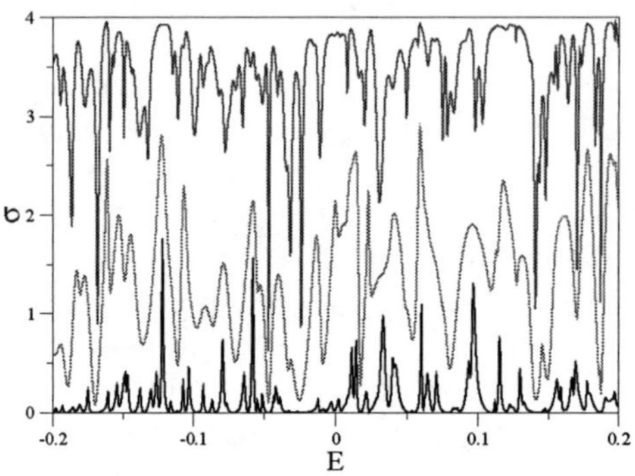

FIGURE 3. Elastic cross-section versus energy E for different strength of coupling to the continuum: for weak, $\kappa = 0.07$, strong, $\kappa = 0.7$, and very strong, $\kappa = 7.0$, for $M = 10$. Other parameters are the same as in Fig.1.

Fluctuations of cross sections. The fluctuations of both elastic and inelastic cross sections strongly depend on the coupling to the continuum, see examples in Fig. 3. According to the standard Ericson theory, in the region of strongly overlapping resonances, $\kappa \approx 1$, the variance of fluctuations of both elastic and inelastic cross sections can be expressed via the average cross sections. Specifically, for equal channels one gets,

$$\mathrm{Var}(\sigma) = \langle\sigma_{\rm fl}\rangle\,(2\langle\sigma_{\rm dir}\rangle + \langle\sigma_{\rm fl}\rangle) = \langle\sigma_{\rm fl}\rangle^2. \tag{13}$$

The latter equality is written for the absence of direct processes, that is our case. Thus, one can expect the relation, $\mathrm{Var}(\sigma)/\langle\sigma_{\rm fl}\rangle^2 = 1$, for a large number of channels. Our data confirm this relation for inelastic cross sections, independently of the interaction strength λ, and for any number of channels.

As for the elastic cross sections, the expression (13) emerges for a quite large number of channels, $M \geq 25$. This fact of slow convergence to the limiting value was also found analytically in Refs. [31, 32] for the GOE case, any number of channels and any coupling strength with the continuum. However, simple expressions were derived only for $MT \gg 1$. Our data reveal a clear dependence on λ that is stronger for the non-normalized variance $\mathrm{Var}(\sigma)$ than for the normalized variance $\mathrm{Var}(\sigma)/\langle\sigma_{\rm fl}\rangle^2$. This effect was explained in Ref. [1].

Another well known prediction of Ericson theory is that for strongly overlapping resonances both inelastic and elastic cross sections have the exponential distribution,

$$P(x) = e^{-x}, \quad x = \frac{\sigma_{\rm fl}}{\langle\sigma_{\rm fl}\rangle}. \tag{14}$$

Indeed, our data for inelastic cross sections show good agreement with this prediction, independently of the interaction strength λ and for large number of channels, $M \geq 10$. For the elastic cross sections, the exponential distribution arises for larger number of channels, $M \geq 25$, with a weak dependence on λ. In general, one can say that the exponential form of the fluctuations is a good approximation in the region $\kappa \approx 1$.

Correlation functions. The correlation function of cross sections is defined as

$$C(\varepsilon) = \langle\sigma(E)\sigma(E+\varepsilon)\rangle - \langle\sigma(E)\rangle^2. \tag{15}$$

By assuming the Gaussian form of distribution for fluctuating amplitudes $\mathcal{T}_{\rm fl}$, one can obtain that the correlation function has a Lorentzian form, with the correlation length equal to the average width, $l_\sigma = \langle\Gamma\rangle$. In our model we have found that $l_\sigma \approx l_S$ for a large number of channels in the region $\kappa \approx 1$, and for any interaction strength λ. Here, l_S is the correlation length obtained from the correlation function of the scattering matrix. On the contrary, for smaller M, our data show that $l_S < l_\sigma$, and this difference grows for the weaker interaction between the particles.

For large M, the correlation functions have, indeed, the Lorentzian form for all λ. Meanwhile, for a small number of channels the correlation function is not Lorentzian, in agreement with the results of [33]. It is important to stress that for any M, the correlation length is different from the average width, apart from the region of small κ. Instead, for a large number of channels, the correlation length is determined by the Weisskopf relation, see [34] and references therein,

$$\frac{l}{D} = \frac{MT}{2\pi} = \frac{M}{2\pi}\frac{4\kappa}{(1+\kappa)^2}. \tag{16}$$

The Weisskopf relation (16) is confirmed by our data in Fig. 2, it has been also derived in Ref. [3] for small values of the ratio $m = M/N$, in the overlapping regime for the TBRI model with the infinite interaction, as well as in Ref. [4] for the GOE ensemble. The fact that the correlation length is not equal to the average width was recognized long ago, see Refs. [35] and [5].

Conductance fluctuations

Our model can be also used to study conductance and its fluctuations for quantum dots with interacting electrons. In such an application one can treat $M/2$ channels as the *left channels* corresponding to incoming electron waves, and other $M/2$ channels, as the *right channels* corresponding to outgoing waves. Then, one can define the conductance G in a standard way,

$$G = \sum_{a=1}^{M/2} \sum_{b=M/2+1}^{M} \sigma^{ab} \tag{17}$$

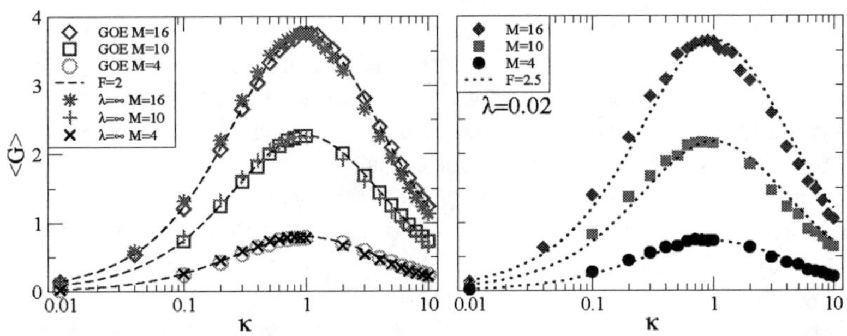

FIGURE 4. Average conductance versus κ for the GOE and very strong interaction, $\lambda \to \infty$ (left panel), and for a weak interaction, $\lambda = 0.02$ (right panel) for different number of channels, $M = 4, 10, 16$. Dashed curves correspond to the analytical expression (18) with $F = 2.0$, (left panel), and dotted curves with $F = 2.5$, (right panel). The data are obtained for $n = 6$ particles, $m = 12$ orbitals, and $M = 4, 10, 16$ channels.

Therefore, the properties of the conductance are entirely determined by the inelastic cross-sections, $b \neq a$. In our case of equivalent channels the average conductance reads,

$$\langle G \rangle = \frac{M^2}{4}\langle \sigma^{ab} \rangle = \frac{M^2}{4}\frac{T}{F+M-1} \to \frac{MT}{4} \tag{18}$$

where $T = 4\kappa/(1+\kappa)^2$ and the last expression is valid for $M \gg 1$. As one can see, the influence of the inter-particle interaction is due to the enhancement factor F only. This factor ranges from $F = 3$ for the absence of the interaction, $\lambda = 0$, to $F = 2$ for a very strong interaction that can be modeled by the GOE ensemble [36]. Since currently there is no theory relating the enhancement factor to the degree of chaos in the internal dynamics, we used this factor as the fitting parameter. In Fig. 4 we report our numerical data obtained for the average conductance in dependence on the coupling strength κ for different values of M. As expected, for a very strong interaction (absence of the mean field part, $d_0 = 0$, or, the same, $\lambda \to \infty$) the results are very close to those obtained for the GOE Hamiltonian H, and there is a good agreement with the expression (17) with the theoretical value $F = 2$. On the other hand, for a relatively weak interaction, $\lambda = 0.02$, (below the chaos threshold in a closed model) the data can be described by the same expression with $F = 2.5$ that is in the middle of the region between the Poisson and Wigner-Dyson statistics for the energy spacings. As one can see, for this intermediate case the correspondence between the relation (18) and data is also good. Note, that for a very large number of channels the influence of the internal dynamics disappears.

Our main interest was in the variance, $\mathrm{Var}(G)$, of the conductance when we change the energy E around the center of the energy spectrum. In Fig. 5 we report the data for the variance for the GOE case (left part) and for a relatively weak interaction $\lambda = 0.02$ in dependence on the coupling parameter κ and number M of channels. First observation concerns a clear evidence that for a weak coupling, $\kappa \approx 0.1$, the role of the interaction should not be neglected. This fact is in contrast with the average conductance for which the influence of the internal chaos seems to be much less in comparison with the conductance fluctuations.

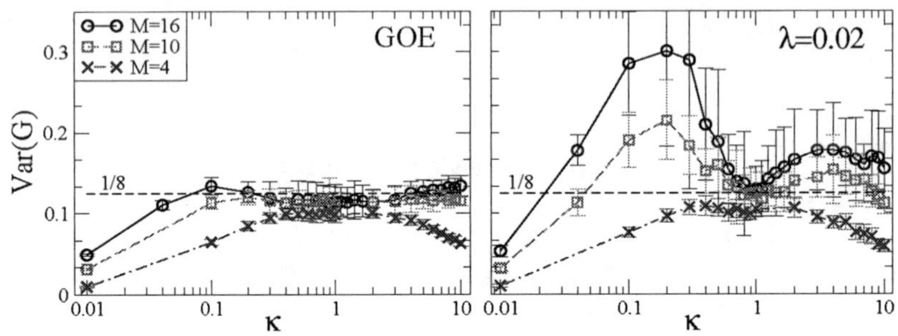

FIGURE 5. Variance of the conductance versus κ for the GOE (left panel), and for $\lambda = 0.02$, (right panel) . Circles, squares and crosses stand for $M = 16, 10, 4$ channels, respectively. The data are given for $n = 6$ and $m = 12$.

Second important observation is that at the transitional point, $\kappa \approx 1$ (known in the solid state applications as "perfect coupling"), practically there is no influence of the inter-particle interaction. This fact is quite instructive, it manifests an important role of the superradiance when the (strong) interaction between the particles is mainly due to the continuum. Also, the dependence of fluctuations on the number of channels at the transitional region is relatively weak (and disappears with an increase of M). In the theory of conductance fluctuations it is known that for a perfect coupling the variance takes the universal value in the limit of a very large number of channels. This value for time-invariant systems is either 2/15 or 1/8 depending on the type of transport (ballistic or diffusive, both for a bulk scattering, see, e.g. [37]). It is quite instructive that our data confirm the value 1/8 for the GOE at $\kappa = 1$ with a very good accuracy. Specifically, the data also reveal small corrections due to the finite values of M, that reduce the variance to that of slightly below 1/8, in agreement with the analytical results obtained in Ref. [37]). Also, the data indicate the value 2/15 for the TBRI model with a moderate interaction between the particles.

We have performed a detailed analysis of the emergence of the universal value 1/8. We should point out that one has to take into account the correlations between different cross sections in the expression for the variance of the conductance. Otherwise, one gets the incorrect result,

$$\mathrm{Var}(G) = \frac{M^2}{4}\mathrm{Var}(\sigma) = \frac{M^2}{4}\left(\frac{T}{F+M-1}\right)^2 \to \frac{T^2}{4}, \tag{19}$$

where the last expression is obtained for $M \gg 1$. Thus, for the perfect coupling, we would have $\mathrm{Var}(G) = 1/4$, which is twice the correct value.

In order to reveal the role of correlations between different cross sections, instead of Eq. (19) we write the relation that takes into account all the correlations,

$$\mathrm{Var}(G) = \frac{M^2}{4}\left(\frac{T}{F+M-1}\right)^2 + N^* C_\Sigma + (N_c - N^*) C_\Lambda, \tag{20}$$

where $N_c = L(L-1)$, $N^* = L(M-2)$, $L = M^2/4$. Here the terms C_Λ and C_Σ stand for the correlation functions,

$$C = \overline{\langle \sigma^{ab}\sigma^{a'b'} \rangle} - \overline{\langle \sigma^{ab} \rangle \langle \sigma^{a'b'} \rangle} \tag{21}$$

with one common index (either $a = a'$ or $b = b'$) for the $\Sigma-$correlations, and with no common indexes (both $a \neq a'$ and $b \neq b'$) for the $\Lambda-$correlations. The bar in Eq. (21) represents the average over different possible combinations of a, b, a', b', and $\langle ... \rangle$ indicate the average over energy. Our analysis shows that for $M \gg 1$ one obtains, $C_\Sigma \approx -M^{-3}$ and $C_\Lambda \approx 2M^{-4}$. Therefore, in the limit of large number of channels, one can write, in the correspondence with the structure of Eq.(20),

$$\mathrm{Var}(G) = \frac{1}{4} - \frac{1}{4} + \frac{1}{8} = \frac{1}{8}. \tag{22}$$

This result clearly demonstrates the crucial role of correlations for the variance of the conductance. These correlations are neglected in the traditional theory of Ericson fluctuations.

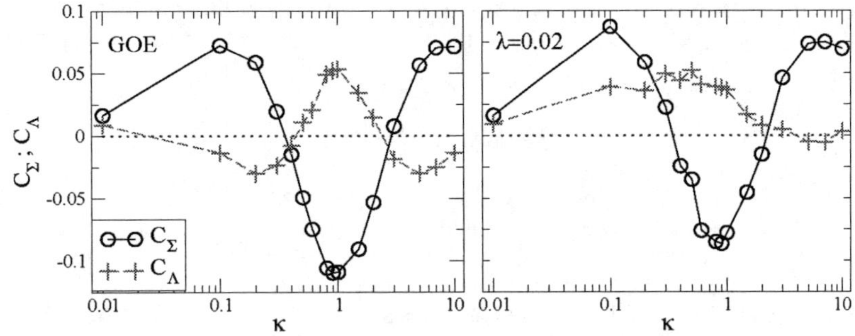

FIGURE 6. Dependence of average correlations between cross sections on the coupling κ for the GOE case (left panel) and for a weak interaction with $\lambda = 0.02$ (right panel). Circles stand for the C_Σ and for the C_Λ correlations (refer to text for details). Both panels are on the same scale for $n = 6$ and $m = 12$.

Of special interest is the $\kappa-$dependence of the correlations for a non-perfect coupling, in the regions $\kappa < 1$ and $\kappa > 1$. As one can see, the $\Lambda-$ and $\Sigma-$correlations have very different behavior, both for weak and strong intrinsic chaos. The remarkable fact is that, for the perfect coupling with $\kappa = 0$, the $\Lambda-$ and $\Sigma-$correlations have maximal values of opposite signs. These results may be important for the analysis of experimental nuclear or solid state data.

Conclusions

In conclusion, we have studied the statistics of resonance widths and cross sections for a fermion system coupled to open decay channels. We used the effective non-Hermitian Hamiltonian that reflects exact quantum-mechanical features of many-body dynamics in an open interacting system. Main attention was paid to various signatures of the crossover from isolated to overlapping resonances in dependence on the strength of inter-particle interaction modeled by the two-body random matrices in the intrinsic part of the non-Hermitian Hamiltonian.

In the framework of our model we have tested main predictions of the Ericson fluctuation theory. We have shown that the assumption that the fluctuations of the resonance widths become negligible for a large number of channels, is invalid in the overlapping regime. We found that the fluctuations of resonance widths increase along with the coupling to the continuum, resulting in the divergence of the relative fluctuation of the widths (the ratio of the variance to the square of the average width) at the transition point $\kappa = 1$ for any number of channels.

We have also shown that, in agreement with previous studies, the correlation length differs from the average width for any number of channels. On the other hand, the Weisskopf relation (16) that connects the correlation length of the cross section to the transmission coefficient, works, for a large number of channels, at any value of the intrinsic interaction strength λ. In many situations the data show that increase of λ suppresses the fluctuations in the continuum. This can be understood qualitatively as a manifestation of many-body chaos that makes all internal states uniformly mixed [38].

Our results can be applied to any many-fermion system coupled to the continuum of open decay channels. The natural applications first of all should cover neutron resonances in nuclei, where rich statistical material was accumulated but the transitional region from isolated to overlapped resonances was not studied in detail. Other open mesoscopic systems, for example, quantum dots and quantum wires, should be analyzed as well in the crossover region.

It was our special interest to study the mesoscopic fluctuations of the conductance in the dependence on the coupling strength and degree of intrinsic chaos. In particular, we have analyzed how the correlations between different cross sections influence the variance of the conductance fluctuations. Our data manifest that these correlations determine the universal value 1/8 (see, for example, [39, 37]) for the conductance variance in the perfect coupling regime, $\kappa = 1$, and for very strong interaction between particles. Thus, we demonstrate that the Ericson theory of cross-section fluctuation that neglects these correlations, is insufficient for the description of the mesoscopic fluctuations.

ACKNOWLEDGMENTS

We acknowledge useful discussion with T. Gorin, Y. Fyodorov, D. Savin, and V. Sokolov. The work was supported by the NSF grants PHY-0244453 and PHY-0555366. The work by G.P.B. was carried out under the auspices of the National Nuclear Security Administration of the U.S. Department of Energy at Los Alamos National Laboratory under Contract No. DE-AC52-06NA25396. S.S. acknowledges the financial support from the Leverhulme Trust.

REFERENCES

1. G.L. Celardo, F.M. Izrailev, V.G. Zelevinsky, and G.P. Berman, *Phys. Rev. E* **76**, 031119 (2007); *Phys. Lett. B* **659**, 170 (2008).
2. C. Mahaux and H.A. Weidenmüller, *Shell Model Approach to Nuclear Reactions* (North Holland, Amsterdam, 1969).
3. D. Agassi, H.A. Weidenmüller, and G. Mantzouranis, *Phys. Rep.* **3**, 145 (1975).
4. J.J.M. Verbaarschot, H.A. Weidenmüller, and M.R. Zirnbauer, *Phys. Rep.* **129**, 367 (1985).
5. T.A. Brody, J. Flores, J.B. French, P.A. Mello, A. Pandey, and S.S.M. Wong, *Rev. Mod. Phys.* **53**, 385 (1981).
6. T. Guhr, A. Müller-Groeling, and H.A. Weidenmüller, *Phys. Rep.* **299**, 189 (1998).
7. P.A. Mello and N. Kumar, *Quantum Transport in Mesoscopic Systems* (Oxford University Press, Oxford, 2004).
8. P.A. Moldauer, *Phys. Rev.* **157**, 907 (1967); *Phys. Rev. Lett.* **23**, 708 (1969).
9. M. Simonius, *Phys. Lett. B* **52**, 279 (1974).
10. P. Kleinwächter and I. Rotter, *Phys. Rev. C* **32**, 1742 (1985).
11. V.V. Sokolov and V.G. Zelevinsky. *Phys. Lett. B* **202**, 10 (1988).
12. V.V. Sokolov and V.G. Zelevinsky, *Nucl. Phys.* **A504**, 562 (1989).
13. V.V. Sokolov and V.G. Zelevinsky, *Ann. Phys. (N.Y.)* **216**, 323 (1992).
14. R.H. Dicke, *Phys. Rev.* **93**, 99 (1954).
15. F.M. Izrailev, D. Sacher and V.V. Sokolov, *Phys. Rev. E* **49**, 130 (1994).
16. Y.V.Fyodorov and H.-J.Sommers, *J. Math. Phys.* **38**, 1918 (1997); H.-J. Sommers, Y.V.Fyodorov, and M.Titov, *J. Phys. A: Math. Gen.* **32**, L77 (1999).
17. I. Rotter, *Rep. Prog. Phys.* **54**, 635 (1991).
18. A. Volya and V. Zelevinsky, *Phys. Rev. Lett.* **94**, 052501 (2005).
19. A. Volya and V. Zelevinsky, in *Nuclei and Mesoscopic Physics*, ed. V. Zelevinsky, AIP Conference Proceedings **777**, 2005, p. 229.
20. A. Volya and V. Zelevinsky, *Phys. Rev. C* **74**, 064314 (2006).
21. V.V. Flambaum and F.M. Izrailev, *Phys. Rev. E* **56**, 5144 (1997).
22. V. Zelevinsky and A. Volya, *Phys. Rep.* **391**, 311 (2004).
23. F. Haake, F.M. Izrailev, N. Lehmann, D. Saher and H.-J. Sommers, *Z. Phys.* **B88**, 359 (1992).
24. M. Simonius, *Phys. Lett. B* **52**, 279 (1974).
25. P.A. Moldauer, *Phys. Rev.* **171**, 1164 (1968).
26. S. Mizutori and V.G. Zelevinsky, *Z. Phys.* **A346**, 1 (1993).
27. T. Ericson, *Ann. Phys.* **23**, 390 (1963); *Phys. Rev. Lett.* **5**, 430 (1960).
28. D. Brink and R. Stephen, *Phys. Lett.* **5**, 77 (1963).
29. T. Ericson and T. Mayer-Kuckuk, *Ann. Rev. Nucl. Sci.* **16**, 183 (1966).
30. A. Muller and H.L. Harney, *Phys. Rev. C* **33**, 1228 (1987).
31. E.D. Davis and D. Boose, *Phys. Lett. B* **211**, 379 (1988).
32. E.D. Davis and D. Boose, *Z. Phys.* **A332**, 427 (1989).
33. F.M. Dittes, H.L. Harney and A. Muller, *Phys. Rev. A* **45**, 701 (1992).
34. N.Lehmann, D.Saher, V.V. Sokolov and H.-J. Sommers, *Nucl. Phys.* **A582**, 223 (1995).
35. P.A. Moldauer, *Phys. Rev. C* **11**, 426 (1975).
36. T. Gorin and T.H. Seligman, *Phys. Rev. E* **65**, 26214 (2002).
37. C.W.J. Beenakker, *Rev. Mod. Phys.* **69**, 731 (1997).
38. V. Zelevinsky, B.A. Brown, N. Frazier, and M. Horoi, *Phys. Rep.* **276**, 85 (1996).
39. S. Ida, H. A. Weidenmüller and J. A. Zuk, *Phys. Rev. Lett.* **64**, 583 (1990); *Ann. Phys. (N.Y.)* **200**, 219 (1990).

Decay Rates Statistics of Unstable Classically Chaotic Systems

Valentin V. Sokolov

Budker Institute of Nuclear Physics, Novosibirsk, Russia
Center for Nonlinear and Complex Systems, Università degli Studi dell'Insubria, Via Valleggio 11, 22100 Como, Italy

Abstract. Decay law of a complicated unstable state formed in a high energy collision is described by the Fourier transform $K(t)$ of the two-point correlation function of the scattering matrix. Although each constituent resonance state decays exponentially the decay of a state composed of a large number $N \gg 1$ of such interfering resonances is not, generally, exponential. We introduce the decay rates distribution function $w(\Gamma)$ by representing the decay law in the form of the mean-weighted decay exponent $K(t) = \int_0^\infty d\Gamma e^{-\Gamma t} w(\Gamma)$. In the framework of the random matrix approach we investigate the properties of the distribution function $w(\Gamma)$ and its relation to the more conventional statistics of the decay widths. The latter is not in fact conclusive as concerns the evolution at the times shorter than the characteristic Heisenberg time. Exact analytical consideration is presented for the case of systems without time reversal symmetry.

Keywords: Chaotic scattering, Decay rates, Resonance widths
PACS: 05.45.Mt, 03.65.Nk, 05.45.+b, 24.60.-v

INTRODUCTION

The temporal aspect of chaotic resonance scattering is an interesting and important issue which repeatedly attracted much attention starting from the seminal works of Wigner and Smith [1, 2]. Later on different problems concerning the duration of resonance collisions were posed and discussed in detail. Distribution of resonance widths on the one hand and statistics of partial and proper delay times on the other have been investigated in a number of papers [3–16].

It is generally known that a single quasi-stationary state formed by the time $t = 0$ in a collision with a given energy E decays afterwards exponentially with the decay rate Γ_r which is given by the imaginary part of the resonance complex energy $\mathscr{E}_r = E_r - \frac{i}{2}\Gamma_r$. In accordance with the Bohr energy-time uncertainty principle this rate defines the width of the Breit-Wigner resonant curve.

However at high enough collision energy E a very complicated unstable state is formed which is a superposition of many, $N \gg 1$, interfering resonance contributions whose spectrum is given by the N complex eigenvalues $\mathscr{E}_r = E_r - \frac{i}{2}\Gamma_r$ of $N \times N$ matrix of the non-Hermitian effective Hamiltonian $\hat{\mathscr{H}} = \hat{H} - \frac{i}{2}\hat{W}$, $\hat{W} = \hat{A}\hat{A}^\dagger$. The resonances, generally, overlap so that their mean width is larger than the mean level spacing, $\langle\Gamma\rangle > \Delta$. The evolution of such an unstable configuration formed via an incoming channel a and decaying onto an outgoing channel b is described [4, 7] by the Fourier transform

$$K^{ba}(\tau) = \frac{1}{2\pi}\int_{-\infty}^{\infty} d\omega\, e^{i\omega\tau} C^{ba}(\omega) \tag{1}$$

of the two-point S-matrix correlation function

$$C^{ba}(\omega = t_H\varepsilon) = \frac{\langle S^{ba}(E+\varepsilon/2) S^{ba*}(E-\varepsilon/2)\rangle_{conn}}{T^b T^a}; \qquad b \neq a \tag{2}$$

Here $T^c = 1 - |S^{cc}|^2$ is the transmission coefficient in the channel c and $t_H = 2\pi/\Delta$ is the Heisenberg (Weisskopf) time. It is convenient to measure the time in the units of this characteristic time interval, $\tau = t/t_H$. Correspondingly, the dimensionless energy shift $\omega = \varepsilon t_H$ is introduced in the Eqs. (1, 2). To avoid complications irrelevant in the present context we restrict for a while our consideration to inelastic collisions only.

Contrary to the case of an isolated quasi-stationary state, the function $K^{ba}(\tau)$ does not, generally, decays exponentially and a connection to the widths of the constituting resonances is not immediately seen. More than that, an opinion has not once been expressed that the very notion of the resonance widths becomes irrelevant when resonances strongly

CP995, *Nuclei and Mesoscopic Physics - WNMP 2007*, edited by P. Danielewicz, P. Piecuch, and V. Zelevinsky

overlap. Nevertheless, intuitively, one would expect that decay properties of a complicated unstable state should somehow depend on the distribution of the individual resonance poles of the scattering amplitudes in the complex energy plane. Could then a non-exponential decay results from an incoherent mixture of exponential decays with different rates? If and how such a mixture is related to the statistics of the widths of resonances? We address these questions below. They are answered explicitly by the example of the chaotic systems without time-reversal symmetry. A rigorous analytical solution is found in this case.

DECAY RATES DISTRIBUTION FUNCTION

The function $C^{ba}(\omega)$ is analytical in the lower part of the complex ω plane and, obviously, satisfies the condition $C^{ba}(-\omega) = C^{ba*}(\omega)$. As a result its Fourier transform $K^{ba}(\tau) \equiv 0$ when $\tau < 0$ and is real when $\tau > 0$. More than that, independently of the number M of the reaction channels, [17]

$$K^{ba}(0) = 1 \text{ and } K^{ba}(\tau) > 0 \text{ when } \tau > 0. \tag{3}$$

Notice at last that the mean reaction cross section is expressed as

$$\langle\sigma^{ba}\rangle = T^b T^a C^{ba}(0) = T^b T^a \int_0^\infty d\tau K^{ba}(\tau). \tag{4}$$

Such properties suggest the Laplace representation for the decay law function

$$K^{ba}(\tau) = \int_0^\infty d\gamma w^{ba}(\gamma) e^{-\gamma\tau}. \tag{5}$$

(In the ordinary units $\gamma = \Gamma t_H = \frac{2\pi}{\Delta}\Gamma$ where Γ is the decay rate.) In view of the Eq. (3) the function w^{ba} is normalized to unity,

$$\int_0^\infty d\gamma w^{ba}(\gamma) = K^{ba}(0) = 1. \tag{6}$$

It follows from the Eqs. (1, 5) that

$$C^{ba}(\omega) = \int_0^\infty d\gamma \frac{w^{ba}(\gamma)}{\gamma + i\omega} \tag{7}$$

so that the correlation function is not, generally, single-valued in the complex ω plane. Finally, according to the Eqs. (4, 7),

$$\langle\sigma^{ba}\rangle = T^b T^a \int_0^\infty d\gamma \frac{w^{ba}(\gamma)}{\gamma}. \tag{8}$$

From this point on we suppose that all channels are statistically equivalent, $T^b = T^a = T$, so that the functions $C(\omega)$, $K(\tau)$, $w(\gamma)$ do not depend on the channel indices.

THE SEMICLASSICAL ASYMPTOTIC EXPANSION

Let us consider first the semiclassical limit: $N \gg 1$, $M \gg 1$ but the ratio $m = M/N$, is finite though small, $m \ll 1$, (this condition is physically justified). The function $w(\gamma)$ can easily be extracted from the results reported in: [5, 9] (GOE)

$$C(\omega) = \frac{1}{MT}\frac{\gamma(\omega)}{\gamma(\omega) + i\omega} \Rightarrow \frac{1}{\gamma_W + i\omega} - \frac{2T}{(\gamma_W + i\omega)^2} + \frac{MT^2}{(\gamma_W + i\omega)^3} + \ldots \tag{9}$$

where $\gamma(\omega)$ is a slow varying function and $\gamma(0) \equiv \gamma_W = MT = t_H\Gamma_W$ where $\Gamma_W = \frac{\Delta}{2\pi}MT$ is the Weisskopf width. The asymptotic series in the r.h.s which is obtained by expanding near the pole $\omega = i\gamma_W$ coincides with that derived first in a different manner in: [18]. The first term corresponds to the Hauser-Feshbach approximation and implies the purely exponential decay $e^{-\gamma_W\tau}$ when the subsequent ones account for deviations. In the limit $N, M \to \infty$, $0 < m \ll 1$ the Weisskopf width equals to the empty gap between the real energy axis and the cloud of the poles corresponding to contributing resonances [9]. The width Γ_W is in this case the smallest possible width.

It immediately follows from the expansion (9) that

$$w(\gamma) = \delta(\gamma - \gamma_W) - 2T\,\delta'(\gamma - \gamma_W) + \frac{MT^2}{2}\,\delta''(\gamma - \gamma_W) + \ldots. \qquad (10)$$

The function $w(\gamma)$ thus obtained is neither smooth nor positive definite. We will see below that the expansion (9) is in this respect somewhat misleading. Nevertheless it works well in certain cases. In particular, for the mean cross section we get from the Eqs. (8, 10)

$$\langle\sigma\rangle^{(GOE)} = \frac{T}{M}\left(1 - \frac{1}{M} + \ldots\right). \qquad (11)$$

Similar calculation yields in the GUE-case $\langle\sigma\rangle^{(GUE)} = \frac{T}{M}$.

Notice that the difference $\langle\sigma\rangle^{(GUE)} - \langle\sigma\rangle^{(GOE)} = \frac{T}{M^2}$ determines the "weak localization" in the quantum transport.

DECAY RATES VERSUS WIDTHS STATISTICS

Heuristic" arguments have been adduced in [19] which assume a simple connection between the function $w(\gamma)$ and the distribution $\rho_M(\gamma)$ of the resonance widths

$$w_M(\gamma) = A\,\gamma^2 \rho_M(\gamma) \qquad (12)$$

with some normalization constant A. A formula of such a kind has been derived a little later in [8]. It was shown, however, that the relation supposed can reproduce only the long-time asymptotic behavior. Only the most long-lived resonances survive by this time, so they can be expected to obey the χ^2_M widths statistics.

$$\rho_M(\gamma) = \frac{1}{(\beta\frac{M}{2})!}\,\frac{1}{\gamma}\left(\beta\frac{\gamma}{2T}\right)^{\beta\frac{M}{2}} e^{-\beta\frac{\gamma}{2T}}, \quad \langle\gamma\rangle = MT = \gamma_W,\ \beta = 1, 2. \qquad (13)$$

Here the parameter β marks the Dyson's symmetry class: GOE ($\beta = 1$) or GUE ($\beta = 2$) and the subscript M indicates explicitly the number of open channels. Notice that in our dimensionless units $\Gamma/\langle\Gamma\rangle = \gamma/MT$ which means that $\langle\gamma\rangle = MT = \gamma_W$. In the case of isolated resonances the Weisskopf width defines the mean rather than the minimal width. The formal extending the relation (12) to all values of γ yields the result

$$K_M(\tau) = AM\left(M + \frac{\beta}{2}\right)\frac{T^2}{\left(1 + \frac{2T}{\beta}\tau\right)^{\beta\frac{M}{2}+2}} \rightarrow \frac{const.}{\tau^{\beta\frac{M}{2}+2}} \qquad (14)$$

which reproduces rightly the power law asymptotic behavior [5] though the proper calculation of the constant is beyond the validity of this approximation. The constant A can be chosen to quantitatively fit the correct asymptotics. However at shorter times the found in such a way expression disagree with the actual decay law, the discrepancy being the stronger the closer is the transmission coefficient T to its maximal value 1 (see [8]). There is no way to reconcile all necessary conditions which must me satisfied by the decay function $K_M(\tau)$ by means of only one matching parameter A. There are two possible reasons why the supposed relation (12) fails: i. the χ^2_M distribution is not valid when resonances overlap [11]; ii. the naive derivation is wrong. We will clarify below this guess-work by the example of the systems with broken time-reversal symmetry which are described by the Gaussian unitary ensemble (GUE) of random Hamiltonians.

SYSTEMS WITH NO TIME-REVERSAL SYMMETRY

General consideration

The GUE two-point correlation function derived in: [20] reduces in the case of equivalent channels to

$$\begin{array}{c} C_M(\omega) = \int_0^1 d\lambda \int_0^\infty d\lambda_1 e^{-i\omega(\lambda+\lambda_1)}\,\frac{1}{\lambda+\lambda_1}\,\frac{(1-T\lambda)^{M-2}}{(1+T\lambda_1)^{M+2}} \times \\ \left\{\left(\frac{2}{T} - 1\right)(1 - T\lambda)(1 + T\lambda_1) - \frac{1-T}{T}\left[(1 - T\lambda) + (1 + T\lambda_1)\right]\right\}. \end{array} \qquad (15)$$

Making use of the following simple identity

$$\frac{1}{\lambda_1+\lambda} = \int_0^\infty d\gamma e^{-\gamma(\lambda_1+\lambda)} \tag{16}$$

we factorize the integrations over λ and λ_1. On the next step we use a second identity

$$\int_0^\infty d\lambda_1 \frac{e^{-(\gamma+i\omega)\lambda_1}}{(1+T\lambda_1)^{M+1+\mu}} = \frac{1}{(M+\mu)!}\int_0^\infty d\eta \frac{\eta^{M+\mu}\, e^{-\eta}}{T\eta+\gamma+i\omega},\ \mu=0,1. \tag{17}$$

The correlation function (15) is thus reduced to a linear combination of terms like

$$c_M^{\mu\mu'}(\omega) = \frac{1}{(M+\mu)!}\int_0^\infty d\gamma \int_0^\infty d\eta\, \eta^{M+\mu} e^{-\eta} \int_0^1 d\lambda (1-T\lambda)^{M-1-\mu'} e^{-\gamma\lambda} \frac{e^{-i\omega\lambda}}{T\eta+\gamma+i\omega}. \tag{18}$$

After a chain of simple transformations of the integration variables the Fourier transform of the latter function reduces finally to

$$\begin{aligned} k_M^{\mu\mu'}(\tau) = \tfrac{1}{T^{2+\mu+\mu'}(M+\mu)!} \textstyle\int_0^\infty d\gamma e^{-\gamma\tau} \int_0^\gamma d\gamma' \gamma'^{(\mu+\mu')} \times \\ \textstyle\int_{\frac{1-T}{T}\gamma'}^{\gamma/T} d\nu\, \Theta\left(\tau - \frac{1}{T} + \frac{\nu}{\gamma}\right) \nu^{M-1-\mu'} e^{-\nu}. \end{aligned} \tag{19}$$

The symbol Θ stands for the step function. All successive integrations in this expression can be carried out explicitly [21].

To simplify further consideration we restrict ourselves to the case of the perfect coupling to the continuum, $T=1$, when the naive consideration is the least satisfactory. Only the term with $\mu=\mu'=0$ remains in this limit and

$$C_M(\omega) = \frac{1}{M!}\int_0^\infty d\gamma \int_0^\infty d\gamma' \gamma'^M e^{-\gamma'} \int_0^1 d\lambda\, (1-\lambda)^{M-1} e^{-\gamma\lambda} \frac{e^{-i\omega\lambda}}{\gamma+\gamma'+i\omega}. \tag{20}$$

Notice that when $T=1$ the formula (20) and so all consequent results are valid also for the elastic collisions including the purely elastic process with one open channel $M=1$.

The found representation enables us to evaluate the mean cross section in the following simple and elegant way:

$$\langle\sigma\rangle = C_M(0) = \frac{1}{M!}\int_0^\infty d\gamma F_M(\gamma)\Phi_{M-1}(\gamma) = \frac{1}{M}\int_0^\infty d\gamma e^{-\gamma} F_0(\gamma) = \frac{1}{M}. \tag{21}$$

The third equality follows from the fact that the functions

$$F_M(\gamma) = \int_0^\infty d\eta\, \eta^M \frac{e^{-\gamma\eta}}{1+\eta}, \qquad \Phi_{M-1}(\gamma) = \int_0^\gamma d\zeta\, (\gamma-\zeta)^{M-1} e^{-\zeta} \tag{22}$$

which appear in this equation obey the simple recursions

$$F_M(\gamma) = -\frac{d}{d\gamma}F_{M-1}(\gamma), \qquad \frac{d}{d\gamma}\Phi_M(\gamma) = M\Phi_{M-1}(\gamma). \tag{23}$$

The decay function looks now as

$$K_M(\tau) = \int_0^\infty d\gamma e^{-\gamma\tau} \frac{1}{M!}\int_0^\gamma d\gamma' \int_0^{\gamma'} d\nu \Theta(\tau - 1 + \nu/\gamma')\, \nu^{M-1} e^{-\nu}. \tag{24}$$

The later consideration depends on whether the time τ exceeds the Heisenberg time $\tau_H = 1$ or not.

Long-time asymptotics

After the Heisnberg time, $\tau > 1$, the step function in the Eq. (24) equals to one in the whole integration region and the decay function gets the required form

$$K_M(\tau > 1) = \int_0^\infty d\gamma e^{-\gamma\tau} w_>(M,\gamma) \tag{25}$$

with the decay rates distribution function given by

$$\begin{array}{c} w_>(M,\gamma)=\frac{1}{M!}\int_0^\gamma d\gamma'\,\varphi_{M-1}(\gamma'),\\ \varphi_{M-1}(\gamma)=\int_0^\gamma d\nu\,\nu^{M-1}e^{-\nu}=(M-1)!-\Gamma(M,\gamma)\end{array} \tag{26}$$

where $\Gamma(M,\gamma)$ is the incomplete Γ function

$$\Gamma(M,\gamma)=(M-1)!\,e^{-\gamma}\sum_{m=0}^{M-1}\frac{\gamma^m}{m!}. \tag{27}$$

The next integration over γ' yields finally

$$\begin{array}{c} w_>(M,\gamma)=\frac{\Gamma(M+1,\gamma)-\gamma\Gamma(M,\gamma)}{M!}+\gamma\frac{1}{M}-1=\widetilde{w}_M(\gamma)-\left(1-\gamma\frac{1}{M}\right),\\ \widetilde{w}_M(\gamma)=e^{-\gamma}\sum_{m=0}^{M-1}\left(1-\frac{m}{M}\right)\frac{\gamma^m}{m!}\equiv e^{-\gamma}P_{(M-1)}(\gamma).\end{array} \tag{28}$$

As expected, this distribution is smooth and positive definite. Notice that, due to subtraction of the last two terms, the distribution $w_>(M,\gamma)$ vanishes as γ^2 when $\gamma\to 0$.

At last the final integration over γ yields

$$K_M(\tau>1)=\int_0^\infty d\gamma e^{-\gamma\tau}w_>(M,\gamma)=\frac{1}{M}\frac{1}{\tau^2(1+\tau)^M},\quad K_M(\tau\to\infty)=\frac{\langle\sigma\rangle}{\tau^{(M+2)}} \tag{29}$$

thus fixing not only the power but also the constant of asymptotic power behavior.

The found results are closely connected to statistics of the decay width. Indeed the widths distribution function in the case of M equivalent channels, arbitrary degree of resonance overlapping and perfect coupling to the continuum is currently well known to be [11]

$$\rho_M(\gamma)=\frac{1}{(M-1)!\,\gamma^2}\int_0^\gamma d\nu\,\nu^M e^{-\nu} \tag{30}$$

and has nothing to do with the χ^2_M-distribution (13). It is immediately seen that

$$\varphi_{M-1}(\gamma)=(M-2)!\,\gamma^2\rho_{M-1}(\gamma) \tag{31}$$

(see the Eq. (26)). This results in the following connection between the two distributions

$$\begin{array}{c} w_>(M,\gamma)=\frac{1}{M(M-1)}\int_0^\gamma d\gamma'\,\gamma'^2\rho_{M-1}(\gamma')=\\ \frac{1}{M^2}\left[\gamma^2\rho_M(\gamma)+\int_0^\gamma d\gamma'\,\gamma'^2\rho_M(\gamma')\right]=\frac{1}{M^2}\left(\frac{d}{d\gamma}+1\right)\int_0^\gamma d\gamma'\,\gamma'^2\rho_M(\gamma').\end{array} \tag{32}$$

The extra term which appears in this relation is missing in the naive formula (12).

Short-time behavior

At the times shorter than τ_H, $\underline{\tau<1}$, an additional term arises,

$$\int_0^\gamma d\nu\,\Theta(\tau-1+\nu/\gamma)\,\nu^{M-1}e^{-\nu}=\varphi_{M-1}(\gamma)-\int_0^{(1-\tau)\gamma}d\nu\,\nu^{M-1}e^{-\nu}, \tag{33}$$

which depends on the time. As a consequence the decay law contains along with the smoothly distributed exponential contributions an additional polynomial term

$$\begin{array}{c} K_M(\tau<1)=\int_0^\infty d\gamma e^{-\gamma\tau}\widetilde{w}_M(\gamma)+\frac{1}{M}\frac{1-(1-\tau)^M}{\tau^2}-\frac{1}{\tau}=\\ \int_0^\infty d\gamma e^{-\gamma\tau}\widetilde{w}_M(\gamma)-\frac{1}{M}\sum_{m=0}^{M-2}(-1)^m\begin{pmatrix}M\\ m+2\end{pmatrix}\tau^m.\end{array} \tag{34}$$

Formally, we still can present this formula in the form

$$K_M(\tau < 1) = \int_0^\infty d\gamma e^{-\gamma\tau} w_<(M,\gamma) \tag{35}$$

with a "distribution density" $w_<(M,\gamma) = \widetilde{w}_M(\gamma) + w_M^{(sing.)}(\gamma)$. The price paid is the singular nature of the additional weight function

$$w_M^{(sing.)}(\gamma) = -\frac{1}{M}\sum_{m=0}^{M-2}(-1)^m \left(\begin{array}{c} M \\ m+2 \end{array}\right) \frac{d^m}{d\gamma^m}\delta(\gamma), \quad M \geqslant 2\,. \tag{36}$$

This singular "distribution" differs in two important respects from that extracted from the Verbaarschot's asymptotic expansion: first, it contains only a finite sum of derivatives of the δ-function and, second, all singular terms are concentrated near $\gamma = 0$ rather than $\gamma = \gamma_W$.

We should stress that the both regular and singular parts play, generally, equally important roles. In particular,

$$K_M(0) = \frac{M+1}{2} - \frac{M-1}{2} = 1\,. \tag{37}$$

The first contribution comes from the regular and the second from the singular parts of the "distribution" $w_<(M,\gamma)$. They are of the same order of magnitude when $M \gg 1$.

Integration over γ gives now

$$K_M(\tau < 1) = \frac{1}{M}\frac{1-(1-\tau^2)^M}{\tau^2(1+\tau)^M}, \quad K_M(0) = 1\,. \tag{38}$$

Combining all found results we arrive finally at

$$K_M(\tau) = \frac{1}{M}\frac{1}{\tau^2(1+\tau)^M}\left[1-\Theta(1-\tau)(1-\tau^2)^M\right], \quad 0 \leqslant \tau < \infty\,. \tag{39}$$

This function satisfies all necessary conditions and is continuous though not analytical in the point $\tau = \tau_W = 1$.

In the semiclassical limit $M \gg 1$ the characteristic decay time $\tau_W = 1/\gamma_W = 1/M \ll 1$ is much shorter than the Heisenberg time. The polynomial term is therefore of principal importance. It becames obvious from the following equivalent presentation of the Eq. (38)

$$\begin{array}{c} K_M(\tau < 1) = \frac{1}{M\tau^2}\left[e^{-M\ln(1+\tau)} - e^{M\ln(1-\tau)}\right] = \\ e^{-M(\tau+O(\tau^3))}\,\frac{2\sinh\left[\frac{M}{2}(\tau^2+O(\tau^4))\right]}{M\tau^2} \approx e^{-M\tau} = e^{-\gamma_W\tau}\,. \end{array} \tag{40}$$

So we come to conclusion that the exponential semiclassical decay with the characteristic Weisskopf's decay rate γ_W cannot be directly traced to the statistics of resonance widths. The omitted terms which describe deviation from the purely exponential law become significant after the time $\tau_q \sim 1/\sqrt{M} \gg \tau_W$ [22, 23].

Mean cross section

The mean cross section can be expressed in two equivalent forms

$$\begin{array}{c} \langle\sigma\rangle = \frac{1}{M} = \int_0^1 d\tau K_M(\tau) + \int_1^\infty d\tau K_M(\tau) = \\ \int_0^\infty \frac{d\gamma}{\gamma} e^{-\gamma}\left[P_{(M-1)}(\gamma) - 1 + \frac{\gamma}{M}\right] + \int_0^\infty w_M^{(sing.)}(\gamma)\,\frac{1-e^{-\gamma}}{\gamma}\,. \end{array} \tag{41}$$

The long-time contribution ($\tau > 1$) rapidly decreases when the number of channels M grows. Even for $M = 1,2$ it amounts only to 30% and 10% respectively. When $M \gg 1$ the following asymptotic expansion

$$\int_1^\infty d\tau K_M(\tau) = K_M(1)\frac{2}{M}\int_0^{M/2} d\xi\, e^{-(\xi+3\xi^2/2M+\ldots)} \approx \frac{2}{M^2}e^{-M\ln 2} \tag{42}$$

shows that this contribution diminishes very fast. The mean cross section is defined by the short time evolution.

On the other hand, the contributions of the regular and singular parts of the decay rates distribution are equally important. Indeed, contribution of the regular part equals

$$\int_0^\infty \frac{d\gamma}{\gamma} e^{-\gamma}\left[P_{(M-1)}(\gamma) - 1 + \frac{\gamma}{M}\right] = \frac{1}{M} + \sum_{m=2}^{M} \frac{1}{m}. \tag{43}$$

The logarithmically growing extra term is perfectly compensated by the contribution of the singular part.

CONCLUSIONS

The decay law $K(t)$ of a complicated unstable state of a classically chaotic system can be formally presented in the form of a Laplace integral as a weighted mean value of the decay exponents. By this the notion of the decay rates distribution is introduced irrelative of statistics of the resonance poles of the scattering amplitudes. The connection of this distribution with statistics of the resonance widths is then investigated. Exact analytical solution is found in the case of systems with broken time-reversal symmetry. It is demonstrated that only the long-time $\tau \gg \tau_H = 1$ asymptotic behavior of the decay law is governed by the statistics of the resonance widths. For the shorter times $\tau < \tau_H = 1$ the contributions of the connected to the widths statistics smooth part and the singular part of the "distribution" $w_<(M,\gamma)$ are equally important. In particular, the approximately exponential semiclassical decay with the characteristic Weisskopf's decay rate γ_W results from interrelation of the both of them.

ACKNOWLEDGEMENTS

I am very grateful to D.V. Savin for his interest to this work, plentiful discussions and advice. I am also thankful to Y. Fyodorov for countenance. At last I greately appreciate discussions of the found results with V. Zelevinsky and A. Richter. The financial support from the RAS Joint scientific program "Nonlinear dynamics and solitons" is acknowledged with thanks.

REFERENCES

1. E.P. Wigner and L. Eisenbud, **72** 29 (1947). E.P. Wigner, Phys. Rev. **98** 145 (1955).
2. F.T. Smith, Phys. Rev. **118** 349 (1960); **119** 2098 (1960).
3. E. Porter and R.G. Thomas, **104** 483 (1956).
4. V.L. Lyuboshitz, Phys. Lett. B **72** 41 (1977); Pis. ZhETF. **28** 32 (1978)(Sov.JEPT **28** 30 (1978)); Yad. Fiz. **27** 948 (1978)(Sov. J. Nucl. Phys. **27** 502 (1978)).
5. C.H. Lewenkopf and H.A. Weidenmueller, Ann. Phys (N.Y.) **212** 53 (1991).
6. B. Eckhard, Chaos **3**, 613 (1991).
7. F.M. Dittes, H.L. Harney, and A. Müller, Phys. Rev. A Phys. Rev. A **45**, 701 (1992).
8. F.M. Izrailev, D. Saher, V.V. Sokolov, Phys. Rev. E **49** 130 (1994).
9. N. Lehmann, D.V. Savin, V.V. Sokolov, H.-J. Sommers, Physica D **86** 572 (1995).
10. V.F. Gopar, P.A. , and Büttiker, Phys. Rev. Lett. **77** 3005 (1996).
11. Y.V. Fyodorov and H.-J. Sommers, JETP Lett. **63** 1026 (1996); J. Math. Phys. **38** 1918 (1997).
12. Y.V. Fyodorov, D.V. Savin, and H.-J. Sommers, Phys. Rev. E **55** R4857 (1997).
13. P.W. Brouwer, K.M. Frahm, and C.W.J. Beenakker, Phys. Rev. Lett. **78** 4737 (1997).
14. T. Kottos and Smilansky, Phys. Rev. Lett. **85** 968 (2000).
15. H.-J. Sommers, D.V. Savin, and V.V. Sokolov, Phys. Rev. Lett. **87** 094101 (2001).
16. T. Kottos and M. Weiss, Phys. Rev. Lett. **89** 056401 (2002).
17. Gorin T., Seligman T.H. Phys. Rev. E **65**, 026214, (2002).
18. J.J.M. Verbaarschot, Ann. Phys.(N.Y.) **168** 362 (1986).
19. F.-M. Dittes, H.L. Harney, I. Rotter, Phys. Lett. A **153** 451 (1991); F.-M. Dittes, Phys. Rep. **339** 215 (2000).
20. Y.V. Fyodorov, D.V. Savin, H.-J. Sommers, Journ. Phys. A: Math. Gen. **38** 10731 (2005).
21. Explicit calculation of the (GUE) decay function $K(t)$ has been fulfilled by prof. A. Richter and his coworkers. Privete communication.
22. G. Casati, G. Maspero, D.L. Shepelyansky, Phys.Rev. E **56** R6233 (1997).
23. D.V. Savin, V.V. Sokolov, Phys. Rev. E **56** R4911 (1997); K. Frahm, Phys. Rev. **56** R6237 (1997).

Doorway States and the Super-Radiant Mechanism

N.Auerbach

School of Physics and Astronomy, Tel Aviv University, Tel Aviv , 69978, Israel, and Department of Physics and Astronomy, Michigan State University, East Lansing, MI 48824, USA

Abstract. A new approach to physics of doorway states in nuclear reactions is developed. It is argued that the coupling of intrinsic states to the continuum through special doorway state(s) may create the situation similar to the optical super-radiance, when the segregation of broad (short-lived) resonances and trapped (long-lived) states occurs. Necessary conditions are discussed, along with applications to isobaric analog states, single-particle resonances, giant resonances, and fission through a double-humped barrier. This pattern can be observed in other mesoscopic systems as well.

Keywords: doorway states, nuclear reactions, resonances, super-radiance

INTRODUCTION

In the early 1950s the possibility of forming a "super-radiant" (SR) state in a gas of atoms confined to a volume of a size smaller than the wave length of radiation was suggested by Dicke [1]. In the absence of a direct interaction, the atoms are coupled through their common radiation field. This indirect interaction through the continuum leads to a redistribution of lifetimes among unstable intrinsic states. A rapidly decaying SR state is created at the expense of the rest of the states of the system that are "robbed" of their decay probability and become narrow.

This mechanism has a general origin, and analogous phenomena should appear in many quantum systems when quasi-bound states are strongly coupled through common decay channels. The SR approach has been used in many different fields. For example, it was applied in chemistry [2, 3], atomic physics [4], condensed matter physics [5, 6], intermediate energy nuclear physics [7, 8], in particle physics [9], in the theory of nuclear reactions [6, 10, 11]. In the next section we will develop the SR formalism for some simple situations in order to arrive in a heuristic fashion to the idea of a super-radiant state and its connection to the concept of a doorway.

THE EFFECTIVE HAMILTONIAN

Following the projection formalism of Feshbach [7, 12], let us divide the Hilbert space of nuclear states into two parts, the $\{Q\}$-subspace involving very complicated many-body states $|q\rangle$, and the subspace $\{P\}$ of open channels $|c\rangle$. We use the notations Q and P for the corresponding projection operators onto the above subspaces. The total wave function of the system,

$$|\Psi\rangle = Q|\Psi\rangle + P|\Psi\rangle, \tag{1}$$

satisfies the stationary Schrödinger equation

$$H|\Psi\rangle = E|\Psi\rangle \tag{2}$$

that can be written as a set of coupled equations,

$$(E - H_{QQ})Q|\Psi\rangle = H_{QP}P|\Psi\rangle \tag{3}$$

and

$$(E - H_{PP})P|\Psi\rangle = H_{PQ}Q|\Psi\rangle, \tag{4}$$

where we use the notations $H_{AB} = AHB$. Eliminating the part $P|\Psi\rangle$, we come to the equation in the Q-space,

$$\left(E - \tilde{H}_{QQ}\right)Q|\Psi\rangle = 0 \tag{5}$$

with the effective Hamiltonian

$$\tilde{H}_{QQ} = H_{QQ} + H_{QP}\frac{1}{E^{(+)} - H_{PP}}H_{PQ}. \tag{6}$$

CP995, *Nuclei and Mesoscopic Physics - WNMP 2007*, edited by P. Danielewicz, P. Piecuch, and V. Zelevinsky

Here $E^{(+)} = E + i0$ contains the infinitesimal imaginary term ensuring right asymptotic conditions for the continuum wave functions.

The second term of the effective Hamiltonian [6] contains a real and imaginary part of the propagator

$$G^{(+)}(E) = \frac{1}{E^{(+)} - H_{PP}} \tag{7}$$

emerging from the principal value and the delta-function $\delta(E - H_{PP})$ (on-shell contributions from channels c open at energy E), respectively. The imaginary part, $-(i/2)W$, of the effective Hamiltonian is given by

$$W = 2\pi \sum_c H_{QP}|c\rangle \langle c|H_{PQ}. \tag{8}$$

Thus, the effective Hamiltonian [7] in Q-space is non-Hermitian,

$$\tilde{H}_{QQ} = \overline{H}_{QQ} - \frac{i}{2}W, \tag{9}$$

where $\overline{H}_{QQ}$ is a symmetric and real matrix that includes, apart from the original Hamiltonian in the Q-space, H_{QQ}, the principal value contribution of the PQ-coupling. The second part is anti-Hermitian. The eigenvalues of $\tilde{H}$, $\tilde{E} = E - (i/2)\Gamma$, are complex poles of the scattering matrix corresponding to the resonances in the cross sections.

To demonstrate in a simple way the role of the anti-Hermitian term we assume that only one channel is open. Then the matrix W, Eq. [9], has a completely separable form,

$$\langle q|W|q\rangle = 2\pi A_q^c A_q^{c*}, \tag{10}$$

where

$$A_q^c = \langle q|H_{QP}|c\rangle. \tag{11}$$

The rank of the matrix W is 1, so that all the eigenvalues of this matrix are zero except one that has the value equal to the trace of the matrix:

$$\Gamma_s = \sum_q \langle q|W|q\rangle = 2\pi \sum_q |A_q^c|^2 \equiv \sum_q \Gamma_q^{\uparrow} \tag{12}$$

with $\Gamma_q^{\uparrow}$ denoting the escape width of the individual levels before the -matrix is diagonalized.

The special unstable state with width Γ_s is often referred to as the *super-radiant* (SR), in analogy to the Dicke coherent state [1] of a set of two-level atoms coupled through the common radiation field. Here the coherence is generated by the common decay channel. The stable states are *trapped* and decoupled from the continuum. In the more general case of N intrinsic states and N_c open channels with $N_c \ll N$ the super-radiant mechanism survives if the mean level spacing D of internal states and their decay widths obey:

$$\kappa = \frac{\Gamma^{\uparrow}}{D} > 1. \tag{13}$$

DOORWAYS

It is often the case that only a subset of intrinsic states $\{Q\}$ connects directly to the $\{P\}$ space of channels. The rest of states in $\{Q\}$ will connect to $\{P\}$ only when they obtain admixtures of these selected states of the first type. The special states directly coupled to continuum are the *doorways*, $|d\rangle$. They form the doorway subspace $\{\mathscr{D}\}$ within $\{Q\}$, and the corresponding projection operator will be denoted here as $\mathscr{D}$. The remaining states in $\{Q\}$ will be denoted as $|\tilde{q}\rangle$ and the subspace as $\{\tilde{Q}\}$.

The full Hamiltonian can be decomposed the following way:

$$H = \left(H_{\tilde{Q}\tilde{Q}} + H_{\mathscr{D}\mathscr{D}} + H_{\tilde{Q}\mathscr{D}} + H_{\mathscr{D}\tilde{Q}}\right) \\ + \left(H_{PP} + H_{\mathscr{D}P} + H_{P\mathscr{D}}\right). \tag{14}$$

Note that the terms $H_{P\tilde{Q}}$ and $H_{\tilde{Q}P}$ are missing because in accordance with the doorway hypothesis they are very small. Also note that diagonalizing the operator in the upper line of (14) would give back the states $|q\rangle$ with the components of $|d\rangle$ mixed with $|\tilde{q}\rangle$ states. The two last items in the above equation couple the doorway states, and therefore all the $|q\rangle$ states to the open channels.

THE CASE OF A SINGLE DOORWAY

We start our discussion with the case when there is only one important doorway state $|d\rangle$. The matrix elements of the effective operator W in the intrinsic space are now given by

$$\langle q|W|q'\rangle = 2\pi \sum_{c=1}^{N_c} \langle q|H_{\mathscr{D}P}|c\rangle\langle c|H_{P\mathscr{D}}|q'\rangle \tag{15}$$

with the doorway assumption:

$$\langle q|H_{\mathscr{D}P}|c\rangle = \langle q|d\rangle\langle d|H_{\mathscr{D}P}|c\rangle, \tag{16}$$

where $\langle q|d\rangle$ is the amplitude of the admixture of the doorway into the $|q\rangle$ state. Eq. (16) becomes:

$$\langle q|W|q'\rangle = 2\pi\langle q|d\rangle\langle d|q'\rangle \sum_{c} |\langle d|H_{\mathscr{D}P}|c\rangle|^2; \tag{17}$$

again we have separable matrix elements of the matrix W, this time irrespective of the number of open channels N_c.

As discussed above, the criterion of validity of the SR mechanism is that the average spacing between the levels in $\{Q\}$ is smaller than the decay width of such a state "before" the SR mechanism takes effect. This can be expressed , in the case of doorways, the following way. Consider the *spreading width*, $\Gamma_d^{\downarrow}$, of the doorway state representing the fragmentation of $|d\rangle$ into compound states $|\tilde{q}\rangle$. If N_q is the number of compound states in the interval covered by the spreading width, their average energy spacing is

$$\overline{D} \approx \frac{\Gamma_d^{\downarrow}}{N_q}. \tag{18}$$

Before the SR mechanism is turned on, the average decay width of a typical $|q\rangle$ state is:

$$\Gamma_q^{\downarrow} = 2\pi \sum_{c} |\langle q|H_{QP}|c\rangle|^2 \tag{19}$$

that can be estimated as

$$\Gamma_q^{\uparrow} = \frac{\Gamma_s}{N_q}. \tag{20}$$

Thus:

$$\frac{\Gamma_q^{\uparrow}}{D_q} \approx \frac{\Gamma_s}{\Gamma_d^{\downarrow}} \approx \frac{\Gamma_d^{\uparrow}}{\Gamma_d^{\downarrow}}. \tag{21}$$

We conclude that the requirement for the SR doorway mechanism to be valid can be formulated as:

$$\frac{\Gamma_d^{\uparrow}}{\Gamma_d^{\downarrow}} > 1. \tag{22}$$

EXAMPLES

Isobaric Analog States

The isobaric analog state (IAS), $|A\rangle$, is obtained from the parent state $|\pi\rangle$ with certain isospin T [13] by acting with the isospin lowering operator T_- that changes a neutron into a proton:

$$|A\rangle = \text{const} \cdot T_-|\pi\rangle. \tag{23}$$

In a compound nucleus, the IAS is surrounded by many compound states $|q\rangle$ of lower isospin $T_< = T-1$. The Coulomb interaction violates the isospin symmetry fragmenting the strength of the IAS over many states $|q\rangle$ giving rise to the spreading width $\Gamma_A^{\downarrow}$ of the IAS. If located above thresholds, the IAS can also decay into several continuum channels that gives rise to the decay width $\Gamma_A^{\uparrow}$. In medium and heavy mass nuclei the condition in Eq. (22) is satisfied. For example in ^{208}Pb the spreading width is about 80 keV while the escape width is about 160 keV [13], thus $\Gamma_A^{\uparrow}/\Gamma_A^{\downarrow} \approx 2$. The SR mechanism is therefore relevant to this case providing a straightforward explanation why the IAS appears as a single resonance with the decay width given by that of $|A\rangle$:

$$\Gamma_A^{\uparrow} = 2\pi\, |\langle A|H_{QP}|P\rangle|^2. \tag{24}$$

Single-particle resonance

Let us consider a single-particle state $|\phi_{sp}\rangle$ that belongs to the space $\{Q\}$ but has energy embedded in the continuum (above the decay threshold). Such a single-particle state couples directly to the continuum channels, often through a one-body mean field potential V. Due to the residual interaction, the single-particle state will be spread over complicated compound states. At low energy above threshold, the main coupling of the compound states $|q\rangle$ to the continuum channels takes place via the single-particle component $\langle\phi_{sp}|q\rangle$. Thus, the state serves as a doorway and therefore our discussion of the single doorway case applies to this situation.

This mechanism will produce a state with a large single-particle width given by

$$\Gamma_{sp} = 2\pi \sum_{c} \left|\langle\phi_{sp}|V|c\rangle\right|^2. \tag{25}$$

This width should be identified with the width of single-particle resonances known in the framework of the optical potential in Feshbach's theory [12]. The narrow neutron resonances enveloped by a single-particle resonances represent the trapped states.

Giant resonances

Giant resonances in nuclei (or atomic clusters) present another example of similar physics [14, 15]. Usually, the giant resonances are discussed in terms of $p-h$ configurations with identical spin-parity quantum numbers. The coherent residual interactions form a correlated state that carries much of the transition strength of a corresponding multipole operator. The giant resonances are mostly located in the particle continuum decaying via particle emission to the ground and excited states in the daughter nucleus.

Since the $p-h$ giant resonance $|G\rangle$ is surrounded usually by a dense spectrum of $2p-2h$ and more complex configurations, the residual strong interaction will mix $|G\rangle$ with this background. Each of the resulting states denoted as $|b\rangle$ will contain the admixture $\langle G|b\rangle$ of the giant resonance. The mixed states $|b\rangle$ couple to the continuum. If we assume that the dominant coupling is that through the admixture of the giant state, then $|G\rangle$ serves as a doorway, and the matrix W is separable, given by:

$$\langle b|W|b'\rangle = \langle b|G\rangle\langle G|b'\rangle \sum_{c=1}^{N_c} \langle G|V|c\rangle\langle c|V|G\rangle. \tag{26}$$

The matrix W is again of rank one and the non-zero eigenvalue is

$$\Gamma_G^{\uparrow} = 2\pi \sum_{c} |\langle G|V|c\rangle|^2. \tag{27}$$

This holds under the assumption that the energy spread of the background states $|b\rangle$ is small compared to their decay width:

$$\Gamma_b^{\uparrow} = 2\pi \sum_{c} |\langle b|V|c\rangle|^2. \tag{28}$$

As before, this condition can be expressed in this case as:

$$\frac{\Gamma_G^{\uparrow}}{\Gamma_G^{\downarrow}} > 1, \tag{29}$$

the spreading width of the giant resonance is smaller than its total decay width. When this condition is not satisfied, so that the energy intervals between the $2p-2h$ states are larger than their decay widths $\Gamma_b^{\uparrow}$, the situation of a single decay peak for the giant resonance might not hold. Still one could expect some bunching of background states into groups and spacings within the group smaller than their decay widths $\Gamma_b^{\uparrow}$. Each such group then can be treated separately using the SR mechanism and appear as a single peak in the decay (or excitation) curve. Then one will observe intermediate structure resonances in the GR energy domain.

In many continuum RPA calculations it was found that high lying giant resonances have substantial decay widths [15] larger than the spreading widths. For example the isovector monopole resonances calculated in heavy nuclei had decay widths of the order of 8-10 MeV, which is larger than the spreading width which is of the order of several MeV. In these cases the condition (29) for super-radiance is satisfied.

Double humped potentials

As a function of a collective coordinate, double (or even triple) humped potentials are often found, especially in heavy nuclei. For example, such structure in the dependence of collective energy on the quadrupole deformation parameter β_2 corresponds to normal-deformed and super-deformed wells. The first potential is usually deeper, and the decay of intrinsic states $|q\rangle$ in this potential is slow because they have to penetrate the double barrier when decaying to the continuum. The states defined in the first well can couple to a few states $|d\rangle$ in the shallower second potential that couples directly to the continuum through a single outer barrier. These states $|d\rangle$ serve as doorways for the decay of states $|q\rangle$ in the first potential well. Similar to the previous examples, the matrix W can be written in the form:

$$\langle q|W|q'\rangle = 2\pi \sum_{cdd'} \langle q|d\rangle\langle d|V|c\rangle\langle c|V|d'\rangle\langle d'|q'\rangle. \tag{30}$$

If the number of doorways $|d\rangle$ is small compared to the number of states $|q\rangle$, one should expect, as a result of diagonalization of the above matrix, several broader states with the rest of the states being narrow.

An outstanding example are the fission isomers [16] in several heavy nuclei, such as ^{240}Pu, ^{237}Np. The fission cross section in the neutron induced reaction on ^{240}Pu shows the structures with larger widths spaced several hundred eV apart, whereas the average spacing between the compound resonances is of the order of a few eV. This phenomenon is described as a direct consequence of a double humped potential in the fission process of a compound nucleus. The energy spacing between the few excited (usually, collective – rotational or vibrational) states in the second shallower potential are larger than between the compound states in the deeper well. When the compound states in the deeper well are in the energy vicinity of the states in the second well, they couple forming mixed states which, in turn, couple to the fission channel via the admixture of the state from the second well.

The states in the second minimum are the doorways $|d\rangle$ for the decay of compound states into the open channels, in this case the fission ones, $|f\rangle$. The matrix W for the compound states $|q\rangle$ in the vicinity of the doorway $|f\rangle$ is:

$$\langle q|W|q'\rangle = \langle q|d\rangle\langle d|q'\rangle \sum_{f} |\langle d|V|f\rangle|^2. \tag{31}$$

Diagonalizing this W matrix, one finds a state with a considerably wider decay width than that of a typical compound state but of course smaller than a single-particle decay width. The situation repeats itself for each of the states in the second well. As a result, one will observe intermediate structure in the fission cross section.

CONCLUSIONS

We have considered few examples of well known nuclear physics phenomena and described the formation of doorways and emergence of intermediate structure in the framework of the SR mechanism in these phenomena. Because of the generic nature of this mechanism it would be interesting to extend this approach to other mesoscopic systems. In the physics of cold atoms [17] in a trap, the coupling of the internal states to the continuum via the Feshbach resonance influences profoundly the system affecting not only the imaginary part of the effective Hamiltonian but also the real part. This enables one to manipulate the properties of the system by changing the sign and/or the strength of the two-body interaction. The Feshbach resonance in this case serves as a doorway.

The traditional theory of intermediate structure [18, 19] considers usually statistical sequences in time of multi-step particle interactions with possible pre-compound (or pre-equilibrium) decay to the continuum. The present formulation emphasizes complementary aspects of underlying physics, namely those that follow from the strict quantum-mechanical description of complicated many-body dynamics generated by the mean field and residual interactions. The approach based on the effective non-Hermitian Hamiltonian predicts the existence of different scales in the energy dependence of observed cross sections. These scales are defined by the intrinsic dynamics that combines collective (coherent) and chaotic (incoherent) features. The new collectivity responsible for the emergence of the doorways and related intermediate structure appears, under certain physical conditions, due to the factorized nature of corresponding terms in the effective Hamiltonian. This separability is related to the properties of unitarity of the scattering matrix. In this way we obtain a supplementary description of many-body dynamics.

ACKNOWLEDGEMENTS

This work was done in collaboration with V. Zelevinsky; the work was supported by the National Superconducting Cyclotron Laboratory (Michigan State University) and by a grant from the Binational Science Foundation USA-Israel.

REFERENCES

1. R.H. Dicke, *Phys. Rev.* **93**, 99 (1954).
2. V.B. Pavlov-Verevkin, *Phys. Lett. A* **129**, 168 (1988).
3. F. Remacle, M. Munster, V.B. Pavlov-Verevkin, and M. Desouter-Lecomte, *Phys. Lett. A* **145**, 265 (1990).
4. V.V. Flambaum, A.A. Gribakina, and G.F. Gribakin, *Phys. Rev. A* lbf 54, 2066 (1996).
5. K. Pichugin, H. Schanz, and P. Łeba, *Phys. Rev. E* **64**, 056227 2001).
6. A. Volya and V. Zelevinsky, in: V. Zelevinsky (Ed.), *Nuclei and Mesoscopic Physics* (AIP Conference Proceedings, vol. 777, 2005) p. 229.
7. N. Auerbach, *Phys. Rev. C* **50**, 1606 (1994).
8. N. Auerbach and V. Zelevinsky, *Phys. Rev. C* **65**, 034601 (2002).
9. N. Auerbach, V. Zelevinsky, and A. Volya, Phys. Lett. B **590** (2004) 45.
10. V.V. Sokolov and V.G. Zelevinsky, *Nucl. Phys.* **A504**, 562 (1989).
11. I. Rotter, *Rep. Prog. Phys.* **54**, 635 (1991).
12. H. Feshbach, *Ann. Phys.* (N.Y.) **5**, 357 (1958); H. Feshbach, *Ann. Phys.* (N.Y.) **19**, 287 (1962).
13. N. Auerbach, J. Hüfner, A.K. Kerman, and C.M. Shakin, *Rev. Mod. Phys.* **44**, 48 (1972).
14. P.F. Bortignon, A. Bracco, and R.A. Broglia, *Giant Resonances: Nuclear Structure at Finite Temperature* (Harwood Academic, Amsterdam, 1998).
15. N.Auerbach and A.Klein, *Nucl. Phys.* **A375**, 77 (1983).
16. A. Bohr and B.R. Mottelson, *Nuclear Structure*, vol. 2 (Benjamin, New York, 1989).
17. H. Smith and C.J. Pethick, *Bose-Einstein Condensation in Dilute Gases* (Cambridge Univ. Press, Cambridge, 2001).
18. R.M. Wood, R.R. Borchers, and H.H. Barschall, *Nucl. Phys.* **71**, 529 (1965).
19. H. Feshbach, A. Kerman, and S. Koonin, *Ann. Phys.* (N.Y.) **125** 429 (1980).

Time-dependent Green's functions approach to nuclear reactions

Arnau Rios[*] and Pawel Danielewicz[*,†]

[*]*National Superconducting Cyclotron Laboratory and Department of Physics and Astronomy, Michigan State University, East Lansing, MI 48824-1321, USA*
[†]*Kavli Institute for Theoretical Physics, University of California, Santa Barbara, CA 93106-4030, USA*

Abstract. Nonequilibrium Green's functions represent underutilized means of studying evolution of quantum many-body systems. In view of a rising computer power, an effort is underway to apply the Green's functions to the dynamics of central nuclear reactions. As the first step, mean-field evolution for the density matrix for colliding slabs is studied in one dimension. Strategy to extend the dynamics to correlations is described.

Keywords: Green's functions, Kadanoff-Baym equations, time-dependent Hartree-Fock, nuclear reactions
PACS: 24.10.-i, 24.10.Cn

INTRODUCTION

Nuclear reactions represent active means of studying nuclei, nuclear interactions and nuclear medium. Central reactions disturb nuclei more than do the peripheral ones and can lead to nuclear fusion, multifragmentation and dramatic compression of nuclear matter, depending on incident energy. Due to the increasing many-body nature of the central reactions with higher energies and due to the corresponding population of a multitude of final states, the description of those reactions has relied on the time domain and has remained primarily semiclassical. Because of their semiclassical nature, the descriptions for central reactions have then remained somewhat disconnected from the descriptions employed for nuclear structure, peripheral reactions or giant nuclear excitations. The coarseness of the central-reaction approaches, associated to their semiclassical nature, is not easy to quantify and this has proven to be a difficulty in linking correctly the observables from the reactions to the properties of nuclear systems under the special conditions created in those reactions.

Historically, the only practical approach to central collisions, that can be considered quantal, has been the Time-Dependent Hartree-Fock (TDHF) method [1] and its extensions [2, 3]. The starting point for TDHF is the mean-field description for nuclei, with individual wave-function components propagated in time according to Schrödinger-like single-particle equations of motion. The validity of TDHF relies on the presumption that correlations play a negligible role in the dynamics. However, with the weakening of Pauli principle as the excitation energy increases in actual reactions, correlations can lead to a fast thermalization of the population of single-particle states as well as to enhanced stopping. Such phenomena are completely absent in any TDHF-like description. In view of the issues above, it would seem important to develop a quantal approach to central reaction dynamics including the effects of correlations beyond the mean-field. These correlations are also known to play a significant role in the initial state [4].

The time-dependent Green's functions formalism [5] is a natural candidate for the sought theory. On one hand, equilibrium Green's functions can describe the microscopic and macroscopic properties of stationary strongly interacting many-body system, accommodating the effects of correlations [4]. On the other hand, an extension of the approach to the time-dependent domain, by means of the Kadanoff-Baym (KB) equations, can also incorporate the effect of correlations onto the time evolution [5]. To date, the application of the nonequilibrium formalism to nonuniform quantum many-body systems has been relatively scarce, but this can be in large part attributed to an involved numerical effort which should become of a lesser concern as the computational power grows. Besides central reactions, the approach could be used to study the response of nuclei to external time-dependent fields [3].

A very important feature of the Green's functions formalism is the fact that, within certain approximations for the self-energies, the evolution of the observables is conserving, *i.e.* the main conservation laws (energy, momentum, angular momentum, etc) are preserved during the time evolution [6, 7]. In addition, the propagator formalism naturally takes into account the quantal properties of the problem. The non-local features of quantum mechanics are for instance reflected into the off-diagonal components of the Green's functions in real space representation. Little is known about

CP995, *Nuclei and Mesoscopic Physics - WNMP 2007,* edited by P. Danielewicz, P. Piecuch, and V. Zelevinsky

the structure of these components and their importance in the time-evolution of quantum many-body systems.

KADANOFF-BAYM EQUATIONS

The KB equations generally describe the evolution, under rather liberal assumptions, of expectation values of the products of two single-particle annihilation and creation operators, at different time arguments, i.e. 2-point Wightman functions. In the context of nonequilibrium theory, these are called the single-particle Green's functions. At equal time-arguments, the expectation values yield the single-particle density matrix. For realistic modeling of reactions, a nonuniform three-dimensional (3D) system needs to be studied. First, however, we shall restrict to the less involved one-dimensional (1D) case to get acquainted with the effect that correlations might induce on the time evolution of the system. Here, in addition, we shall assume in the following that the initial state of the system is uncorrelated, i.e. describable by mean-field theory. This approximation will be eventually relaxed, leading to a modification of the KB equations which we shall again not treat in here [8].

The initial state of the systems is specified in terms of an A-body density operator $\hat{\rho}_0$ at time t_0. The one-body Green's functions are then defined as expectation values, with respect to $\hat{\rho}_0$, of products of Heisenberg-picture creation $\hat{a}^\dagger(x,t)$ and destruction $\hat{a}(x,t)$ operators:

$$\mathscr{G}^<(x_1,t_1;x_2,t_2) = i\langle \hat{a}^\dagger(x_2,t_2)\hat{a}(x_1,t_1)\rangle\,, \tag{1}$$

$$\mathscr{G}^>(x_1,t_1;x_2,t_2) = -i\langle \hat{a}(x_1,t_1)\hat{a}^\dagger(x_2,t_2)\rangle\,. \tag{2}$$

Up to a factor, the time-diagonal Green's function $\mathscr{G}^<$ reduces to the one-body density matrix. The KB equations, governing the evolution of Green's functions in their arguments,

$$\left\{ i\hbar\frac{\partial}{\partial t_1} + \frac{\hbar^2}{2m}\frac{\partial^2}{\partial x_1^2}\right\}\mathscr{G}^{\lessgtr}(\mathbf{1}\mathbf{1}') = \int \mathrm{d}\bar{\mathbf{r}}_1 \Sigma_{HF}(\mathbf{1}\bar{\mathbf{1}})\mathscr{G}^{\lessgtr}(\bar{\mathbf{1}}\mathbf{1}') + \int_{t_0}^{t_1}\mathrm{d}\bar{\mathbf{1}}\,\Sigma^+(\mathbf{1}\bar{\mathbf{1}})\mathscr{G}^{\lessgtr}(\bar{\mathbf{1}}\mathbf{1}') + \int_{t_0}^{t_{1'}}\mathrm{d}\bar{\mathbf{1}}\,\Sigma^{\lessgtr}(\mathbf{1}\bar{\mathbf{1}})\mathscr{G}^-(\bar{\mathbf{1}}\mathbf{1}'), \tag{3}$$

$$\left\{ -i\hbar\frac{\partial}{\partial t_{1'}} + \frac{\hbar^2}{2m}\frac{\partial^2}{\partial x_{1'}^2}\right\}\mathscr{G}^{\lessgtr}(\mathbf{1}\mathbf{1}') = \int \mathrm{d}\bar{\mathbf{r}}_1 \mathscr{G}^{\lessgtr}(\mathbf{1}\bar{\mathbf{1}})\Sigma_{HF}(\bar{\mathbf{1}}\mathbf{1}') + \int_{t_0}^{t_1}\mathrm{d}\bar{\mathbf{1}}\,\mathscr{G}^+(\mathbf{1}\bar{\mathbf{1}})\Sigma^{\lessgtr}(\bar{\mathbf{1}}\mathbf{1}') + \int_{t_0}^{t_{1'}}\mathrm{d}\bar{\mathbf{1}}\,\mathscr{G}^{\lessgtr}(\mathbf{1}\bar{\mathbf{1}})\Sigma^-(\bar{\mathbf{1}}\mathbf{1}'), \tag{4}$$

follow from considerations of the equations of motion for creation and destruction operators. In the above, the simplified notation $\mathbf{1} = (x_1,t_1)$ has been introduced and the retarded and advanced functions are defined according to:

$$F^\pm(\mathbf{1},\mathbf{2}) = F^\delta(\mathbf{1},\mathbf{2}) \pm \Theta[\pm(t_1-t_2)]\left[F^>(\mathbf{1},\mathbf{2}) - F^<(\mathbf{1},\mathbf{2})\right]\,, \tag{5}$$

with F^δ standing for a possible singular contribution at $t_1 = t_2$. A generalized self-energy $\Sigma(\mathbf{1},\mathbf{2})$ may be introduced [8] which accounts for all the interaction effects of the system, including the instantaneous Hartree-Fock contribution, $\Sigma_{HF}(\mathbf{1},\mathbf{2})$, as well as the $\Sigma^{\lessgtr}$ terms, generated by correlation effects that go beyond the mean-field.

The complex integro-differential KB equations have to be solved in a self-consistent way, since the self-energies depend on the Green's functions that are being solved for. This self-consistency simultaneously guarantees a microscopic consistency of the theory. In particular, for certain many-body approximations to the self-energy (those of the conserving type), one can show that the time evolution induced by the self-consistent KB equations preserves the conservation laws obeyed by the system as a whole [7, 8]. These conservation laws can therefore be used as tests for the numerical implementation of the equations.

An attractive feature of the KB equations is their generality. The time evolution induced by Eqs. (3) and (4) can easily describe different types of correlations, given in terms of different approximations to the self-energy. In addition, the equations do not focus on any particular physical system and can therefore be used to study different many-body problems. So far, the KB equations have been used to study the time evolution of uniform nuclear matter [9, 10] and the uniform electron gas [11], as well as inhomogeneous atomic and molecular systems [12].

The KB equations account for the so-called memory effects. From Eq. (3), one can easily see that that the Green's function $\mathscr{G}^<$ at times t_1 and $t_{1'}$ depends on the Green's functions and self-energies at all the previous times t, $t_1 > t > t_0$ and $t_{1'} > t > t_0$, via the time integrals on the r.h.s of the equation. In consequence, to find a solution of the equations one must keep track of all the previous time-steps. A strategy to attack this problem, before addressed in [9] and [13], is discussed further on.

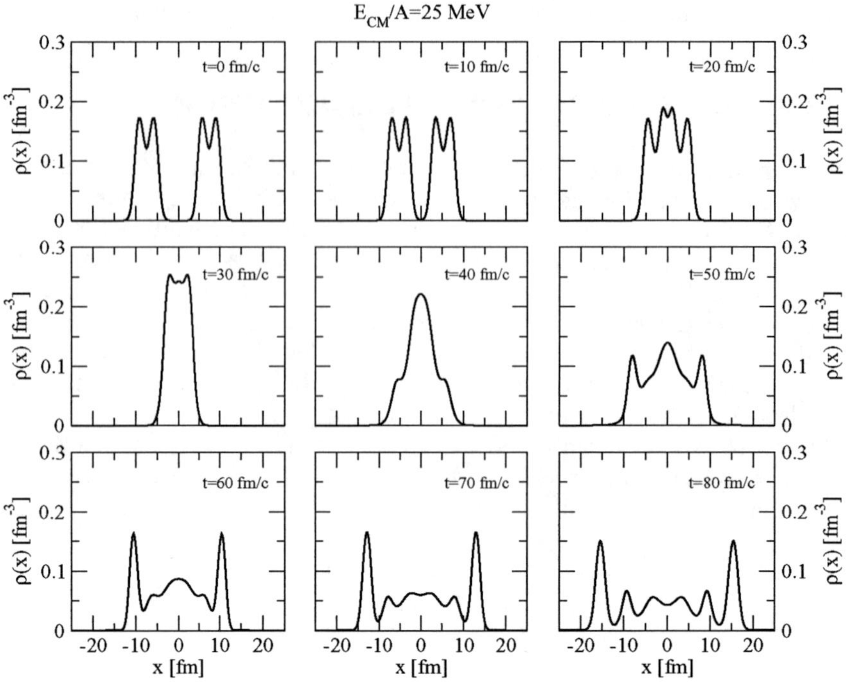

FIGURE 1. Time evolution of density for slabs colliding in the mean-field approximation.

Finally, it is important to note that the KB equations are quantal and, in particular, respect the non-local features of quantum mechanics. This is directly reflected in the fact that the evolution, even in the 1D case, depends on two space and two time variables, (x_1, t_1) and $(x_{1'}, t_{1'})$. This doubling of variables might be a disadvantage in terms of computational cost, especially compared to other time-dependent approaches which do not share this feature. We believe, however, that the increase in computational cost (if any) might be compensated by the advantages of describing the system in terms of the Wightman's functions.

MEAN-FIELD APPROXIMATION

The rather involved calculations associated with the KB equations simplify when one deals only with the Hartree-Fock part of the self-energy. The evolution of the correlation functions is then given in terms of the mean-field and there are no memory effects involved in the calculations. The KB equations reduce in that case to the TDHF equations for the evolution of single-particle density matrix for the system. This simple framework can be used as a starting point for the studies of a time-dependent approach formulated in terms of the density matrix, rather than in terms of single-particle wave-functions, as would have been standard for a TDHF approach.

Since this is meant as a first qualitative example, we have chosen a rather simple local density-dependent mean-field of the Skyrme-type:

$$\Sigma_{HF}(1\bar{1}) = \delta(t_1 - t_{1'})\,\delta(x_1 - x_{1'})\left[\frac{3}{4}\,t_0\,n(x_1,t) + \frac{2+\sigma}{16}\,t_3\,[n(x_1,t)]^{(\sigma+1)}\right], \tag{6}$$

with the parameters t_0, t_3 and σ fitted to reproduce the saturation properties of infinite nuclear matter. The local density $n(x,t)$ corresponds actually to a 3D system and it is related to the 1D density by a scaling factor, $n(x,t) = -i\xi\,\mathscr{G}(x,t;x_{1'} = x, t_{1'} = t)$, where $\xi = (4n_0^2)^{1/3}$ and $n_0 = 0.16\,\text{fm}^{-3}$ is the density of normal nuclear matter. In the above, we assume a uniformity of the 3D system in the perpendicular y and z directions, as well as a spin-isospin degeneracy of $\nu = 4$.

Since the Hartree-Fock interaction is local, one can implement the time evolution in a rather straightforward way by means of the so-called Split Operator Method (SOM) [14]. This is based on the fact that, for infinitesimally short time-steps Δt, the evolution of the density matrix is formally given in terms of the single-particle kinetic K and mean-field

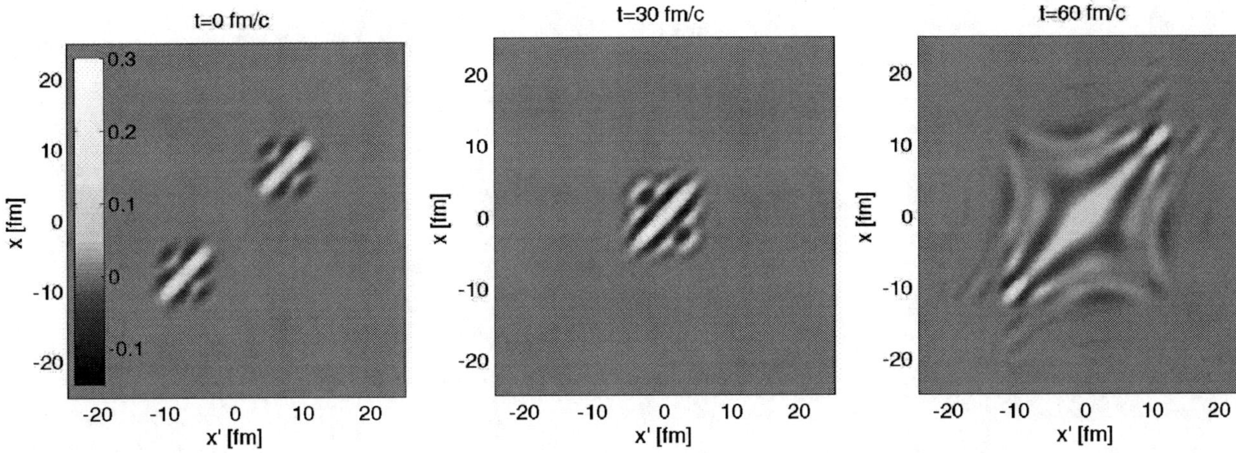

FIGURE 2. Intensity plots of the real part of $\mathscr{G}^<$ as a function of the x and x' variables. The left, center and right panels represent, respectively, the initial state of the system; the state at $t = 30\,fm/c$ when slabs interpenetrate and the state at $t = 60\,fm/c$ when the slabs move away from each other. Note that the white tones correspond to positive values of the function and the darker ones (dark blue on-line) to negative values.

U operators as:

$$\mathscr{G}(x,t+\Delta t;x',t'+\Delta t') = \mathrm{e}^{-i\frac{(\hat{K}+\hat{U})\Delta t}{\hbar}}\,\mathscr{G}(x,t;x',t')\,\mathrm{e}^{i\frac{(\hat{K}+\hat{U})\Delta t'}{\hbar}}\,. \tag{7}$$

For short times, one can use the decomposition $\mathrm{e}^{i\frac{(\hat{K}+\hat{U})\Delta t}{\hbar}} = \mathrm{e}^{i\frac{\hat{U}\Delta t}{2\hbar}}\mathrm{e}^{i\frac{\hat{K}\Delta t}{\hbar}}\mathrm{e}^{i\frac{\hat{U}\Delta t}{2\hbar}} + O(\Delta t^3)$ to divide the exponential into mean-field and kinetic terms. Since the mean-field is diagonal in real space and the kinetic part is diagonal in momentum space, we can trivially compute the impact of these two exponentials by applying them separately to the real and the momentum space density matrices. By using fast Fourier transforms, we can easily switch from one representation to the other and therefore implement the time evolution in a rather efficient way.

It is worthwhile to comment on the preparation of the initial mean-field density matrix. In order to avoid spurious oscillations in various variables for slabs representing the colliding nuclei, it is convenient to start the time-dependent calculation from a density matrix that is the ground state for the mean-field. To obtain this ground state, we have started from a 1D harmonic oscillator density matrix embedded in a harmonic oscillator potential. Then, following Eq. (7), we have very slowly switched off the harmonic oscillator potential, while switching on, at the same time, the mean-field. When this process is carried on adiabatically, the ground state of the self-consistent mean-field density-matrix is reached at the end of the evolution. After this ground state propagator is obtained, we can repeatedly apply the algorithm of Eq. (7) to the time evolution of the superposition of boosted ground-state density matrices. This in turn gives access to the evolution of one-body observables and total energy.

In Fig. 1, we show the time evolution for a system where two shells of 1D harmonic oscillators have been filled for each slab. The slabs have been evolved adiabatically to self-consistency and then boosted by transforming the density matrices with the operators $\mathrm{e}^{\pm iPx}$. The momentum P corresponded to the center-of-mass energy of 25 MeV per particle. It is observed that the two initially separated density distributions eventually overlap and interpenetrate each other. As the system flies apart, a low density region is left behind in which some light fragments seem to be forming. This could be naively associated with some form of multifragmentation process.

One can observe, already in this very crude model, that changing collision energy leads to the appearance of qualitatively different physical processes. At high energies, the slabs interpenetrate each other and break apart, as already commented. At low collision energies, one can observe the fusion of the slabs as well as the subsequent evolution of the excited state. These results are in qualitative agreement with the findings of the time evolution of 1D slabs within the TDHF formalism in terms of wave-functions [15].

The time-diagonal density matrix depends on two position variables, x and x'. In Fig. 2. we show intensity plots for the real part of $\mathscr{G}^<$ at three different stages of the collision, specifically at times $t = 0$, 30 and 60 fm/c. These plots give an insight on the structure of the Green's functions away from the $x = x'$ diagonal. Thus, the initial $t = 0$ density matrices show some spread in direction perpendicular to the diagonal, an effect which can be understood in terms of the wave-functions in the structure of the matrix. For one, the support of the matrix is constrained by the support of

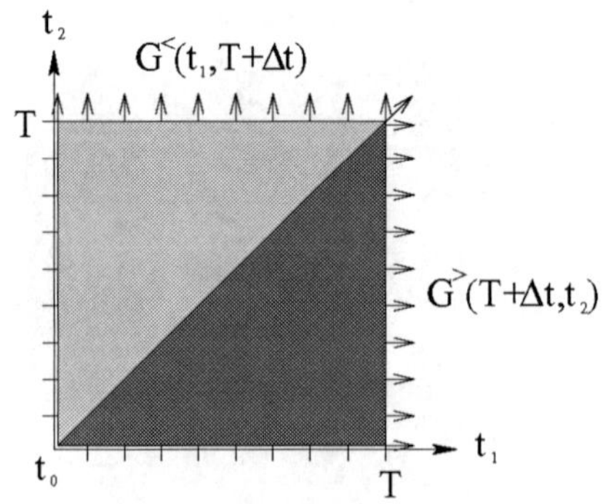

FIGURE 3. Strategy for solving the Kadanoff-Baym equations in the two time arguments.

the wave-functions. This yields square outlines where the values of the matrix can be significant. Second, the wave-functions signs interplay, producing a band structure along the diagonal. In more dimensions, phases would matter. The changing signs and/or phases are generally associated with the changing momentum content in the wave-functions. The bigger the momentum spread, the more narrow the spread in $|x-x'|$. The spread remains similar in the slab collision when the slabs interpenetrate at $t = 30\,\mathrm{fm/c}$. However, the situation qualitatively changes after the slabs collide and begin to move away from each. On top of the original structure, the support for the matrix $\mathscr{G}^<$ expands, as illustrated for $t = 60\,\mathrm{fm/c}$. A larger but fainter square of support emerges, with two corners at the location of slabs on the diagonal and two fainter cross-corners persisting into the future. While the matrix values still exhibit sharp maxima along the diagonal, the wider region of significant values reflects the fragmentation of original wave-functions. Nucleons from the originally populated states can end up in either of the residues and in the space in-between. The portions of the wave-functions maintain a phase relationship that would be important for the interference if the portions were ever to recombine.

In the late of the nuclear reactions, though, systems progress through expansions and decays and nearly never recombine. Thus, following nuclear interpenetration, we expect to be able to neglect the matrix elements far away from the $x = x'$ diagonal without affecting the evolution of the elements close to the axis. In fact, in our preliminary tests, we find virtually no effect on the evolution of density on the $x = x'$ axis when ignoring elements farther away from the axis than those which are significant for a single slab. Ignoring the elements amounts to the assumption of decoherence over farther-away distances.

CORRELATED APPROXIMATION

When integrating the KB equations beyond the mean-field approximation, the SOM method alone is not sufficient. Even in a 1D case, the solution of the KB equations becomes a relatively involved process. At present, we are working on the numerical implementation of the equations for nuclear reactions, following closely the implementations already existing for uniform matter [9, 13] and generalizing them to the non uniform case. In this first implementation, we will work within the direct Born approximation for the self-energy. This is the simplest conserving approximation that includes effects beyond the mean-field. In the momentum representation, the collisional self-energies are:

$$\Sigma^{\lessgtr}(p_a,t_a;p_b,t_b) = \int_{-\infty}^{\infty}\frac{\mathrm{d}p_1}{2\pi}\int_{-\infty}^{\infty}\frac{\mathrm{d}p_2}{2\pi} V(p_a-p_1)V(p_b-p_2)\mathscr{G}^{\lessgtr}(p_1,t_a;p_2,t_b)\Pi^{\lessgtr}(p_a-p_1,t_a;p_b-p_2,t_b) \tag{8}$$

where the function $\Pi^{\lessgtr}$ is a convolution integral of two Green's functions:

$$\Pi^{\lessgtr}(p_a,t_a;p_b,t_b) = \int_{-\infty}^{\infty}\frac{\mathrm{d}p_1}{2\pi}\int_{-\infty}^{\infty}\frac{\mathrm{d}p_2}{2\pi}\mathscr{G}^{\lessgtr}(p_1,t_a;p_2,t_b)\mathscr{G}^{\gtrless}(p_2-p_b,t_b;p_1-p_a,t_a)\,. \tag{9}$$

In those initial studies, we plan to use a gaussian two-body interaction, $V(p) = V_0\,\mathrm{e}^{-\frac{\eta^2p^2}{4}}$, with parameters chosen to reproduce semiquantitatively the free-space nucleon-nucleon (NN) scattering cross section. This should give the first hint on the effect that the inclusion of NN collisions in the formalism has on the time evolution of the different properties of the system. Since the Born self-energies are given in terms of convolution integrals, they can be efficiently computed using fast-Fourier transform techniques.

To optimally implement the time evolution given by the KB equations, symmetries should be exploited to minimize some of the computational expenses involved in the calculations. A particularly useful symmetry is given by the relation:

$$[F^{\lessgtr}(\mathbf{1},\mathbf{2})]^* = F^{\lessgtr}(\mathbf{2},\mathbf{1}), \tag{10}$$

which is fulfilled by both the Green's functions and self-energies. Therefore, one can store each of the Green's functions in one of the half-planes of the (t_1,t_2) variables and access the other half by using Eq. (10). This further suggests the implementation of the time evolution following the square structure of Fig. 3. Let us assume that, starting from t_0, the Green's functions have been determined for $T \geq t_1, t_2 \geq t_0$ and let us denote the r.h.s. of (3) as $I^{\lessgtr}(\mathbf{1},\mathbf{1}')$. To the second order in Δt, the Green's function values at $T+\Delta t$ can be then found from

$$\mathscr{G}^>(p_1,T+\Delta t;p_2,t_2) \simeq e^{i\frac{\varepsilon(p_1)\Delta t}{\hbar}}\mathscr{G}^>(p_1,T;p_2,t_2) + \frac{1}{2}\left[I^>(p_1,T;p_2,t_2) + I^>(p_1,T+\Delta t;p_2,t_2)\right]\frac{e^{i\frac{\varepsilon(p_1)\Delta t}{\hbar}}-1}{\varepsilon(p_1)}, \tag{11}$$

with $\varepsilon(p_1) = \hbar^2 p_1^2/2m$. An analogous equation holds for $\mathscr{G}^<$. A combination of Eqs. (3) and (4) yields an algorithm for evolving the functions along the diagonal in Fig. 3. Advancing of the functions needs to be repeated to achieve self-consistency as new function values appear on the r.h.s. as well, e.g. in (11).

We expect a general drop in function values [9] for increasing difference $|t_1 - t_2|$. Similarly, we expect a drop in values for increasing $|x_1 - x_2|$ or $|p_1 - p_2|$, see Fig. 2, limiting the size of region within the Green's function matrix, around the diagonal, that needs to be remembered during evolution. The bookkeeping of values is the particularly costly aspect of the numerical implementation of the KB equations.

In 1D already, we expect to gain meaningful insights into how the inclusion of correlations makes the dynamics different from the mean-field dynamics or from semiclassical dynamics with nucleon-nucleon collisions. The inclusion of correlations may cause turn-on problems early on in the evolution and we intend to test the adiabatic switching on of those correlations. Useful for the 2D and 3D calculations should be an experience with treating the far-off elements of the Green's functions and even the strategies for discretizing the functions.

ACKNOWLEDGMENTS

Discussions with Dennis Lacroix are gratefully acknowledged. This work was partially supported by the National Science Foundation, under Grant Nos. PHY-0555893 and PHY-0551164.

REFERENCES

1. J. W. Negele, *Reviews of Modern Physics* **54**, 913 (1982).
2. M. Tohyama, *Physical Review C* **36**, 187 (1987).
3. D. Lacroix, S. Ayik, and P. Chomaz, *Progress in Particle and Nuclear Physics* **52**, 497 (2004).
4. W. Dickhoff, and D. van Neck, *Many-body theory exposed!*, World Scientific Publishing, London, 2005.
5. L. P. Kadanoff, and G. Baym, *Quantum Statistical Mechanics*, Benjamin, New York, 1962.
6. G. Baym, and L. P. Kadanoff, *Physical Review* **124**, 287 (1961).
7. G. Baym, *Physical Review* **127**, 1391 (1962).
8. P. Danielewicz, *Annals of Physics* **152**, 239 (1984).
9. P. Danielewicz, *Annals of Physics* **152**, 305 (1984).
10. H. S. Köhler, *Physical Review C* **51**, 3232 (1995).
11. N. H. Kwong, M. Bonitz, R. Binder, and H. S. Köhler, *Physica Status Solidi B* **206**, 197 (1998).
12. N. E. Dahlen, and R. van Leeuwen, *Physical Review Letters* **98**, 153004 (2007).
13. H. S. Köhler, N. H. Kwong, and H. A. Yousif, *Computer Physics Communications* **123**, 123 (1999).
14. M. D. Feit, J. A. Fleck, and A. Steiger, *Journal of Computational Physics* **47**, 412 (1982).
15. P. Bonche, S. Koonin, and J. W. Negele, *Physical Review C* **13**, 1226 (1976).

Quantum Transport in Mesoscopic Semiconductor Devices: Vortices, Flows and Atomistic Effects

John R. Barker

Nanoelectronics Research Centre
Department of Electronics and Electrical Engineering
University of Glasgow, Glasgow G12 8LT, UK

Abstract. Exciting developments in nanometer scale semiconductor MOS technology are briefly reviewed. The physics of such devices may be understood by applying 2D and 3D non-equilibrium Green function theory to their simulation. Although the current-voltage characteristics show excellent transistor behaviour the underlying transport is strictly quantum mechanical. This will be illustrated by showing the self-consistent energy resolved current flows and charge densities for typical target devices particularly double-gate MOS devices and wrap-round gate silicon nanowire MOS devices. The existence of quantized vortices even at very high temperature will be discussed and the effects of phase de-coherence. Scattering processes are particularly interesting in such devices and many novel interference phenomena and atomistic scattering effects occur including polarisation effects arising from the close proximity of source and drain and gate regions. These small devices comprise in effect finite many-body systems that are driven far from equilibrium. Each device has a different microscopic structure and environment at the atomic level and this leads to variability in the macroscopic device parameters. The full modelling of such systems is one of the major challenges to present day technology. At present the field is at a watershed where future modelling will require atomic level descriptions of the devices thus making contact with molecular electronics and novel media such as carbon nanotubes.

Keywords: semiconductor devices, silicon, Green functions, quantum transport, vortices.

1. INTRODUCTION

The modern semiconductor electronics industry is mostly built on silicon technology. It drives forward by scaling: the continued decrease in size of the basic MOSFET transistor and their interconnects: this underpins Moore's Law - the increase in components per chip, the increase in transistor speed, and a myriad dependent system parameters and economic advantages follow an exponential growth law. For nearly 50 years, scaling has worked by building transistor architectures which are integratable into increasingly complex systems and where the underlying device physics is invariant with respect to the scaling. The industry is facing a crisis in the 21st century because present transistor designs cannot be scaled beyond the 22 nm node. There has been a frenzy of activity hoping to provide a successor technology. Historically, this has occurred before. In the late 1970s, the heat death of silicon NMOS technology was widely mooted, successor non-silicon technologies popped up and were as quickly struck down by harsh realities. Changing the architecture from NMOS to CMOS brought a new lease of life to silicon technology which continued apace; first striking below the 1.0 μm channel length scale, until today we are on the brink of 18 nm scale channels. Recently, power dissipation problems to do with leakage (tunnelling) into substrate and via the thin oxide into the gate and high resistance effects in the contact regions have been held back by changing both the architecture and the clock frequencies of the devices. Substrates are replaced by insulating material (or nothing!) to remove substrate leakage. The most advanced mainstream device today is the silicon CMOS transistor used in Intel's 45 nm node (18 nm channel) device which uses raised source, drain and mushroom gate to reduce resistances, it uses high- k dielectric oxide instead of SiO_2 giving the same equivalent electrostatic thickness but with a real thickness that is sufficiently large to kill-off gate

CP995, *Nuclei and Mesoscopic Physics - WNMP 2007*, edited by P. Danielewicz, P. Piecuch, and V. Zelevinsky

leakage by tunnelling. Beyond this scale there are problems: according to the International Semiconductor Technology Road Map [1] (2005 extract shown in Figure 1a), there are no solutions yet for scaling to the 2020 point where the road map looks at the 14 nm node (10 nm channel length). These problems are partly to do with fabrication and partly to do with physics. Already with 50 nm channels the underlying physics is changing. For example, at these scales the discrete nature of the dopant distributions becomes apparent (continuous doping models begin to fail), the transport is strictly quantum mechanical and quantum effects are now routinely taken into account in device modelling.

International Technology Roadmap for Semiconductors

Year	2005	2010	2015	2020
MPU Half Pitch *(nm)*	90	45	25	14
MPU Gate Length *(nm)*	32	18	10	6

2005 edition

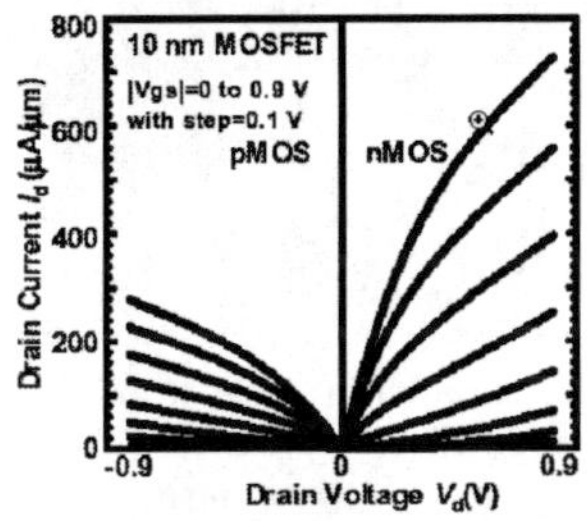

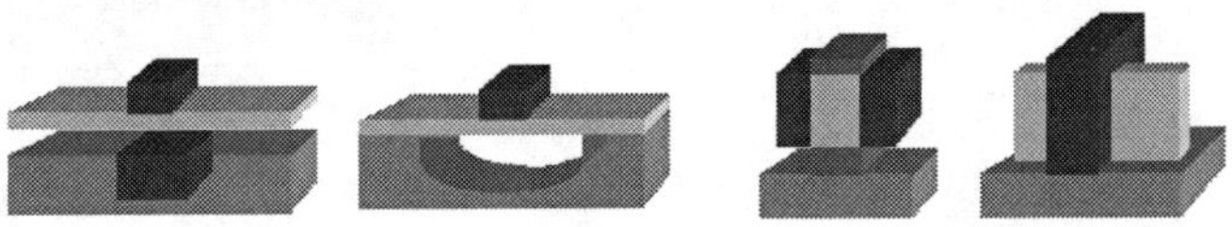

Figure 1:(a) Extract from 2005 Semiconductor Road Map; (b) example of a 6 nm device; (c) alternative transistor architectures.

Several examples of silicon FETs have been demonstrated at channel lengths below 10 nm (Figure 1b gives an example), and a range of prospective new architectures have been proposed and indeed tested (see Figure 1c); but no consensus yet exists on what is the correct choice. It is the opinion of this author that the most promising future devices are silicon nanowire MOSFETs with completely wrap-round gates [2]. These bear a striking resemblance and scale to carbon nanotube transistors (CNT); but the silicon nanowire devices have definite routes to fabrication and integration into large systems.

2. MESOSCOPIC REGIME

The scale below about 50 nm is termed the mesoscopic regime because it lies between the macro-scale that is dominated by bulk material properties and the atomic scale which is dominated by microscopic properties. It is particularly challenging to model this regime. In Figure 2a we sketch the typical architecture of a bulk MOSFET. The source and drain regions are conveniently modelled by *continuum* doping distributions as is the gate region and the channel region. The electrostatics are accurately determined from these continuous charge distributions and the mobile charge distributions of the carriers during device operation. Relatively simple drift-diffusion models describe the current-voltage characteristics very well, provided they are parameterised accurately: something that may be calibrated by detailed semi-classical ensemble Monte Carlo computations. But from Figure 2b, we can see that around the 22 nm channel length scale, there are less than 50 silicon atoms along the channel and the continuum doping model is broken. Looking towards the end of the road map (Figure 2c) we see that a 4 nm channel device is comparable to a large molecules the individual dopant atoms are fully resolved: the atomistic regime. On this scale the device is a strongly coupled (via Coulomb forces) highly degenerate three-dimensional molecular-scale object: it cannot be described by simple bulk properties, nor by semi-classical transport theory since the channel is smaller than the mean free path.. It is dominated by its enclosing surfaces and interfaces and its local structure is highly inhomogeneous.

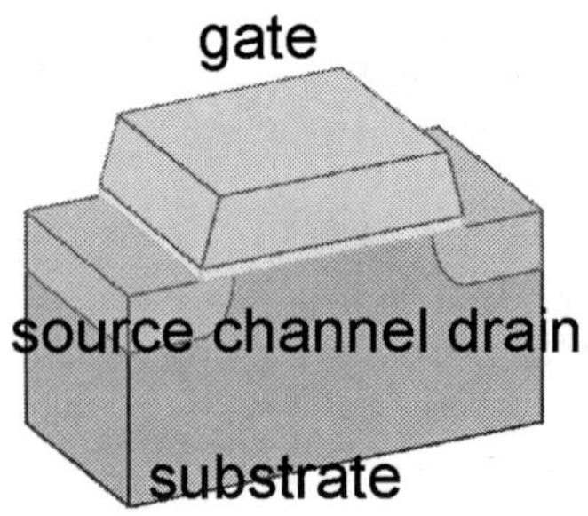

Fig. 2a. The current approach to semiconductor device simulation assumes continuous ionised dopant charge and smooth boundaries and interfaces.

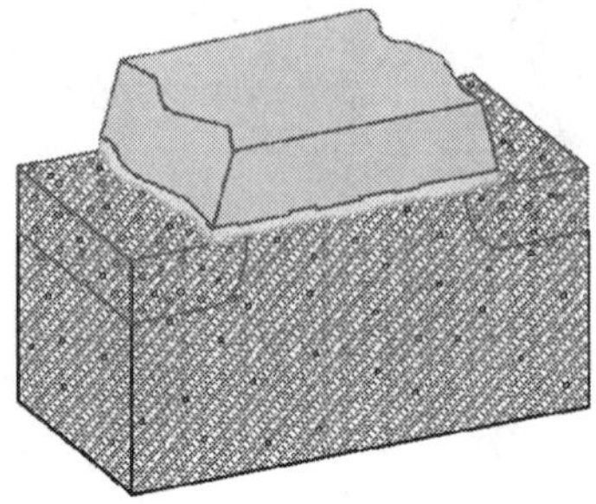

Fig.2b. Sketch of a 22 nm channel MOSFET expected in mass production in 2008. There are less than 50 Si atoms along channel. Random discrete dopants, atomic scale interface roughness and line edge roughness introduce significant parameter fluctuations.

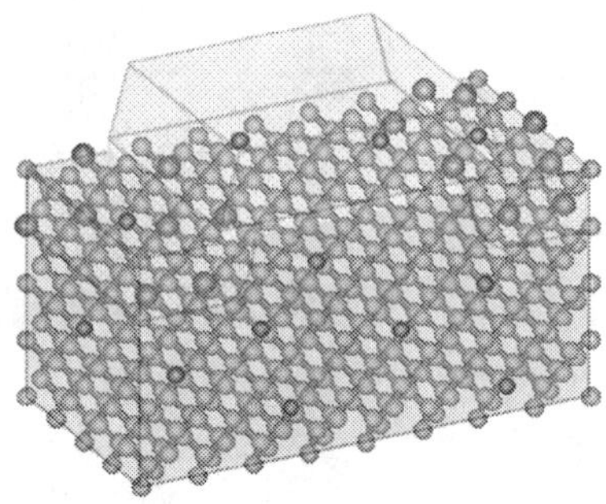

Fig. 2c. Sketch of a 4 nm channel MOSFET expected in mass production in 2023. There are less than 10 Si atoms along the channel. The size of the device becomes smaller than the size of a large molecule.

3. ATOMISTIC ISSUES: MICROVARIABILITY

From a device modelling point of view, the big new issue with the mesoscopic regime is the problem of microvariability: devices which are fabricated to the same *macroscopic* specification (doping densities, feature sizes and scales, edge and interface definition) will have different device characteristics (such as threshold voltage, channel conductance, current drive) because the discrete nature of the dopants and the random variability in edge/interface definition means that each device is *microscopically* different and these microscopic variations cause large differences between the performances of individual devices [3-6]. Figure 3 illustrates the computed large variance in threshold voltage for an atomistic MOSFET.

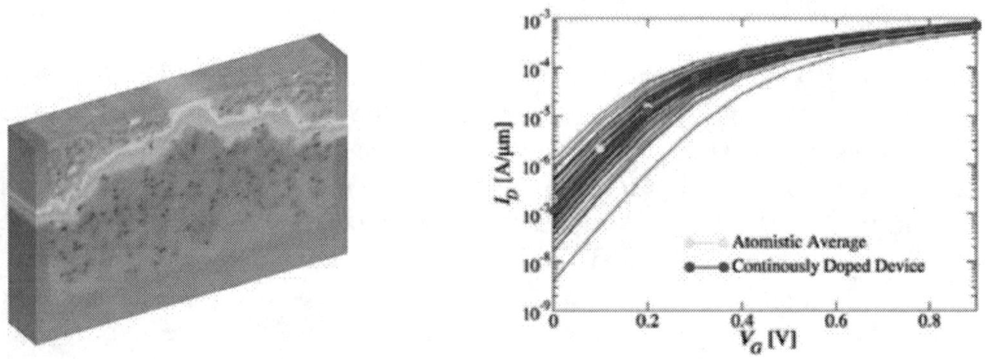

Figure 3: (a) atomistic MOSFET (b) Ensemble of Current-Voltage characteristics

There are many sources of microvariability sketched in Table 1.

Source of microvariability	Example (atomistic picture)	Idealised (non-atomistic)
Discrete dopants: different spatial micro-configurations at a given doping concentration rather than uniform continuum doping. Discrete dopants in channel shown as circles. Problem acute in channel, but also Occurs in source/drain and oxide.	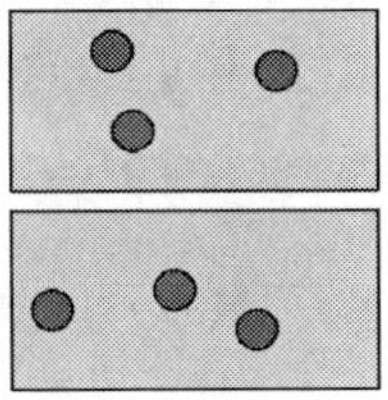	
Discrete dopants: Number fluctuations Particularly acute in source and drain Near channel interface	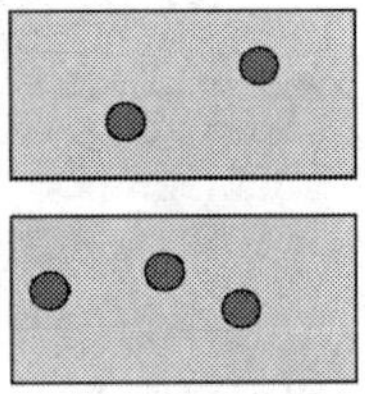	
Unintentional doping Particularly acute in channel.		
Source/Drain random extensions		
Body-thickness variations	Channel-substrate width variations	
Surface/interface roughness: Example shows double gate device with two oxide-channel interfaces	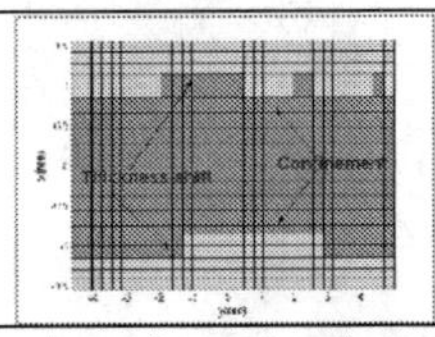	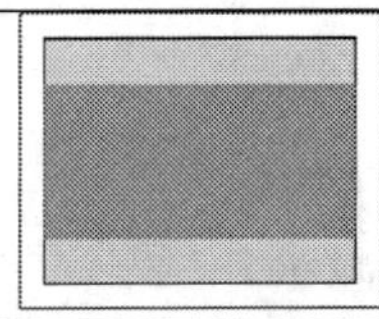
Oxide thickness variations	Similar to interface roughness but includes gate edge variations	Constant thickness
Polyphase gate oxide	Early examples of high k dielectrics	Uniform phase in silicon dioxide
Line edge roughness		
Vortex formation	Quasi-localised charge regions due to Atomistic effects and abrupt boundary changes: dynamic effect	

Table 1: Sources of microvariability in semiconductor devices

4. PHYSICAL ISSUES BELOW 50 nm

Below 50 nm scales the major mesoscopic effects are: (i) size quantization, (ii) quasi-ballistic flows; (iii) quantum transport not classical transport-tunnelling in some regions; (iv) low-dimensional effects (loss of bulk properties, particularly in band structure-see Fig. 4); (v) atomistic effects: microvariability a major issue; (vi) very inhomogeneous and very high local electric fields (leading to hot electron phenomena); (vii) very high doping levels (in source and drain) $N_D>10^{18}$-10^{21} cm^{-3} – high degeneracy, strong screening near and in source and drain; surface to volume ratio is very high: dominant effects from surface roughness. To describe the transport in mesoscopic systems requires a quantum transport theory that can handle spatial inhomogeneities and strong fields. In particular *non-perturbative* methods are required for surface roughness and for scattering on atomistic impurities. In the latter case we cannot define a self-energy using the Kohn-Luttinger *ansatz* because in sufficiently small regions, or more correctly for sufficiently few impurities, the scattering is not self-averaging [7-11].

For example, Fig 5 shows the computed structure factor :

$$K(\mathbf{q}) = |\sum_{j=1}^{N_I} e^{-i\mathbf{q}\cdot\mathbf{r}_i}| / N_I \quad \rightarrow \delta_{\mathbf{q}.0} \ (for \ N_I \rightarrow \infty) \tag{1}$$

for impurities in a small rectangular region. Self-averaging does not occur when $N_I < 1000$.

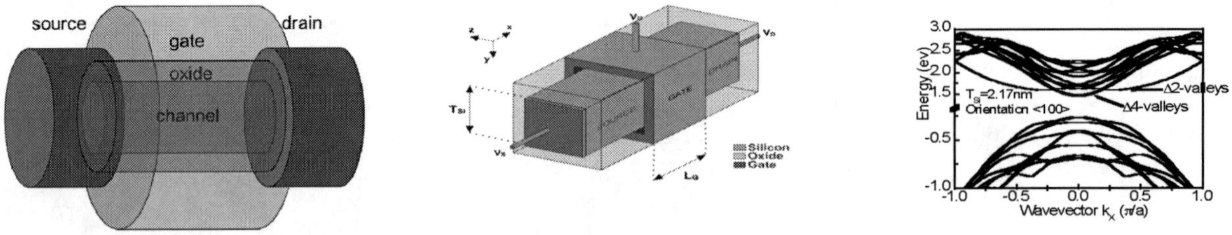

Fig. 4a Cylindrical Nanowire FET Fig. 4b Rectangular Nanowire FET Fig. 4c Tight Binding Band structure

Figure 4: (a)-(b) Nanowire MOSFETs; (c) band structure for device in Fig. 4b.

A detailed discussion is given in [7-11]. Self-averaging still occurs for phonon scattering, but it is less important in channels that are smaller than the inelastic/elastic mean free paths for phonon scattering. Strong inelastic scattering does occur in the contact regions, particularly in the drain.

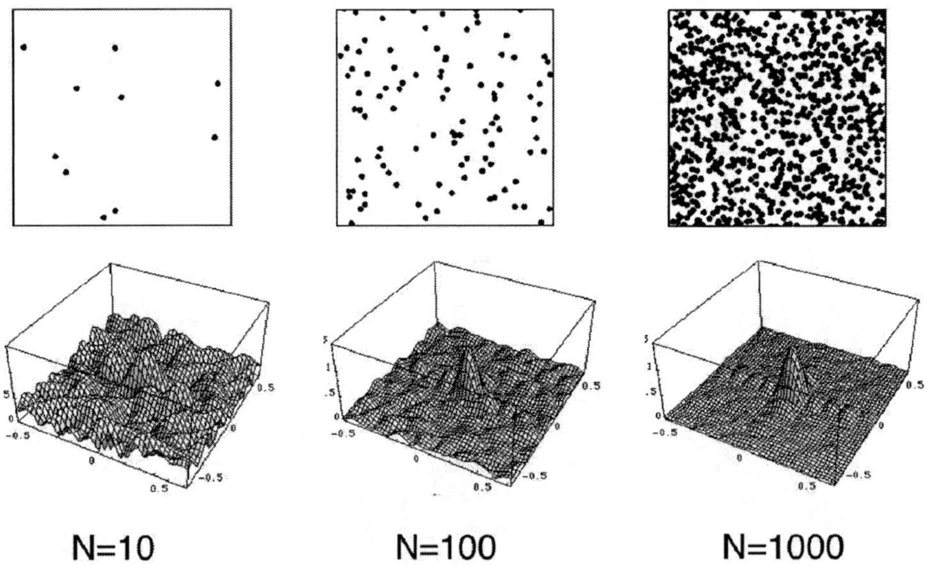

Figure 5: Computed structure factor for N impurities in 2D square region.

5. THE NON-EQUILIBRIUM GREEN FUNCTION APPROACH

The best quantum transport formalism that meets the needs expressed in the previous section has been known since the early 1960s: the thermodynamic Green function formalism introduced by Kelydysh [12]. The basic one-electron Green functions are two-point functions of space-time $G(\mathbf{x},t;\mathbf{x}',t')$ ordered along the Keldysh contour. The complete formalism is daunting from a computational viewpoint since devices comprise strongly-driven, inhomogeneous open many-body systems. However, using projection operator methods the simulation domain for the one-electron Green functions may be reduced to a finite volume containing relevant parts of the source, drain and interface to the oxide and all the channel region [13]. The cost is the introduction of additional self-energy terms near the input and output regions that describe the coupling to the open external regions (including the gate and substrate if leakage currents are important). Most modelling uses 1D spatial dependence (suitable for time-domain) or 2D /3D steady-state Green functions because of the large computational complexity. Details of a fully self-consistent 3D steady state Green function scheme are given in [14]. For quantum ballistic flows (no inelastic scattering) the steady state device response with the aid of the energy-resolved retarded Green function $G^R = [(E+i\eta)\mathbf{1} - H]^{-1}$ (operator form) where H is the full Hamiltonian. Referring to Fig. 6, the Green function may be projected into different regions :

$$G_R = \begin{bmatrix} G_{SS} & G_{SC} & G_{SD} \\ G_{CS} & G_{CC} & G_{CD} \\ G_{DS} & G_{DC} & G_{DD} \end{bmatrix} = \begin{bmatrix} (E+i\eta)1 - H_S & \Lambda_{SC} & 0 \\ \Lambda_{CS} & (E+i\eta)1 - H_C & \Lambda_{CD} \\ 0 & \Lambda_{DC} & (E+i\eta)1 - H_D \end{bmatrix}^{-1} \quad (2)$$

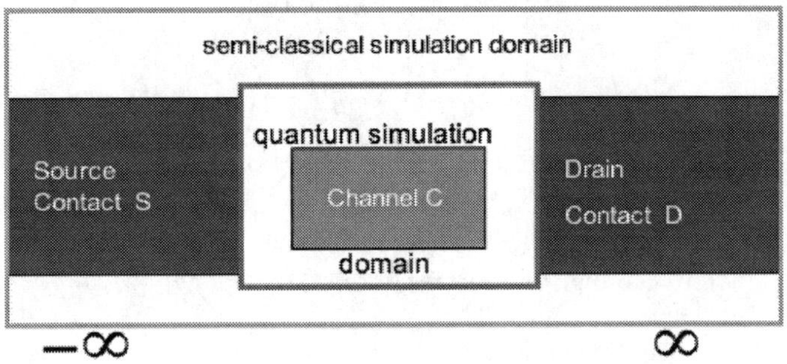

Figure 6: Simulation domains

$$G_{CC} = [(E+i\eta) - H_C - \Sigma_C]^{-1} \quad (3)$$

$$\Sigma_C = \Lambda_{CS} G^0{}_{SS} \Lambda_{SC} + \Lambda_{CD} G_{DD}{}^0 \Lambda_{DC} = \Sigma_{CS} + \Sigma_{CD} \quad (4)$$

$$G^0{}_{SS} = [(E+i\eta) - H_S]^{-1};\ G^0{}_{DD} = [(E+i\eta) - H_D]^{-1} \quad (5)$$

The modern NEGF algorithms are based on calculating G_{CC} where the coupling to the environment is handled by the self-energies Σ_{CS}, Σ_{CD}. The Kelydysh formalism requires four Green functions: G^R, G^A $G^<$, $G^>$ that for steady state are related by:

$$\begin{aligned} &[E\mathbf{1} - H - \Sigma^R]G^R = 1 && \textit{KINETICS} \\ &[E\mathbf{1} - H - \Sigma^R]G^< = \Sigma^< G^A && \textit{STATISTICS} \\ &[E\mathbf{1} - H - \Sigma^R]G^> = \Sigma^> G^A && \textit{STATISTICS} \end{aligned} \quad (6)$$

Numerically, the full self-consistent Green function requires the inversion of a $N_x \times N_y \times N_z$ dimensional matrix at each iterative cycle together with solving Poisson's equation for the total electrostatic field. Special algorithms have been developed that reduce the problem to N_z inversions of $N_x \times N_y$ dimensional matrices. The atomistic impurities and surface roughness are placed non-perturbatively into the Hamiltonian H. Inelastic scattering may be treated but is computationally very expensive. The charge density and current density (energy resolved) are

$$n(E;\mathbf{r}) = -2iG^<(E;\mathbf{r},\mathbf{r})/\pi;\quad \mathbf{J}(E;\mathbf{r}) = -e\frac{\hbar}{4\pi}\mathbf{m}^{*-1} \bullet (\nabla - \nabla')G^<(E;\mathbf{r},\mathbf{r}')|_{r=r'} \quad (7)$$

6. DENSITIES AND FLOWS

If we define the velocity field $\mathbf{v}(\mathbf{r},t) = \mathbf{j}(\mathbf{r},t)/n(\mathbf{r},t)$, we may use the carrier density and velocity field to describe the quantum hydrodynamical flows in ballistic transport. Figure 7 shows results from a 2D self-consistent NEGF simulation of a double gate device (Fig. 7a) which has 3 atomistic impurities in the channel [11, 16-17]. Fig 7b shows the charge density and Fig 7c shows the potential at a particular set of source-drain bias and gate bias. The impurities provide a set of barriers to the current flow. Figure 8 shows the longitudinal component of current density for two different gate voltages. At the higher gate voltage, the impurity barriers are suppressed sufficiently to allow current flow at higher energies close to the interfaces.

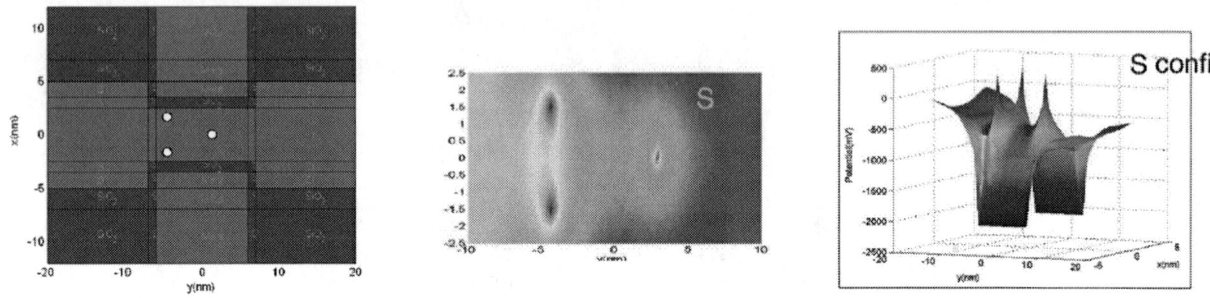

(a) (b) (c)

Figure 7: (a) Double gate device with 3 atomistic impurities (circles) in channel; (b) computed charge distribution in channel; (c) computed potential in channel.

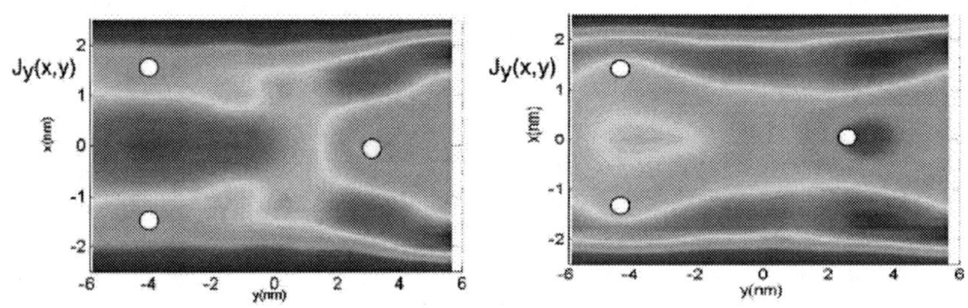

(a) Vgate=1.5 V (b) Vgate=2.4 V

Figure 8: Longitudinal current density for double gate device with 3 atomistic impurities (circles)

The meandering flow bifurcates around the rightmost impurity. This meandering behaviour is very evident when surface roughness is present [18-20]. Fig 9a illustrates the longitudinal current density in an undoped channel of a double gate device with rough interfaces. Fig 9b shows a similar example for a 3D computation of a silicon nanowire device .

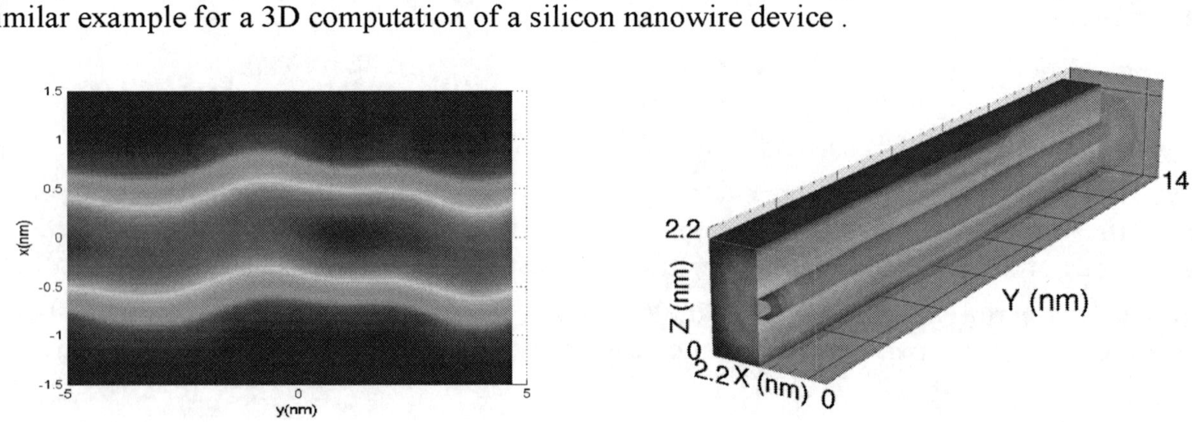

(a) Longitudinal current density in DG device (b) Longitudinal current density in nanowire device

Figure 9: Longitudinal current flow in presence of surface roughness (a) double gate device (b) nanowire device

7. ENERGY RESOLVED

The channel current may be described in terms of a thermal convolution over the energy resolved transmission function using Landauer's formalism [13,15]. The transmission functions for a set of idealised double gate devices with different aspect ratios (source and drain width:: channel width) are shown in Fig. 10 for different gate voltages at source-drain voltage V_D= 0.2V. The quantized conductance is clearly exhibited Figure 11 shows similar self-consistent NEGF computations for the case of 3 atomistic impurities in the channel corresponding to Fig 7-8 . The distortion of the simple transmission is evident.

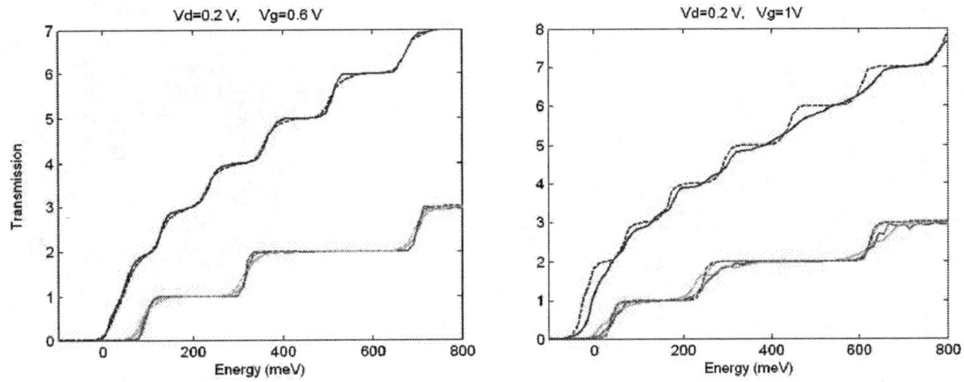

Fig 10. Transmission as function of energy: perfect double gate device, for different aspect ratios

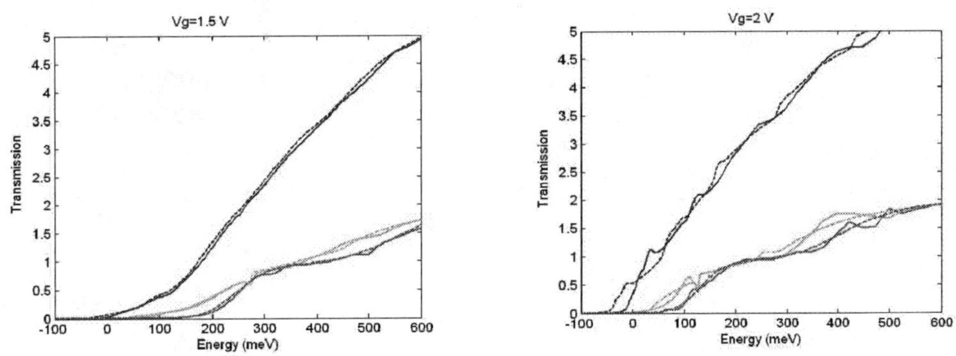

Fig 11. Transmission as function of energy: perfect double gate device, for different aspect ratios; S configuration

8. VORTICES

The current density flows are very simple for steady flows or periodic flows at a given total energy E when the flow topology is determined by nodes of the energy-resolved density (vortex points) and nodes of the energy-velocity field (hyperbolic separation points) where $\mathbf{v}(\mathbf{r},E,t)=\mathbf{J}(\mathbf{r},E,t)/n(\mathbf{r},E,t)$. The full charge density and current involve convolutions of the energy-resolved versions over thermal distributions and may obscure features such as vortices in the energy resolved flows except at low temperatures. However, detailed quantum transport modelling of semiconductor devices [11, 22-26] has predicted the occurrence of quantized vortex formation in the coherent flows of the quantum ballistic transport regime even in the total velocity field and at room temperature. A simple (pure state) example of vortex flows is a non-stationary travelling wave state of a long quantum wire with square cross section of width W formed by superposing two transverse eigenstates with different energies and modulated by longitudinal plane wave states. The transverse flow corresponds to a periodic motion of charge between the walls. The net flow is analogous to the von-Karman vortex street of classical hydrodynamics. The flow is exactly determined from the non-stationary state:

$$\begin{aligned} &|\psi>= c_1 |\psi_1> + c_2 |\psi_2> \\ &<\mathbf{r}|\psi_\alpha>= C\exp[iK_\alpha z - iE_\alpha t]\sin(n_\alpha k_0 x)\sin(n_\alpha k_0 y) \end{aligned} \qquad (8)$$

Figure 12a illustrates the velocity field in the y-z plane for the case: $k_0 = \pi / W$; $K_1 = 3k_0$; $K_2 = 2k_0$; $n_1 = 1; n_2 = 2; E_2 > E_1$. Figure 12b is a still from a computer movie of the moving vortex tube created by the state (8).

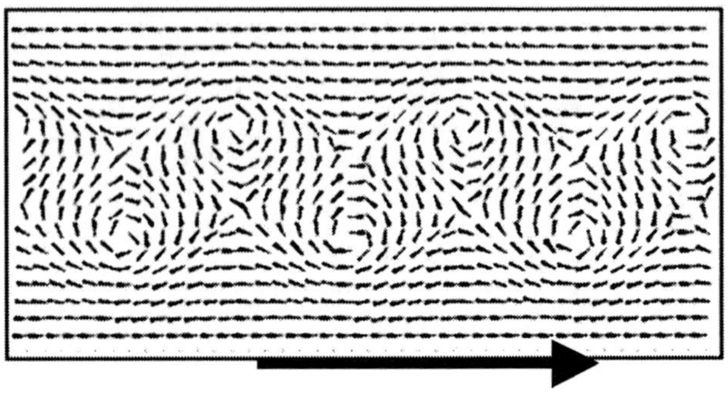

Figure 12a Velocity field

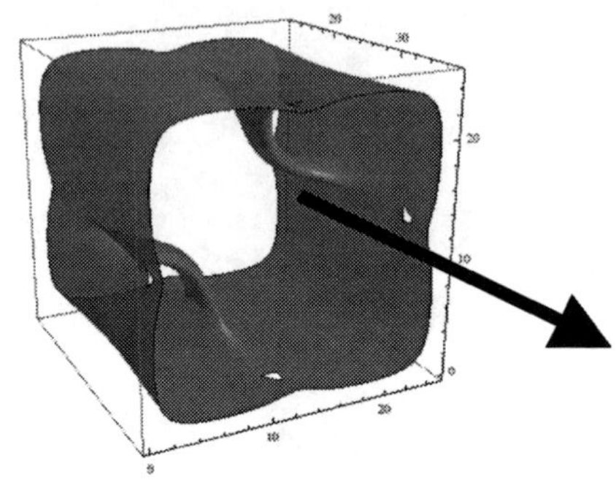

Figure 12b Moving vortex tube

The simplest vortex states for cylindrical quantum wires correspond to angular momentum eigenstates corresponding to circular flow transverse to the nanowire axis. More generally, for stationary open flows at a given energy E, quantized vortices with angular momentum $N_i\hbar$ occur due to the attractive quantum potential that arises near a density node at $\mathbf{x}_i$ with spatial dependence $\sim (\mathbf{r}-\mathbf{x}_i)^{2N_i}$, where N_i is an integer.

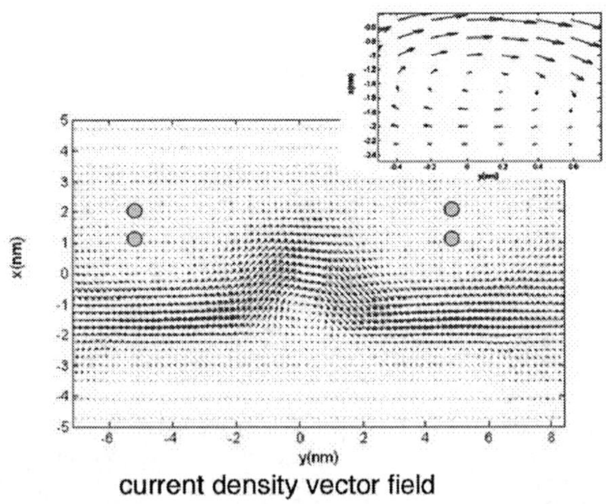

Figure 13a: Total current density vector field for a double gate device with 4 atomistic impurities in the channel (circles) showing vortex flow embedded in meandering open flow.

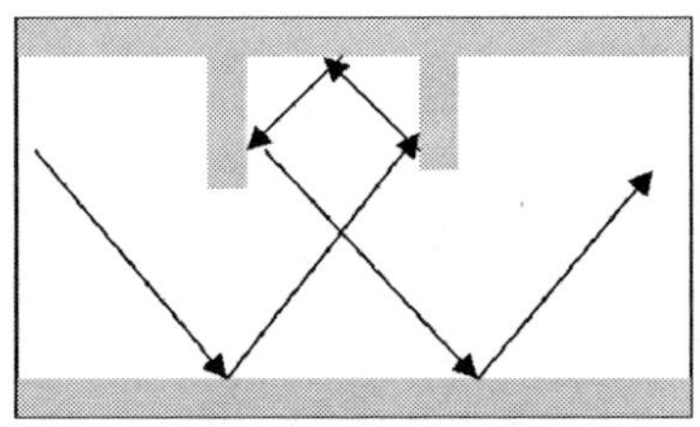

Figure 13b: classical trajectories

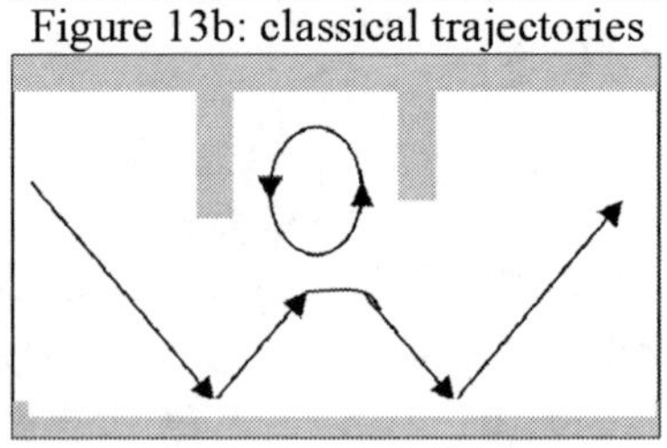

Figure 13c: quantum (Bohm) trajectories

Quantum transport models of ballistic transport in devices show that vortices exist in a wide range of forms many of which are topologically identical to classical vortices [2-24]. Of course, several factors conspire to hide their contributions to current flow in real devices. First, their existence requires that the device regions are smaller enough to avoid de-coherence, which is easily achieved in small quantum wires or coupled quantum dot arrays. De-coherence properties are discussed in [25]. More importantly, the states of a device are not pure states but instead are mixed states comprising thermal admixtures of pure open states. Thus the net flow may show no vortex structure. Nevertheless, we find that given the right conditions, macroscopic vortices are predicted in the high temperature ($\geq$ 300K) transport in small transistor structures. Figure 13a shows such an example from a self-consistent NEGF simulation study of double-gate field effect transistors . Here the macro-vortex arises due to angular momentum transfer between the flow and atomistic impurities arranged in a cavity-like configuration. The vortex formation may be explained easily within the Bohm interpretation of quantum mechanics (see Fig. 13 b,c and [22]). Using energy-resolved NEGF analysis we find that vortex flows frequently occur near atomistic impurities and sudden boundary changes such as surface roughness patterns. The energy-resolved current flow comprises a meandering open flow plus localised vortex flows which act to block the available open flow volume. The net effect of vortices is to reduce the channel conductance at any temperature. They are thus an important part (but hidden) of mesoscopic transistor behaviour.

9. CONCLUDING REMARKS

New challenges exist for the theory and modelling of mesoscopic devices as they approach molecular scales and incorporate new materials. Strong cross-disciplinary input is likely from computational chemistry and many-body physics. Atomic level descriptions of interfaces and their defects is underway via density-functional formalisms. The Green function method maps easily into atomic basis representations but the continuum space version has useful properties for describing many-electron scattering, screening, correlation and polarisation effects (image charges) in the contact regions and gate and for Coulomb blockade effects. The heat dissipation problem and de-coherence issues are addressable in the contact regions but no satisfactory quantum simulations have yet been performed. Many of the techniques described here applicable to exotic electronic materials such as graphene, carbon nanotubes and molecular devices.

ACKNOWLEDGMENT

This work has been supported by grants from the UK Engineering and Physical Sciences Research Council.

REFERENCES

1. *International Technology Road Map for Semiconductors.* Available: http://public.itrs.net/
2. J. Goldberger, A. I. Hochbaum, R. Fan, and P.Yang, Nano Letters, **6**, 973-977, (2006).
3. J . A. Nixon, J. H. Davies and J. R. Barker, in Physics and fabrication of nanostructures, ed M Read and W. P. Kirk, Academic Press 123-127(1989).
4. J. A. Nixon and J. H. Davies, Phys. Rev. B **41**, 7929 (1990).
5. M. J. Laughton, J. R. Barker, J. A. Nixon and J. H. Davies , Phys. Rev. **B44** 1150-1153 (1991).
6. A. Asenov, A. R. Brown, J. H. Davies, S. Kaya, and G. Slavcheva, *IEEE Trans. Electron Devices*, **50**, 1837, (2003).
7. J.R. Barker, Physica **E19**, 62 (2003).
8. J.R. Barker, J.Computational Electronics **2**, 153, (2003).
9. J. R. Barker, Semiconductor Science and Technology, **19S**, 56, (2004).
10. J.R Barker, Superlattices and Microstructures, **34**, 361, (2004).
11. J R Barker, A Martinez, A Svizhenko, A Anantram and A Asenov, J. Phys.: Conf. Ser. **35,** 233, (2006).
12. L. V. Keldysh, Sov. Phys.-JEPT, **20**, 1018, (1965).
13. R. Lake, G. Klimeck, R. C. Bowen and D. Jovanovic, J. App. Phys. **81**, 7845, (1997).
14. A. Martinez, M. Bescond, J.R. Barker, A. Svizhenko, M. P. Anatram, C. Millar and A. Asenov,

IEEE Transactions Electron Devices,**54**, 2213, (2007).

15. R. Landauer, Phil. Mag. **21**, 863, (1970).
16. A, Martinez, J.R. Barker, A. Svizhenko, M.P. Anantram, and A. Asenov, Springer Proceedings in Physics, **110** (August 15th) (2006).
17. A. Martinez, J.R. Barker , A. Svizhenko, M.P.Anantram, A. Brown, B. Biegel, A. Asenov, Journal of Physics Conf series **38,** 192, (2006).
18. A. Martinez, A. Svizhenko, M.P. Anatram, J.R. Barker, A.R. Brown, A. Asenov The International Electronic Devices Meeting IEDM 2005, IEDM Technical Digest, San Francisco, December, 613, (2005).
19. A. Martinez,J. R. Barker, A.Asenov, A. Svizhenko and M.P. Anatram J.Comput. Electron **6**, 215, (2007).
20. A. Martinez, K. Kalna, J.R. Barker and A. Asenov, Physica E **37**, 168, (2007).
21. A. Martinez, J. R. Barker, A. Svizhenko, M. P. Anantram, and A. Asenov IEEE Transactions on Nanotechnology, **6**, 438, (2007).
22. J. R. Barker , D. K. Ferry, and R. Akis, Superlattices and microstructures, **27**, 319, (2000).
23. J.R. Barker,VLSI Design **13**, 237, (2001).
24. J. R. Barker, Microelectronic Engineering, **63** 223, (2002).
25. J.R. Barker and A. Martinez, J. Computational Electronics, **3**, 401, (2004).
26. J. R. Barker,, Physics of Semiconductor, ed J. Menedez and C.G. Van de Walle, AIP Press, **27**, 1493, (2005).

Self-consistent Hartree-Fock approximation for non-equilibrium electron transport through nanostructures

M. F. Gelin and D. S. Kosov

Department of Chemistry and Biochemistry, University of Maryland, College Park, 20742

Abstract. We present the formulation of self-consistent Hartree-Fock theory for non-equilibrium electron transport in nanostructures. The derivations are performed by our method for direct calculations of asymptotic, non-equilibrium steady state averages. We use asymptotic single-particle density matrix to approximate the molecular Hamiltonian by its Hartree-Fock form. Then we obtain the close system of coupled nonlinear integral equations for the transformation matrix, which diagonalizes the Hartree-Fock Hamiltonian, and asymptotic single particle density matrix.

Keywords: Electron transport, non-equilibrium, electron correlations
PACS: 72.10.-d, 72.10.Bg, 73.63.-b, 73.23.-b

INTRODUCTION

When a molecule bridges the nanogap between two metal surfaces with different chemical potentials, the electric current flows through it. Accurate and reproducible measurements of current-voltage characteristics have been reported for single molecules.[1, 2, 3, 4, 5, 6] Much of the theoretical and computational studies of transport properties of nanostructures are based on Landauer's theory which defines the conductance of the device from the electron transmission probability.[7] The use of non-equilibrium Green's functions (NEGF) in the context of Landauer's theory was pioneered by Caroli et al.[8] The modern electron transport calculations combine NEGF with density functional theory (DFT). [9, 10, 11, 12, 13, 14] There are many diverse practical implementations of NEGF-DFT and it has been applied to various specific physical systems.[15, 16, 17, 18, 19, 20, 21, 22, 23, 24, 25, 26, 27, 28, 11, 29, 30] These NEGF-DFT calculations, if they are properly performed, overestimate current by one order of magnitude [16, 31, 11, 27, 28] There is currently no consensus of whether the discrepancies are due to poorly controlled experiments, bad implementations of the NEGF-DFT scheme, or due to an inadequacy of the entire theoretical concept since there is no rigorous justification for merging DFT and NEGF. Therefore it is important to explore approaches to electron transport calculations which are based on Hartree-Fock method or include electron-electron correlations.[32, 33, 34, 35]

HAMILTONIAN AND ASYMPTOTIC NON-EQUILIBRIUM AVERAGES

The Hamiltonian of the molecular junction consists of two parts: The Hamiltonian for decoupled systems H_0 and the tunneling interaction T

$$H = H_0 + T. \tag{1}$$

The Hamiltonian for the decouples systems is

$$H_0 = \sum_l \varepsilon_l a_l^\dagger a_l + \sum_r \varepsilon_r a_r^\dagger a_r + \underbrace{\sum_m E_m a_m^\dagger a_m + \frac{1}{4} \sum_{m_1 m_2 m_3 m_4} V_{m_1 m_2 m_3 m_4} a_{m_1}^\dagger a_{m_2}^\dagger a_{m_4} a_{m_3}}_{H_M}, \tag{2}$$

where indexes l and r refer to the single-particle states in the left and right electrodes respectively, m refers to the molecular states. We assume that the electron-electron interaction is restricted to the molecular space and $V_{m_1 m_2 m_3 m_4}$ is the antisymmetrized Coulomb integral computed in some basis. The creation $a^\dagger$ and annihilation a operators obey the standard anticommutation relations

$$\left\{a_i^\dagger, a_j\right\} = \delta_{ij}, \qquad \left\{a_i^\dagger, a_j^\dagger\right\} = \{a_i, a_j\} = 0. \tag{3}$$

CP995, *Nuclei and Mesoscopic Physics - WNMP 2007*, edited by P. Danielewicz, P. Piecuch, and V. Zelevinsky

The tunneling interaction couples the electrodes with the molecule

$$T = \sum_{lm} t_{lm}(a_l^\dagger a_m + a_m^\dagger a_l) + \sum_{rm} t_{rm}(a_r^\dagger a_m + a_m^\dagger a_r), \tag{4}$$

The current operator is defined by the continuity equation:

$$J = \frac{d}{dt}N_l = i[H, N_l] = i\sum_{lm} t_{lm}(a_l^\dagger a_m - a_m^\dagger a_l). \tag{5}$$

We assume that at $t < 0$ there is no interaction between the molecule and the electrodes. Therefore the system is in equilibrium at $t < 0$ and it is described by the density matrix

$$\rho_0 = \frac{1}{\mathrm{Tr}[e^{-\beta(H_0 - \mu_L N_L - \mu_R N_R)}]} e^{-\beta(H_0 - \mu_L N_L - \mu_R N_R)}, \tag{6}$$

where N_L (N_R) is the number of particles operator for the left(right) electrode. Then at time $t = 0$ we turn on the electrode-molecule coupling T. The density matrix at any specific time $t > 0$ can be obtained by explicit time evolution of the equilibrium density matrix ρ_0

$$\rho(t) = e^{-iHt}\rho_0 e^{iHt}. \tag{7}$$

If the electrodes are macroscopically large, we can assume the existence of a unique steady state. It implies the existence of the limit $t \to +\infty$ for observable quantities. It means that the asymptotic value of the average of an operator A is uniquely defined and can be computed as

$$\begin{aligned}\overline{A} &= \lim_{t\to+\infty} \mathrm{Tr}[\rho(t)A] = \lim_{t\to+\infty} \mathrm{Tr}\left[e^{-iHt}\rho_0 e^{iHt}A\right] \\ &= \lim_{t\to+\infty} \mathrm{Tr}\left[\rho_0 e^{iHt}Ae^{-iHt}\right] = \lim_{t\to+\infty} \langle e^{iHt}Ae^{-iHt}\rangle.\end{aligned} \tag{8}$$

Here $\langle ...\rangle$ means average over the equilibrium density matrix $\langle ...\rangle = \mathrm{Tr}\,[\rho_0 ...]$. We begin calculation of the limit by the use of the following identity, which is valid for any function $f(t)$:

$$f(t) = f(0) + \int_0^t dt' \frac{df(t')}{dt'}. \tag{9}$$

Applying this identity to the limit (8), we get

$$\overline{A} = \langle A\rangle + \int_0^{+\infty} dt \frac{d}{dt}\langle e^{iHt}Ae^{-iHt}\rangle. \tag{10}$$

The existence of the unique steady state implies the convergence of the integral in Eq.(10), therefore we can introduce the infinitesimal factor η (the factor of 2 is introduced for convenience here):

$$\overline{A} = \langle A\rangle + \int_0^{+\infty} dt e^{-2\eta t} \frac{d}{dt}\langle e^{iHt}Ae^{-iHt}\rangle. \tag{11}$$

The integration by parts yields

$$\overline{A} = 2\eta \int_0^{+\infty} dt e^{-2\eta t}\langle e^{iHt}Ae^{-iHt}\rangle. \tag{12}$$

This expression for the non-equilibrium asymptotic average is very interesting physically since the integral is multiplied by infinitesimal number η. Therefore, only singular terms proportional to $1/\eta$ give non-vanishing contribution from the integral to $\overline{A}$. Using Abel's theorem

$$\lim_{T\to\infty} \frac{1}{T}\int_0^T dt f(t) = \lim_{\eta\to 0} \eta \int_0^{+\infty} dt e^{-\eta t} f(t), \tag{13}$$

we can see that (12) is the time-invariant part of the operator A (the part of the operator which is non-vanishing after averaging over the infinitely long time).

SELF-CONSISTENT NON-EQUILIBRIUM HARTREE-FOCK EQUATIONS FOR CURRENT-CARRYING MOLECULAR SYSTEM

We approximate molecular Hamiltonian by its Hartree-Fock mean field form

$$H_M \approx \sum_m E_m a_m^\dagger a_m + \sum_{m_1 m_2 m_3 m_4} V_{m_1 m_2 m_3 m_4} \overline{n}_{m_4 m_2} a_{m_1}^\dagger a_{m_2}, \tag{14}$$

where $\overline{n}_{m_4 m_2} = \lim_{t\to+\infty} \mathrm{Tr}\left[\rho(t) a_{m_2}^\dagger a_{m_1}\right]$ is the asymptotic non-equilibrium single-particle density matrix. Since the Hamiltonian (14) is quadratic, it can be diagonalized exactly by unitary transformation $\mathbf{D}$ of creation/annihlation operators

$$\sum_m D_{m'm} a_m \to a_{m'}. \tag{15}$$

To diagonilize the Hamiltonian (14), $H_M \to \sum_m \varepsilon_m a_m^\dagger a_m$, the matrix $\mathbf{D}$ should satisfy the following equation

$$\sum_{m_2} \left[E_{m_2} \delta_{m_1 m_2} + \sum_{m_3 m_4} V_{m_1 m_3 m_2 m_4} \overline{n}_{m_4 m_3} \right] D_{m_2 m} = \varepsilon_m D_{m_1 m}. \tag{16}$$

This is the nonlinear eigenvalue problem, since $\overline{n}_{m_4 m_2}$ depends on $\mathbf{D}$ and ε_m. This dependence is not trivial and we will establish it. This transformation (15) brings the Hamiltonian of the decoupled systems to the non-interacting form

$$H_0 \approx \sum_l \varepsilon_l a_l^\dagger a_l + \sum_r \varepsilon_r a_r^\dagger a_r + \sum_m \varepsilon_m a_m^\dagger a_m \tag{17}$$

and the tunneling interaction becomes

$$\begin{aligned} T &= \sum_{lm}[\sum_{m'} t_{lm'} D_{mm'}](a_l^\dagger a_m + a_m^\dagger a_l) + \sum_{rm}[\sum_{m'} t_{rm'} D_{mm'}](a_r^\dagger a_m + a_m^\dagger a_r) \\ &= \sum_{lm} T_{lm}(a_l^\dagger a_m + a_m^\dagger a_l) + \sum_{rm} T_{rm}(a_r^\dagger a_m + a_m^\dagger a_r), \end{aligned} \tag{18}$$

where T_{lm} and T_{rm} are new matrix elements renormalized by the transformation matrix $\mathbf{D}$.

Now, we turn our attention to the calculation of

$$\overline{n}_{mm'} = 2\eta \int_0^{+\infty} dt e^{-2\eta t} \langle e^{iHt} a_{m'}^\dagger a_m e^{-iHt} \rangle. \tag{19}$$

This integral can be further transformed by inserting the delta function $\delta(t-t') = \frac{1}{2\pi}\int_{-\infty}^{+\infty} d\omega e^{i\omega(t-t')}$ leading to the following expression for the asymptotic single particle density

$$\overline{n}_{mm'} = \frac{\eta}{\pi} \int_{-\infty}^{+\infty} d\omega \int_0^{+\infty} \int_0^{+\infty} dt dt' \langle e^{-i(\omega - H - i\eta)t} a_{m'}^\dagger a_m e^{i(\omega - H + i\eta)t'} \rangle. \tag{20}$$

With the use of operator identity (51) the single-particle density matrix becomes

$$\overline{n}_{mm'} = \frac{\eta}{\pi} \int_{-\infty}^{+\infty} d\omega \int_0^{+\infty} \int_0^{+\infty} dt dt' \sum_{ij} \left[e^{-i(\omega - \mathbf{H} - i\eta)t}\right]_{im'} \left[e^{i(\omega - \mathbf{H} + i\eta)t'}\right]_{mj} \langle a_i^\dagger a_j \rangle. \tag{21}$$

Here, the Hamiltonian matrix is

$$\mathbf{H} = \begin{bmatrix} \mathbf{H}_L & \mathbf{T}_{LM} & 0 \\ \mathbf{T}_{ML} & \mathbf{H}_M & \mathbf{T}_{MR} \\ 0 & \mathbf{T}_{RM} & \mathbf{H}_R \end{bmatrix}. \tag{22}$$

Now we take into account that $\langle a_i^\dagger a_j \rangle = \delta_{ij} n_i$, where n_i is the equilibrium Fermi-Dirac occupation number of single-particle level i, and introduce the diagonal matrix of occupation numbers

$$\mathbf{n} = \begin{bmatrix} \mathbf{n}_L & 0 & 0 \\ 0 & \mathbf{n}_M & 0 \\ 0 & 0 & \mathbf{n}_R \end{bmatrix}, \qquad (\mathbf{n})_{ij} = \delta_{ij} n_i. \tag{23}$$

The asymptotic single-particle density matrix becomes

$$\bar{n}_{mm'} = \frac{\eta}{\pi}\int_{-\infty}^{+\infty} d\omega \int_0^{+\infty}\int_0^{+\infty} dt dt' \left[e^{-i(\omega-\mathbf{H}-i\eta)t}\mathbf{n}e^{i(\omega-\mathbf{H}+i\eta)t'}\right]_{mm'} \tag{24}$$

Integrating (24) over t and t', we obtain

$$\bar{n}_{mm'} = \frac{\eta}{\pi}\int_{-\infty}^{+\infty} d\omega \mathrm{Tr}\left[\mathbf{G}^{\dagger}(\omega)\mathbf{n}\mathbf{G}(\omega)\right]_{mm'} \tag{25}$$

Here we define the Green's function for the entire system:

$$\mathbf{G}(\omega) = \begin{bmatrix} \mathbf{G}_{LL}(\omega) & \mathbf{G}_{LM}(\omega) & \mathbf{G}_{LR}(\omega) \\ \mathbf{G}_{ML}(\omega) & \mathbf{G}_{MM}(\omega) & \mathbf{G}_{MR}(\omega) \\ \mathbf{G}_{RL}(\omega) & \mathbf{G}_{RM}(\omega) & \mathbf{G}_{RR}(\omega) \end{bmatrix} = [(\omega+i\eta)\mathbf{I}-\mathbf{H}]^{-1} \tag{26}$$

The use of the Frobenius formula (appendix B) leads to the following matrix elements of the Green's function

$$\mathbf{G}_{MM}(\omega) = [(\omega+i\eta)\mathbf{I}_M - \mathbf{H}_M - \Sigma_L(\omega) - \Sigma_R(\omega)]^{-1}, \tag{27}$$

$$\mathbf{G}_{ML}(\omega) = \mathbf{G}_{MM}(\omega)\mathbf{T}_{ML}\mathbf{g}_L(\omega), \qquad \mathbf{G}_{MR}(\omega) = \mathbf{G}_{MM}(\omega)\mathbf{T}_{MR}\mathbf{g}_R(\omega), \tag{28}$$

$$\mathbf{G}_{LL}(\omega) = \mathbf{g}_L(\omega) + \mathbf{g}_L(\omega)\mathbf{T}_{LM}\mathbf{G}_{MM}(\omega)\mathbf{T}_{ML}\mathbf{g}_L(\omega), \tag{29}$$

$$\mathbf{G}_{RR}(\omega) = \mathbf{g}_R(\omega) + \mathbf{g}_R(\omega)\mathbf{T}_{RM}\mathbf{G}_{MM}(\omega)\mathbf{T}_{MR}\mathbf{g}_R(\omega). \tag{30}$$

Here we introduced Green's functions for the left and right electrodes

$$\mathbf{g}_{L/R}(\omega) = \left[(\omega+i\eta)\mathbf{I}_{L/R} - \mathbf{H}_{L/R}\right]^{-1}, \tag{31}$$

and corresponding self-energies

$$\Sigma_{L/R}(\omega) = \mathbf{T}_{ML/R}\mathbf{g}_{L/R}(\omega)\mathbf{T}_{L/RM}. \tag{32}$$

Using the explicit form of the occupation number matrix (23), we obtain the following expression of single-particle density matrix

$$\bar{n}_{mm'} = \frac{\eta i}{\pi}\int_{-\infty}^{+\infty} d\omega\,[\mathbf{G}^*_{MM}(\omega)\mathbf{n}_M\mathbf{G}_{MM}(\omega)]_{mm'}$$

$$+\frac{\eta i}{\pi}\int_{-\infty}^{+\infty} d\omega\,([\mathbf{G}^*_{ML}(\omega)\mathbf{n}_L\mathbf{G}_{LM}(\omega)]_{mm'} + [\mathbf{G}^*_{MR}(\omega)\mathbf{n}_R\mathbf{G}_{RM}(\omega)]_{mm'}) \tag{33}$$

The first term in this expression, which involves the molecular states, is not singular, and being multiplied by η becomes zero. It indicates that initial electronic distribution on the molecule is not relevant for the steady state single-particle density matrix. Now, we substitute the explicit expressions for matrix elements of the Green's functions (27-30) and perform some straightforward algebraic manipulations. With the help of the following "easy-to-prove" identity

$$\mathbf{g}^*_{L/R} - \mathbf{g}^*_{L/R} = 2i\eta\,\mathbf{g}^*_{L/R}\mathbf{g}_{L/R}, \tag{34}$$

the single particle density matrix (33) becomes

$$\bar{n}_{mm'} = \frac{1}{2\pi i}\int_{-\infty}^{+\infty} d\omega\,[\mathbf{G}_{MM}(\omega)\mathbf{T}_{ML}(\mathbf{g}^*_L(\omega) - \mathbf{g}^*_L(\omega))\mathbf{n}_L\mathbf{T}_{LM}\mathbf{G}^*_{MM}(\omega)$$

$$+\mathbf{G}_{MM}(\omega)\mathbf{T}_{MR}(\mathbf{g}^*_R(\omega) - \mathbf{g}^*_R(\omega))\mathbf{n}_R\mathbf{T}_{RM}\mathbf{G}^*_{MM}(\omega)]_{mm'} \tag{35}$$

Since $(\mathbf{g}^*_R(\omega) - \mathbf{g}_R(\omega))_{rr} \sim \delta(\varepsilon_r - \omega)$ and $(\mathbf{n}_R)_{rr} = n(\varepsilon_r - \mu_R)$ (the same is true for the left electrode) then

$$((\mathbf{g}^*_R(\omega) - \mathbf{g}_R(\omega))\mathbf{n}_R)_{rr} = (\mathbf{g}^*_R(\omega) - \mathbf{g}_R(\omega))_{rr}\,n(\omega - \mu_R), \tag{36}$$

$$((\mathbf{g}^*_L(\omega) - \mathbf{g}_L(\omega))\mathbf{n}_L)_{ll} = (\mathbf{g}^*_L(\omega) - \mathbf{g}_L(\omega))_{ll}\,n(\omega - \mu_L), \tag{37}$$

and we can pull the occupation numbers from the matrix product in (35). Then with the use of the eq.(34) we obtain the following expression for the steady state single-particle density matrix:

$$\bar{n}_{mm'} = \frac{1}{2\pi i}\int_{-\infty}^{+\infty} d\omega\,[n(\omega-\mu_L)\mathbf{G}_{MM}(\omega)(\Sigma^*_L(\omega) - \Sigma_L(\omega))\mathbf{G}^*_{MM}(\omega)$$

$$+n(\omega-\mu_R)\mathbf{G}_{MM}(\omega)(\Sigma^*_R(\omega) - \Sigma_R(\omega))\mathbf{G}^*_{MM}(\omega)]_{mm'} \tag{38}$$

The system of coupled nonlinear integral equations (16) and (38) should be solved simultaneously. Combined together they represent self-consistent Hartree-Fock approximation for current carrying molecular system.

ASYMPTOTIC VALUE FOR ELECTRON CURRENT: LANDAUER FORMULA

Having obtained non-equilibrium single particle density matrix $\overline{n}_{mm'}$ and transformation matrix $\mathbf{D}$ we can compute the asymptotic value of electric current:

$$\overline{J} = 2\eta \int_0^{+\infty} dte^{-2\eta t} \langle e^{iHt} J e^{-iHt} \rangle. \tag{39}$$

Likewise the calculations of asymptotic density matrix, we transform this integral by inserting the delta function $\delta(t-t')$ and by using operator identity (51):

$$\overline{J} = \frac{\eta}{\pi} \int_{-\infty}^{+\infty} d\omega \int_0^{+\infty} \int_0^{+\infty} dtdt' \sum_{ij} \left(e^{-i(\omega - \mathbf{H} - i\eta)t} \mathbf{J} e^{i(\omega - \mathbf{H} + i\eta)t'} \right)_{ij} \langle a_i^\dagger a_j \rangle. \tag{40}$$

The matrix, which represents the current is

$$\mathbf{J} = \begin{bmatrix} 0 & \mathbf{T}_{LM} & 0 \\ -\mathbf{T}_{ML} & 0 & 0 \\ 0 & 0 & 0 \end{bmatrix}. \tag{41}$$

Using the matrix representation of the occupation numbers (23), we obtain

$$\overline{J} = \frac{\eta}{\pi} \int_{-\infty}^{+\infty} d\omega \int_0^{+\infty} \int_0^{+\infty} dtdt' \mathrm{Tr} \left[e^{-i(\omega - \mathbf{H} - i\eta)t} \mathbf{J} e^{i(\omega - \mathbf{H} + i\eta)t'} \mathbf{n} \right]. \tag{42}$$

Integrating (42) over t and t', we obtain

$$\overline{J} = \frac{\eta}{\pi} \int_{-\infty}^{+\infty} d\omega \mathrm{Tr} \left[\mathbf{G}^\dagger(\omega) \mathbf{J} \mathbf{G}(\omega) \mathbf{n} \right], \tag{43}$$

where $\mathbf{G}(\omega)$ is the total Green's function (26). Using the explicit expression of the current and occupation number matrices, we obtain the electron current in the following form

$$\begin{aligned} \overline{J} = & \frac{\eta i}{\pi} \int_{-\infty}^{+\infty} d\omega \mathrm{Tr} \left[\left(\mathbf{G}_{MM}^*(\omega) \mathbf{T}_{ML} \mathbf{G}_{LM}(\omega) - \mathbf{G}_{ML}^*(\omega) \mathbf{T}_{LM} \mathbf{G}_{MM}(\omega) \right) \mathbf{n}_M \right] \\ & + \frac{\eta i}{\pi} \int_{-\infty}^{+\infty} d\omega \mathrm{Tr} \left[\left(\mathbf{G}_{LM}^*(\omega) \mathbf{T}_{ML} \mathbf{G}_{LL}(\omega) - \mathbf{G}_{LL}^*(\omega) \mathbf{T}_{LM} \mathbf{G}_{ML}(\omega) \right) \mathbf{n}_L \right] \\ & + \frac{\eta i}{\pi} \int_{-\infty}^{+\infty} d\omega \mathrm{Tr} \left[\left(\mathbf{G}_{RM}^*(\omega) \mathbf{T}_{ML} \mathbf{G}_{LR}(\omega) - \mathbf{G}_{RL}^*(\omega) \mathbf{T}_{LM} \mathbf{G}_{MR}(\omega) \right) \mathbf{n}_R \right]. \end{aligned} \tag{44}$$

The first term in this expression, which involves the trace over the molecular states is not singular enough, and being multiplied by η becomes zero. The rest of the derivations is very similar to the corresponding derivation for asymptotic single particle density matrix. Therefore we do not repeat it here and merely write the final expression for the steady state electric current

$$\overline{J} = \frac{1}{2\pi} \int_{-\infty}^{+\infty} d\omega (n(\omega - \mu_L) - n(\omega - \mu_R)) \mathrm{Tr} \left[(\Sigma_L(\omega) - \Sigma_L^*(\omega)) \mathbf{G}_{MM}^*(\omega) (\Sigma_R(\omega) - \Sigma_R^*(\omega)) \mathbf{G}_{MM}(\omega) \right]. \tag{45}$$

This expression is the Landauer formula for electric current which is typically obtained via formalism of NEGF.[8]

CONCLUSIONS

In this paper, we discuss the derivation of non-equilibrium, self-consisted Hartree-Fock equations for steady state electron transport in nanostructures. We use asymptotic single-particle density matrix to approximate the molecular Hamiltonian by its Hartree-Fock form. Then we apply our technique for calculations of asymptotic non-equilibrium averages to compute the explicit expression for the single particle density and to close the system of coupled nonlinear integral equations (16) and (38).

ACKNOWLEDGMENTS

This work has been supported by NSF-MRSEC DMR0520471 at the University of Maryland and by the American Chemical Society Petroleum Research Fund (44481-G6).

TRANSFORMATION OF OPERATOR $e^{A}Be^{-A}$ FROM SECOND QUANTIZED FORM TO MATRIX MULIPLICATION

For arbitrary operators A and B we have

$$e^{A}Be^{-A}=B+[A,B]+\frac{1}{2!}[A,[A,B]]+\ldots \tag{46}$$

Suppose, that A and B are quadratic in creation and annihilation operators:

$$A=\sum_{ij}A_{ij}a_i^{\dagger}a_i, \qquad B=\sum_{ij}B_{ij}a_i^{\dagger}a_i. \tag{47}$$

If we assume standard anticommutation relations between $a, a^{\dagger}$, we can easily compute

$$[A,B]=\sum_{ij}([\mathbf{A},\mathbf{B}])_{ij}a_i^{\dagger}a_j, \tag{48}$$

$$[A,[A,B]]=\sum_{ij}([\mathbf{A},[\mathbf{A},\mathbf{B}]])_{ij}a_i^{\dagger}a_j, \tag{49}$$

$$e^{\sum_{ij}A_{ij}a_i^{\dagger}a_i}\sum_{kl}B_{kl}a_k^{\dagger}a_l e^{-\sum_{mn}A_{mn}a_m^{\dagger}a_n}=\sum_{ij}\left(\mathbf{B}+[\mathbf{A},\mathbf{B}]+\frac{1}{2!}[\mathbf{A},[\mathbf{A},\mathbf{B}]]+\ldots\right)_{ij}a_i^{\dagger}a_j, \tag{50}$$

$$e^{\sum_{ij}A_{ij}a_i^{\dagger}a_i}\sum_{kl}B_{kl}a_k^{\dagger}a_l e^{-\sum_{mn}A_{mn}a_m^{\dagger}a_n}=\sum_{ij}\left(e^{\mathbf{A}}\mathbf{B}e^{-\mathbf{A}}\right)_{ij}a_i^{\dagger}a_j. \tag{51}$$

THE USE OF FROBENIUS FORMULA TO COMPUTE GREEN'S FUNCTION

The Frobenius formula allows us to compute the inverse of the block matrix

$$\begin{bmatrix}\mathbf{A} & \mathbf{B}\\ \mathbf{C} & \mathbf{D}\end{bmatrix}^{-1}=\begin{bmatrix}\mathbf{K}^{-1} & -\mathbf{K}^{-1}\mathbf{B}\mathbf{D}^{-1}\\ -\mathbf{D}^{-1}\mathbf{C}\mathbf{K}^{-1} & \mathbf{D}^{-1}+\mathbf{D}^{-1}\mathbf{C}\mathbf{K}^{-1}\mathbf{B}\mathbf{D}^{-1}\end{bmatrix}, \tag{52}$$

where $\mathbf{K}=\mathbf{A}-\mathbf{B}\mathbf{D}^{-1}\mathbf{C}$. If we set

$$\mathbf{A}=(\omega+i\eta)\mathbf{I}_M-\mathbf{H}_M \tag{53}$$

$$\mathbf{D}=\begin{bmatrix}(\omega+i\eta)\mathbf{I}_L-\mathbf{H}_L & 0\\ 0 & (\omega+i\eta)\mathbf{I}_R-\mathbf{H}_R\end{bmatrix} \tag{54}$$

$$\mathbf{B}=\mathbf{C}^{T}=[\mathbf{T}_{LM}\ \mathbf{T}_{RM}]. \tag{55}$$

then the Frobenius formula yields matrix elements of the Green's function (26).

REFERENCES

1. B. Xu, and N. J. Tao, *Science* **301**, 1221–1223 (2003).
2. X. Cui, A. Primak, X. Zarate, J. Tomfohr, O. Sankey, A. L. Moore, T. A. Moore, D. Gust, G. Harris, and S. Lindsay, *Science* **294**, 571 – 574 (2001).
3. M. Elbing, R. Ochs, M. Koentopp, M. Fischer, C. von Hänisch, F. Weigend, F. Evers, H. Weber, and M. Mayor, *PNAS* **102**, 8815–8820 (2005).
4. A. S. Blum, J. G. Kushmerick, D. P. Long, C. H. Patterson, J. C. Yang, J. C. Henderson, Y. X. Yao, J. M. Tour, R. Shashidhar, and B. R. Ratna, *Nature Materials* **4**, 167–172 (2005).
5. N. J. Tao, *J. Mater. Chem.* **15**, 3260–3263 (2005).
6. L. Venkataraman, J. Klare, I. Tam, C. Nuckolls, M. Hybertsen, and M. Steigerwald, *Nano Letters* **6**, 458–462 (2006).
7. R. Landauer, *Philos. Mag.* **21**, 863 (1970).
8. C. Caroli, R. Combesco, P. Nozieres, and D. Saintjam, *J. Phys. C* **4**, 916 (1971).
9. P. A. Derosa, and J. M. Seminario, *J. Phys. Chem. B* **105**, 471–481 (2001).
10. Y. Q. Xue, S. Datta, and M. A. Ratner, *Chem. Phys.* **281**, 151–170 (2002), sp. Iss. SI.
11. F. Evers, F. Weigend, and M. Koentopp, *Phys. Rev. B* **69**, 235411 (2004).
12. M. Brandbyge, J. L. Mozos, P. Ordejon, J. Taylor, and K. Stokbro, *Phys. Rev. B* **65**, 165401 (2002).
13. S. H. Ke, H. U. Baranger, and W. T. Yang, *Phys. Rev. B* **70**, 085410 (2004).
14. Z. Li, and D. S. Kosov, *J. Phys.: Cond. Matt.* **18**, 1347–1358 (2006).
15. Z. Li, and D. S. Kosov, *J. Phys. Chem. B* **110**, 9893–9898 (2006).
16. Z. Li, and D. S. Kosov, *Phys. Rev. B* **76**, 035415 (2007).
17. A. Calzolari, N. Marzari, I. Souza, and M. B. Nardelli, *Phys. Rev. B* **69**, 035108 (2004).
18. H. Chen, J. Q. Lu, J. Wu, R. Note, H. Mizuseki, and Y. Kawazoe, *Phys. Rev. B* **67**, 113408 (2003).
19. P. Damle, A. W. Ghosh, and S. Datta, *Chem. Phys.* **281**, 171–187 (2002), sp. Iss. SI.
20. Y. Fujimoto, and K. Hirose, *Phys. Rev. B* **67**, 195315 (2003).
21. H. Geng, S. W. Yin, K. Q. Chen, and Z. G. Shuai, *J. Phys. Chem. B* **109**, 12304–12308 (2005).
22. K. Hirose, N. Kobayashi, and M. Tsukada, *Phys. Rev. B* **69**, 245412 (2004).
23. S. M. Hou, J. X. Zhang, R. Li, J. Ning, R. S. Han, Z. Y. Shen, X. Y. Zhao, Z. Q. Xue, and Q. Wu, *Nanotechnology* **16**, 239–244 (2005).
24. Y. Hu, Y. Zhu, H. Gao, and H. Guo, *Phys. Rev. Lett.* **95**, 156803 (2005).
25. S. H. Ke, H. U. Baranger, and W. T. Yang, *J. Chem. Phys.* **122**, 074704 (2005).
26. N. Kobayashi, M. Brandbyge, and M. Tsukada, *Phys. Rev. B* **62**, 8430–8437 (2000).
27. Y. Q. Xue, and M. A. Ratner, *Phys. Rev. B* **68**, 115406 (2003).
28. Y. Q. Xue, and M. A. Ratner, *Phys. Rev. B* **68**, 115407 (2003).
29. J. M. Seminario, A. G. Zacarias, and P. A. Derosa, *J. Chem. Phys.* **116**, 1671–1683 (2002).
30. J. M. Seminario, C. De La Cruz, P. A. Derosa, and L. M. Yan, *J. Phys. Chem. B* **108**, 17879–17885 (2004).
31. M. Di Ventra, S. T. Pantelides, and N. D. Lang, *Phys. Rev. Lett.* **84**, 979–982 (2000).
32. T. Shimazaki, H. Maruyama, Y. Asai, and K. Yamashita, *J. Chem. Phys.* **123**, 164111 (2005).
33. T. Shimazaki, Y. Xue, M. A. Ratner, and K. Yamashita, *J. Chem. Phys.* **124**, 114708 (2006).
34. X. Wang, C. D. Spataru, M. S. Hybertsen, and A. J. Millis, *Phys. Rev. B* **77**, 045119 (2008).
35. Y. Dahnovsky, V. G. Zakrzewski, A. Kletsov, and J. V. Ortiz, *J. Chem. Phys.* **123**, 184711 (2005).

Electron correlations and the conductance of a quantum point contact at finite temperature

C. Sloggett*, A.I.Milstein† and O. P. Sushkov*

*School of Physics, University of New South Wales, Sydney 2052, Australia
†Budker Institute of Nuclear Physics, 630090 Novosibirsk, Russia

Abstract.
To describe electron transport through a quantum point contact we formulate and solve a nonlocal kinetic equation for the density matrix of electrons. The electron-electron interaction is treated by perturbation theory with account of higher order terms. At zero temperature, the approach results in the Landauer formula for conductance of a quantum point contact. At finite temperature, the leading correction to the current of correlated electrons scales as temperature squared. The corresponding correction to conductance G is negative and strongly enhanced in the region $0.5\frac{2e^2}{h} \leq G \leq 1.0\frac{2e^2}{h}$. The effect may be relevant to the so called "0.7 structure" of conductance observed experimentally.

Keywords: Quantum point contact, 0.7 structure
PACS: 74.72.Dn, 75.10.Jm, 75.30.Fv, 75.50.Ee

INTRODUCTION

The conductance of a quantum point contact (QPC) - a 1D constriction in a 2D electron gas - has been known to be quantized in units of $G_0 = 2e^2/h$ since 1988 [1, 2]. The observed conductance plateaus can be easily understood in the single-electron picture [3, 4]. Below we measure conductance in units of G_0.

The "0.7 structure" appears on the lowest conductance step as a narrow extra plateau at $G \approx 0.7$.The structure was first observed by Thomas *et al.* in 1996[5] and has been the subject of numerous experiments since [6, 7, 8, 9, 10, 11]. The position of the structure varies from 0.5 to 0.7 depending on the device, and the structure becomes more pronounced as the temperature is increased up to $\sim$ 2K, where thermal smearing becomes significant. It is not clear from experiment whether the structure survives at $T = 0$. In a longitudinal magnetic field, which breaks the spin degeneracy, the structure evolves smoothly to the 0.5 plateau that is expected in the single-electron picture. This demonstrates that the effect is in some way related to electron spin.

Observations of the "0.7 structure" have created much theoretical interest. There have been suggested explanations based on spontaneous magnetization within the contact [12, 13, 14, 15, 16, 17], charge density waves within the contact [18, 19], the Kondo effect [20, 21], and even on electron-phonon scattering in the contact [22]. The problem has also stimulated the development of general scattering theory in the presence of leads [23, 24].

The present study is motivated by the "0.7 structure". However, within perturbation theory, we consider the *generic* problem of the correction to the Landauer formula for the conductance of a QPC. Although it is not clear that perturbation theory is capable of explaining the "0.7 structure", our results are very similar to experimental data.

MODEL

Near the center of the QPC the single particle dynamics are described by the parabolic saddle-point potential $U = -\frac{1}{2}m\omega^2x^2 + \frac{1}{2}m\omega_y^2y^2 + V_g$, see, e.g., Ref. [4]. Here m is the effective mass, V_g is a gate voltage, and typical values corresponding to experiment are $\omega_y/\omega \sim 3$, $\omega_y \sim 4$meV. Throughout the paper we set $\hbar = k_B = 1$. The dynamics in the x-direction are described by different channels corresponding to quantization in the y-direction. The lowest, $n = 0$, channel has longitudinal potential $-\frac{1}{2}m\omega^2x^2$, and hence the transmission probability is [25]

$$t_\varepsilon = \left(1 + e^{-2\pi\varepsilon}\right)^{-1}, \tag{1}$$

CP995, *Nuclei and Mesoscopic Physics - WNMP 2007*, edited by P. Danielewicz, P. Piecuch, and V. Zelevinsky

where $\varepsilon = E/\omega$, E is the energy. At $\varepsilon = 0$ the turning point for the nth channel is $x_t = \pm\sqrt{\frac{2n\omega_y}{m\omega^2}}$. So outside the barrier, from $x_{2D} \sim 2/\sqrt{m\omega}$, many transverse channels are occupied and hence the electron-electron Coulomb interaction is strongly screened. The interaction is unscreened only on the top of the barrier where the electron density is low. Therefore the effective electron-electron interaction in the $n = 0$ channel can be approximated as

$$\frac{e^2}{\kappa|x_1 - x_2|} \to H_{int} = \omega\pi^2 g\delta(\xi_1)\delta(\xi_2) , \tag{2}$$

where e is the electron charge, κ is the dielectric constant, and $\xi = \sqrt{m\omega}x$ is the dimensionless distance. In GaAs, $\kappa \approx 13$, $m \approx 0.07m_e$, and for $\omega \sim 1$meV the dimensionless coupling constant is about unity,

$$g \sim \frac{e^2}{\pi^2\kappa}\sqrt{m/\omega} \sim 1 . \tag{3}$$

It is worth to note that, in our opinion, the same many-body screening that we discussed above explains absence of structures in conductance steps with $n = 1$ and higher. For $n \geq 1$ density of electrons on the top of the potential is always large.

SINGLE-PARTICLE WAVE FUNCTIONS

To represent single-particle wave functions in the $n = 0$ channel we consider a 1D wire of length L with a potential barrier in the middle of the wire. The details of the potential shape are unimportant apart from the parabolic top. Wave functions are normalised according to

$$\int_{-L/2}^{L/2} |\psi_k|^2 dx = 1 . \tag{4}$$

The length L is of the order of the inelastic mean free path in the leads. Away from the barrier the wave function ψ_k is the standard combination of incident, reflected and transmitted waves

$$\psi_k(x) = \begin{cases} \frac{1}{\sqrt{L}}\left(e^{ikx} + R_\varepsilon e^{-ikx}\right) & \xi \ll 0 \\ \frac{1}{\sqrt{L}} T_\varepsilon e^{ikx} & \xi \gg 0 \end{cases} \tag{5}$$

where R_ε and T_ε are the complex reflection and transmission coefficients, and the transmission probability is $t_\varepsilon = |T_\varepsilon|^2$. Near $\xi = 0$ we can consider only the nearby shape of the potential barrier $-\frac{1}{2}m\omega^2x^2 = -\frac{1}{2}\omega\xi^2$, which can be solved to give single-particle wavefunctions of the form (see, e.g., Ref. [25])

$$\psi_k(\xi) \approx \frac{1}{\sqrt{L}}\left(\frac{mv_F^2}{2\omega}\right)^{1/4} \varphi_k(\xi) , \tag{6a}$$

$$\varphi_k(\xi) = \sqrt{\frac{e^{\pi\varepsilon/2}}{\cosh(\pi\varepsilon)}} D_{(i\varepsilon-1/2)}(\sqrt{2}\xi e^{-i\pi/4}), \tag{6b}$$

where v_F is the Fermi velocity far from the barrier and D_ν is the parabolic cylinder function. Eqs. (5) and (6) corresponds to the wave incident from the left, $k \geq 0$. The probability density at the top of the potential given by Eq. 6b,

$$\rho(\varepsilon) = |\varphi_k(0)|^2 = \frac{\pi\exp(\pi\varepsilon/2)}{\sqrt{2}\cosh(\pi\varepsilon)\,|\Gamma(3/4 - i\varepsilon/2)|^2}, \tag{7}$$

is peaked at $\varepsilon \approx 0.2$, see Fig. 1. This results in enhancement of the interaction (2), and in the end leads to all effects considered in this paper.

We stress that Eqs. (6b) and (7) are given by the exact solution for the parabolic barrier. The normalised Eq. (6a) assumes a semiclassical approximation at large distances from the top of the barrier where the shape of the barrier is smoothly changing from parabolic to flat. Note that the peak in Fig. 1 is *not* a resonance behaviour. This is clear from Eq. (1): the transmission probability is a smooth function of energy without a resonant behavior. The peak in Fig. 1 is due to semiclassical slowing and it is unrelated to a resonance or virtual level.

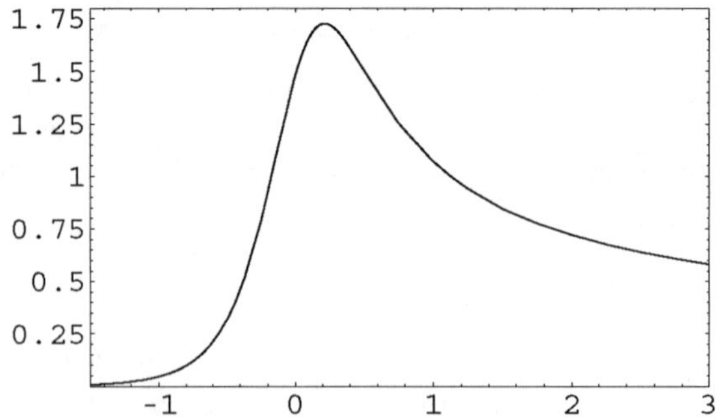

FIGURE 1. The probability density at $\xi = 0$ versus the dimensionless energy ε.

GENERAL EXPRESSION FOR DC ELECTRON CURRENT IN TERMS OF THE DENSITY MATRIX

The operator of electric current $\hat{j}$ is a single particle operator. In the basis of the scattering states (6) matrix elements $\langle k|\hat{j}|k'\rangle$ are nonzero only if $k' = k$ or $k' = -k$.

$$\langle \psi_k|\hat{j}|\psi_k\rangle = e\frac{k}{m}t_\varepsilon \, , \tag{8}$$

$$\langle \psi_{-k}|\hat{j}|\psi_k\rangle = e\frac{k}{2m}(R^*_\varepsilon T_\varepsilon - R_\varepsilon T^*_\varepsilon) = e\frac{k}{m}R^*_\varepsilon T_\varepsilon \, . \tag{9}$$

Here we have used the orthogonality condition, $\langle \psi_{-k}|\psi_k\rangle \propto (R^*_\varepsilon T_\varepsilon + R_\varepsilon T^*_\varepsilon) = 0$. Therefore the operator $\hat{j}$ can be represented as

$$\hat{j} = \frac{e}{L}\sum_{k,\sigma}\frac{k}{m}\left(t_\varepsilon a^\dagger_{k,\sigma}a_{k,\sigma} + R^*_\varepsilon T_\varepsilon a^\dagger_{-k,\sigma}a_{k,\sigma}\right) \, , \tag{10}$$

where $a^\dagger_{k,\sigma}$ is the creation operator of the electron in state (5), (6) with spin σ. The total electric current is given by the trace of the single particle operator with the density matrix,

$$J = \frac{e}{L}\sum_{k,\sigma}\frac{k}{m}t_\varepsilon n_{k,\sigma} \, . \tag{11}$$

Here $n_{k,\sigma} = \langle\langle a^\dagger_{k,\sigma}a_{k,\sigma}\rangle\rangle$ is the usual occupation number. The double brackets denote quantum and Gibbs averaging. We have used the fact that in the basis of stationary single particle states the offdiagonal matrix element of the density matrix is zero, $\langle\langle a^\dagger_{-k,\sigma}a_{k,\sigma}\rangle\rangle = 0$. This treatment of J is valid with or without electron–electron interaction.

LANDAUER FORMULA FOR NONINTERACTING ELECTRONS

Without electron–electron interaction within the contact the steady state is established due to inelastic collisions in the right and left leads. So, there are two Fermi–Dirac distributions n_{0k} for $k > 0$ and for $k < 0$, with different chemical potentials due to the applied voltage V, so that n_{0k} depends on the direction of current flow, which we write $s = k/|k|$. We write n_{0k} instead of n_k to indicate that this is the noninteracting case. Since the applied voltage V is small one can write

$$n_{0k} = n_f + s\frac{eV}{2\omega}n'_f \, , \tag{12}$$

where n_f is the equilibrium Fermi–Dirac distribution, $n'_f = -\frac{\partial n_f}{\partial \varepsilon} \approx \delta(\varepsilon - \mu)$, and μ is the chemical potential in units of ω.

Substitution of (12) into the general Eq. (11) immediately gives the Landauer formula

$$J = 2\frac{e}{L}\int \frac{Ldk}{2\pi}\frac{k}{m}t_\varepsilon n_{0k} = \frac{2e^2}{h}t_\mu V \, , \tag{13}$$

as expected in the noninteracting case. In units of G_0 this just gives $G = t_\mu$.

KINETIC EQUATION FOR OCCUPATION NUMBERS WITH ELECTRON–ELECTRON INTERACTION

At zero temperature the interaction (2) in the leading order gives the usual elastic direct and exchange diagrams for the single-particle scattering amplitude, see Fig. 2. The corresponding corrections renormalise height of the

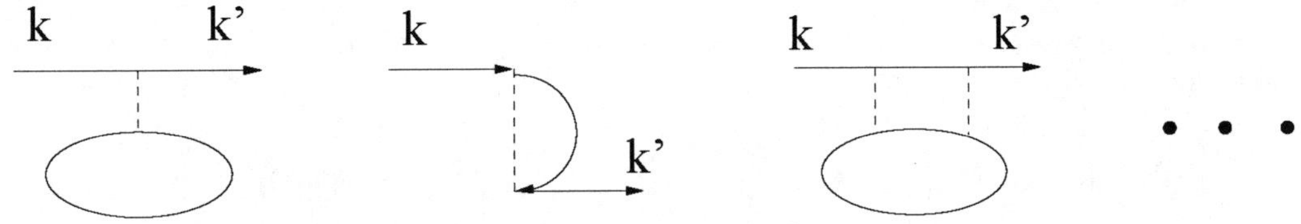

FIGURE 2. Corrections to the single particle potential due to electron–electron interaction. The dashed line shows interaction (2)

single particle potential, renormalise the frequency ω in the potential and even make the effective potential slightly nonparabolic. Nevertheless this does not materially change the profile of the transmission coefficient (1). We have checked numerically that the same is true to higher orders in perturbation theory. Thus the interaction (2) treated perturbatively does not influence the Landauer formula and does not materially change the shape of the transmission coefficient at zero temperature.

At nonzero temperature, the interaction (2) leads to two different effects. The first is related to two-leg diagrams describing elastic scattering and shown in Fig. 2. This effect gives just a weak temperature-dependent renormalisation of the effective potential $-\frac{1}{2}m\omega^2x^2$, similar to the zero-temperature effect discussed above. This does not materially change the profile of the transmission coefficient. The second effect, related to four-leg diagrams describing inelastic scattering, Fig. 3, gives a nontrivial correction to conductance. To account for this inelastic scattering we need to discuss equilibration, for which we formulate a kinetic equation.

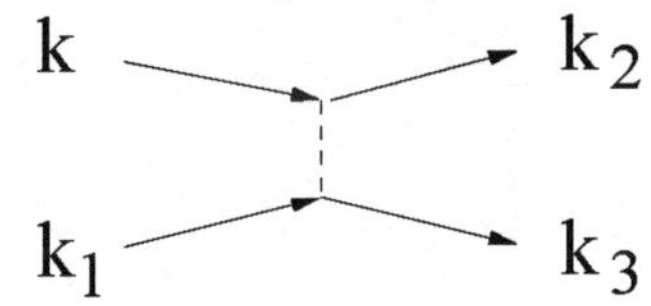

FIGURE 3. Four leg diagram describing inelastic scattering.

Let us first consider the case without electron–electron interaction. The scattering states equilibrate to n_{0k} due to collisions in the leads. We can write the kinetic equation

$$\frac{\partial n_k}{\partial t} = -\frac{n_k - n_{0k}}{\tau} , \tag{14}$$

where τ is the relaxation time in the leads. For *any small* deviation from equilibrium, the occupation numbers must obey this equation simply because the equilibrium density matrix is diagonal in the basis of the scattering states [26], and there is no source of equilibration other than the relaxation effect in the leads. This is a nonlocal kinetic equation, as one cannot in principle apply it independently to any particular point within the QPC. The equation describes a region of the size of the inelastic mean free path L around the contact. This kinetic equation is in principle also valid for finite bias V, but in the present paper we consider only an infinitesimal bias where n_{0k} is given by Eq. (12).

To take into account electron–electron interaction in the QPC, we consider the basic scattering event of Fig. 3. We obtain a collision term in the kinetic equation [27]

$$\begin{aligned}
&\frac{\partial n_k}{\partial t} = -\frac{n_k - n_{0k}}{\tau} + \mathrm{St}(n_k) , \\
&\mathrm{St}(n_k) = 2\pi \int \frac{Ldk_1}{2\pi}\frac{Ldk_2}{2\pi}\frac{Ldk_3}{2\pi} |M_{kk_1k_2k_3}|^2 \left[n_{k_2}n_{k_3}(1-n_k)(1-n_{k_1}) - n_k n_{k_1}(1-n_{k_2})(1-n_{k_3})\right] \\
&\quad \times \delta(E_k + E_{k_1} - E_{k_2} - E_{k_3}) .
\end{aligned} \tag{15}$$

Here $M_{kk_1k_2k_3}$ is the matrix element of the interaction (2). This matrix element corresponds to the real transition between quantum states, $|k,k_1\rangle \rightarrow |k_2,k_3\rangle$, as shown in Fig. 3. Therefore in the matrix element the initial and final wave functions are given by (6). Note that this differs from the Sommerfeld rule, used for the scattering amplitude, where initial wavefunctions are given by Eq. 6, while final wavefunctions are $\psi_k^{(-)}(x) = \psi_{-k}^*(x)$. (For a description of the Sommerfeld rule, see, e.g., [28].)

CONDUCTANCE WITH ELECTRON–ELECTRON INTERACTION

Using occupation numbers from Eq. (12) and expanding the collision integral of Eq. 15 up to the first power in the bias V, we find the integral at $T \ll \omega$,

$$\mathrm{St}(n_k) = -\frac{T^2 eVL^3}{12v_F^3}\delta(E_k - \omega\mu)\sum_{s_1s_2s_3}\left[s+s_1-s_2-s_3\right]\left|M_{kk_1k_2k_3}\right|^2 . \tag{16}$$

All legs in the matrix element are taken at the Fermi surface, so we need only perform summations over the directions s_1, s_2, s_3. Calculating the matrix elements of interaction (2) with the wave functions (6) we find

$$\begin{aligned}\mathrm{St}(n_k) &= -\frac{T^2 eVL^3}{3v_F^3}s\delta(E_k-\omega\mu)\left(|M_{++--}|^2+|M_{+++-}|^2\right)\\ &= -seV\frac{\pi^4 g_e^2}{6}\frac{v_F}{L}\left(\frac{T}{\omega}\right)^2\rho^4(\mu)\delta(E_k-\omega\mu)\,,\end{aligned} \tag{17}$$

where $\rho(\mu)$ is given by (7). Eq. (17) leads to the following steady-state solution of the kinetic equation (15)

$$n_k = n_f + \frac{eV}{2}s\delta(E_k-\omega\mu)\left\{1-\frac{\pi^4 g_e^2}{3}\frac{\tau v_F}{L}\left(\frac{T}{\omega}\right)^2\rho^4(\mu)\right\} . \tag{18}$$

Substitution of this expression into Eq. (11) gives an altered transmission probability (or conductance in units of G_0)

$$t_\mu \rightarrow \bar{t}_\mu = t_\mu\left\{1-\frac{\pi^4 g_e^2}{3}\frac{\tau v_F}{L}\left(\frac{T}{\omega}\right)^2\rho^4(\mu)\right\} . \tag{19}$$

This result is justified only if the second term in brackets is small. The term is due to the current of correlated electrons. It will be most significant within the region $0.5\times 2e^2/h \le G \le 1.0\times 2e^2/h$, since above $G = 0.5$ electrons can flow without tunneling, leading to the peak in $\rho(\mu)$. The length L is of the order of the mean free path in the leads, so it is most natural to assume that the factor $\tau v_F/L$ in (19) is of the order of unity, though we cannot exclude some dependence of the factor on temperature. In the latter case the T^2 dependence of the correlated current will be modified.

A set of plots of $\bar{t}_\mu$ for different temperatures is shown in Fig. 4. Though the result looks quite sensible it is obtained

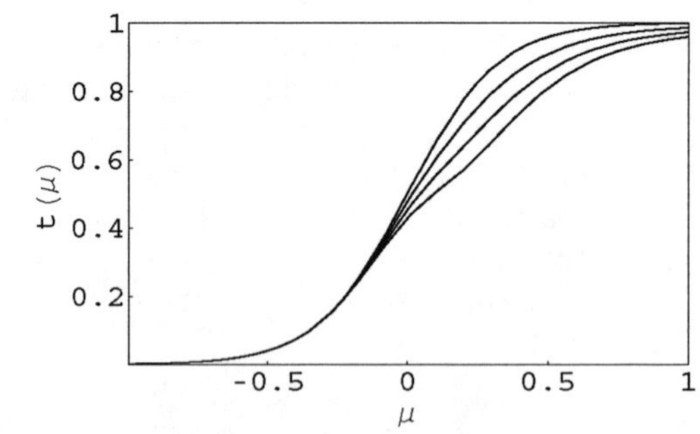

FIGURE 4. Conductance in units of $2e^2/h$ versus μ (chemical potential in units of ω) for different temperatures in the weak coupling limit, $g_e \ll 1$. The uppermost curve corresponds to $g_eT = 0$, while the lowest is $g_eT \approx 0.3$K, assuming $\frac{\tau v_F}{L} \approx 1$.

for $g_e \ll 1$. However, according to our estimate (2), the constant is not small, $g_e \sim 1$, and hence virtual rescattering must be taken into account.

RENORMALIZATION OF THE COUPLING CONSTANT DUE TO RESCATTERING

The leading correction to the Born scattering amplitude Fig. 3 is given by the diagrams shown in Fig. 5. This correction

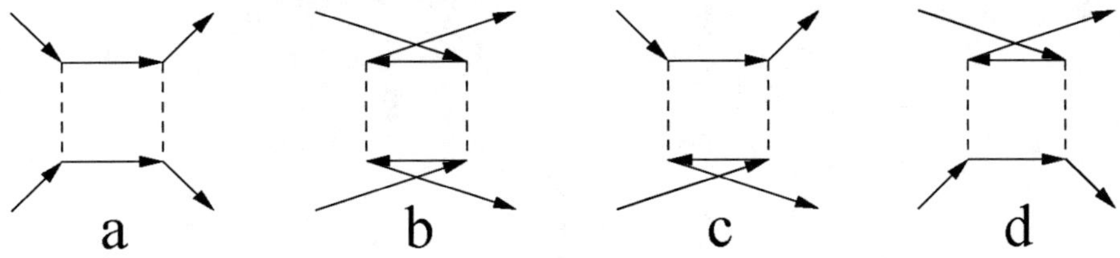

FIGURE 5. The leading correction to the matrix element.

is equivalent to renormalization of the coupling constant, $g_e \to g_e + \delta g_e$, where $\delta g_e(\mu) = 2g_e^2 K(\mu)$ and

$$K(\mu) = \frac{1}{4}\int_{-\infty}^{\infty}\int_{-\infty}^{\infty} \rho(\varepsilon_1)\rho(\varepsilon_2)\left\{\frac{\theta(\varepsilon_1-\mu)\theta(\varepsilon_2-\mu)}{2\mu-\varepsilon_1-\varepsilon_2} + \frac{\theta(\mu-\varepsilon_1)\theta(\mu-\varepsilon_2)}{\varepsilon_1+\varepsilon_2-2\mu} - 2\frac{\theta(\mu-\varepsilon_1)\theta(\varepsilon_2-\mu)}{\varepsilon_1-\varepsilon_2}\right\} d\varepsilon_1 d\varepsilon_2 \,. \quad (20)$$

Here $\theta(y)$ is the step function. The first term in $K(\mu)$ (diagram Fig. 5a) is logarithmically divergent at $\varepsilon_1, \varepsilon_2 \to +\infty$. The divergence is a result of the contact approximation (2). The point is that the δ-functions in Eq. (2) in reality should be replaced by bell-shaped functions with width of the order of the barrier size, $\sim 1/\sqrt{m\omega}$, which is also the screening scale. When the electron wavelength λ is larger than this size one can ignore the finite width and use the contact (δ-function) approximation (2). However, when the energy is large, $\varepsilon \gg 1$, the electron wave function oscillates within the barrier size, and hence the contact approximation strongly overestimates the Coulomb matrix element. This leads to a logarithmic ultraviolet divergence of (20). To fix the problem, we just introduce an ultraviolet cutoff Λ, $\varepsilon_1 + \varepsilon_2 \le \Lambda$. Dependence on the cutoff is weak and we will present all results for $\Lambda = 2$. The integrals in (20) cannot be calculated analytically. However numerical integration is very simple and we present a plot of $K(\mu)$ in Fig. 6.

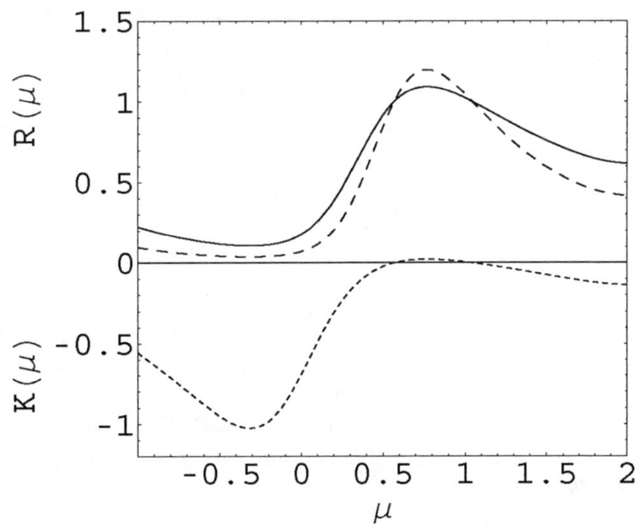

FIGURE 6. Dashed line: the function $K(\mu)$ for the second order correction. Solid line: Brueckner correction factor $R(\mu)$ for $g_e = 1$. Long dashed line: Brueckner correction factor $R(\mu)$ for $g_e = 2$.

Since the coupling constant $g_e \sim 1$, the second order correction alone is not sufficient. However in this regime the Brueckner approximation[29] usually works well, see analysis in Ref. [30]. Since the kernel $K(\mu)$ is independent of external momenta, the Brueckner approximation is equivalent to the summation of a geometrical progression, and hence the renormalised coupling constant g_R is

$$g_R^2 = g_e^2 R(\mu)\,, \quad R(\mu) = \frac{1}{[1-2g_e K(\mu)]^2}\,. \quad (21)$$

Plots of $R(\mu)$ for $g_e = 1$ and $g_e = 2$ are presented in Fig. 6. Fig. 7 shows a set of plots of conductance $\overline{t_\mu}$ for different temperatures, using Eq. (19) with $g_e \to g_R(g_e)$ for $g_e = 1$. The results presented in Figs. 4, 7 are very similar to the

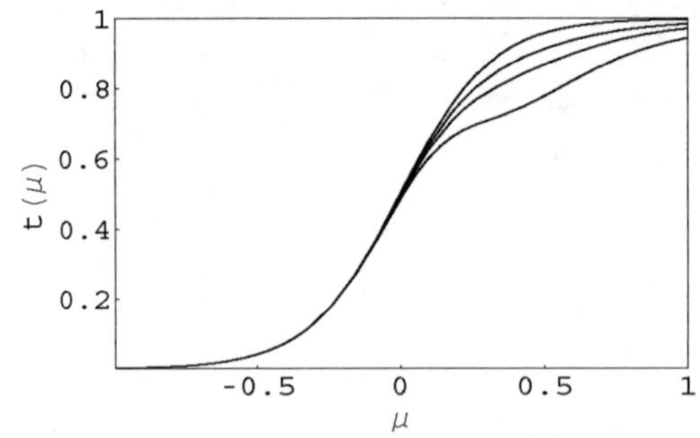

FIGURE 7. Conductance in units of $2e^2/h$ versus μ (chemical potential in units of ω) for different temperatures in the intermediate coupling limit, $g_e = 1$. The uppermost curve corresponds to $T = 0$, while the lowest is $T \approx 0.3$K, assuming $\frac{\tau v_F}{L} \approx 1$.

experimental data on the 0.7 structure. According to our calculation the exact position of the structure depends on the coupling constant g_e: for small g_e it is more like a "0.5 structure" and for $g_e \sim 1$ it is a "0.6–0.7 structure".

LONGITUDINAL MAGNETIC FIELD

The effects we have discussed are due to the interaction between electrons with opposite spins. The interaction between electrons with parallel spins vanishes because the exchange diagram exactly cancels out the direct one for the contact Hamiltonian (2). Therefore under an applied longitudinal magnetic field B we should take $\rho^4(\mu) \to \rho^2(\mu)\rho^2(\mu')$. Here $\mu' = \mu - (2g_s\mu_B)B/\omega$, where g_s is the gyromagnetic ratio and μ_B is the Bohr magneton. Since $\rho(\mu)$ is a peaked function (Fig. 1) a magnetic field $B \sim \omega/(2g_s\mu_B) \sim 10$T effectively switches off the interaction. At the same time, the ordinary spin-splitting of the steps is being switched on. This gives an explanation for the smooth evolution of the 0.7 structure to the 0.5 plateau of non-interacting electrons under a magnetic field. The calculated dependence on magnetic field for $g_e = 1$ and $T = 0.3$K shown in Fig. 8 agrees with experimental data[5] remarkably well.

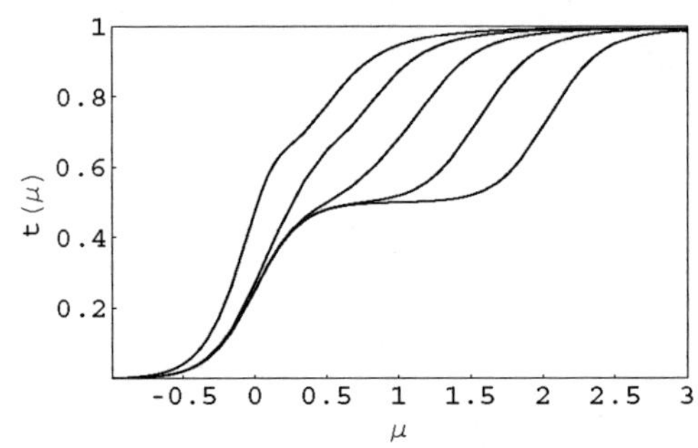

FIGURE 8. Conductance in units of $2e^2/h$ versus μ (chemical potential in units of ω) for $g_e = 1$, $T = 0.3$K and different values of longitudinal magnetic field, $2g_s\mu_B B/\omega = 0, 0.5, 1, 1.5, 2$. The uppermost curve corresponds to $B = 0$, while the lowest is $B \approx 17$T, assuming $\omega = 1meV$ and $g_s = 1$, see Ref. [5].

CONCLUSION

Within perturbation theory, we have considered transport of correlated electrons through a quantum point contact. At zero temperature, the approach results in the usual Landauer formula and the conductance does not show any structures. However at nonzero temperature the electron–electron interaction gives rise to a current of correlated electrons which scales as temperature squared at very low temperatures. The corresponding correction to conductance is negative and strongly enhanced in the region $0.5 \times 2e^2/h \leq G \leq 1.0 \times 2e^2/h$, as seen in Fig. 4 and Fig. 7. We believe that these results are directly relevant to the 0.7 conductance structure. Our model is consistent with the experimental behavior of the 0.7 structure under a magnetic field: a field smoothly “switches off” the effective interaction between electrons, Fig. 8. Effects considered in the present paper have a very simple physical origin: the electron wave function at the barrier and hence the electron–electron interaction is strongly peaked when the transmission coefficient is slightly higher than 0.5.

REFERENCES

1. D. A. Wharam *et al*, J. Phys. C **21**, L209 (1988).
2. B. J. van Wees *et al* Phys. Rev. Lett. **60**, 848 (1988).
3. R. Landauer, Physics Letters A **85**, 91 (1981).
4. M. Büttiker, Phys. Rev. B **41**, 7906 (1990).
5. K. J. Thomas *et al*, Phys. Rev. Lett. **77**, 135 (1996).
6. K. J. Thomas *et al*, Phys. Rev. B **58**, 4846 (1998).
7. A. Kristensen *et al*, Phys. Rev. B **62**, 10950 (2000).
8. D. J. Reilly *et al*, Phys. Rev. B **63**, 121311 (2001).
9. S. M. Cronenwett *et al*, Phys. Rev. Lett. **88**, 226805 (2002).
10. P. Roche *et al*, Phys. Rev. Lett. **93**, 116602 (2004).
11. R. Danneau *et al*, Appl. Phys. Lett. bf 88, 012107 (2006).
12. Chuan-Kui Wang and K.-F. Berggren, Phys. Rev. B **54**, 14257 (1996); Phys. Rev. B **57**, 4552 (1998).
13. L. Calmels and A. Gold, Solid State Commun. **106**, 139 (1998).
14. N. Zabala, M. J. Puska, and R. M. Nieminen, Phys. Rev. Lett. **80**, 3336 (1998).
15. H. Bruus, V. V. Cherepanov, and K. Flensberg, Physica E, **10**, 97 (2001).
16. A. A. Starikov, I. I. Yakimenko, and K.-F. Berggren, Phys. Rev. B **67**, 235319 (2003).
17. D. J. Reilly, Phys. Rev. B **72**, 033309 (2005).
18. B. Spivak and F. Zhou, Phys. Rev. B **61**, 16730 (2000).
19. O. P. Sushkov, Phys. Rev. B **64**, 155319 (2001); **67**, 195318 (2003).
20. Y. Meir, K. Hirose, and N. S. Wingreen, Phys. Rev. Lett. **89**, 196802 (2002); K. Hirose, Y. Meir, and N. S. Wingreen, Phys. Rev. Lett. **90**, 026804 (2003).
21. P. S. Cornaglia and C. A. Balseiro, Europhys. Lett., **67**, 634 (2004).
22. G. Seelig and K. A. Matveev, Phys Rev Lett. **90**, 176804 (2003).
23. R.A. Molina *et al*, Phys. Rev. B 67, 235306 (2003).
24. V. Meden and U. Schollwoeck, Phys. Rev. B **67**, 193303 (2003).
25. A. Bohr and B. Mottelson, *Nuclear structure, vol. 2*, Benjamin, New York, Amsterdam, 1974.
26. M. Büttiker, Phys. Rev. Lett. **65**, 2901 (1990).
27. E. M. Lifshitz and L P. Pitaevskii, *Physical kinetics*, Pergamon Press; Oxford , New York; 1981.
28. L. D. Landau and E. M. Lifshitz, *Quantum Mechanics*, Pergamon Press; Oxford, New York; 1965.
29. K. A. Brueckner and C. A. Levinson, Phys. Rev **97**, 1344 (1955).
30. V. A. Dzuba, V. V. Flambaum, and O. P. Sushkov, Phys. Lett. A **140**, 493 (1989); **141**, 147 (1989).

Scanning Charge Accumulation Probe of Semiconductor Donor Molecules

S. H. Tessmer[a], I. Kuljanishvili[b], C. Piermarocchi[a], T. A. Kaplan[a] and J. F. Harrison[c]

[a]*Department of Physics and Astronomy, Michigan State University, East Lansing, Michigan 48824, USA*
[b]*Department of Physics and Astronomy, Northwestern University, Evanston, Illinois 60208, USA*
[c]*Department of Chemistry, Michigan State University, East Lansing, Michigan 48824, USA*

Abstract. We have developed a scanning probe method that is able to detect individual electrons entering a system of semiconductor donor atoms. We have applied the method to a system of Si donors within a GaAs-AlGaAs heterostructure sample. The data compare well to a model that considers donor molecules, effectively formed by nearest-neighbor silicon atoms.

Keywords: scanning probe microscopy, defects, quantum computing
PACS: 71.22.+I, 71.23.An, 71.55.Eq, 73.20.Hb, 73.21.La, 73.22.-f

The ability to manipulate and probe small numbers of dopant atoms in semiconductors represents an emerging line of research, motivated by the continued miniaturization of semiconductor devices and potential applications where single dopants themselves form the functional part of a device [1,2]. The most well known potential device of this kind is the quantum computer proposed by Bruce Kane in 1998 [3]. In the original Kane device, the quantum bits were defined by nuclear spin states of buried phosphorus dopants in a silicon crystal. Subsequent theoretical work extended these ideas to utilize the electron spin or electron charge state of the donor atoms [4,5]. Although donor properties are well-understood with respect to bulk semiconductors, new questions arise in nanosystems, such as the correct quantum description for small clusters of isolated donors [6]. Moreover, because shallow donors and acceptors are analogous to hydrogen atoms [7], experiments on small numbers of dopants have the potential to be a testing ground for fundamental questions of atomic and molecular physics, such as the maximum negative ionization of a molecule with a given number of positive ions [8]. Resonant Tunneling Spectroscopy (RTS) through isolated dopants has been observed in transport studies [9,10]. In addition, ground breaking RTS work by Andre Geim and coworkers identified resonances due to two closely spaced donors, effectively forming donor molecules [11]. In this paper, we describe our recent scanning-probe study of silicon donors in a gallium-arsenide heterostructure sample [12]. To the best of our knowledge, this study was the first example of single-electron capacitance spectroscopy performed directly with a scanning probe tip [13]. In contrast to the work of Geim et al., our measurements showed discernible peaks attributed to successive electrons entering the molecules.

Fig. 1 introduces the effective-mass approximation developed by Kohn and Luttinger that has proven to be very successful in describing the properties of donors and acceptors inside a host semiconductor [7,14]. With respect to ionized donors, the theory describes the attraction of electrons toward the ion using a hydrogenic potential. A basic assumption is that the donor potential varies slowly compared to the crystal potential, decoupling the long and the short wavelength physics. Compared to natural hydrogen, the mass of the electron is replaced by the effective mass, and the charge is screened by the dielectric properties of the host. As a result, the effective Bohr radius a_0* and Rydberg energy R* are rescaled considerably, as shown in Fig. 1. For example, a_0*=7.0 nm and R*= 8.1 meV for Si donors in $Al_{0.3}Ga_{0.7}As$ [15]; these value differ considerably from natural hydrogen, for which a_0=0.53 Ǎ and R=13.6 eV.

CP995, *Nuclei and Mesoscopic Physics - WNMP 2007*, edited by P. Danielewicz, P. Piecuch, and V. Zelevinsky

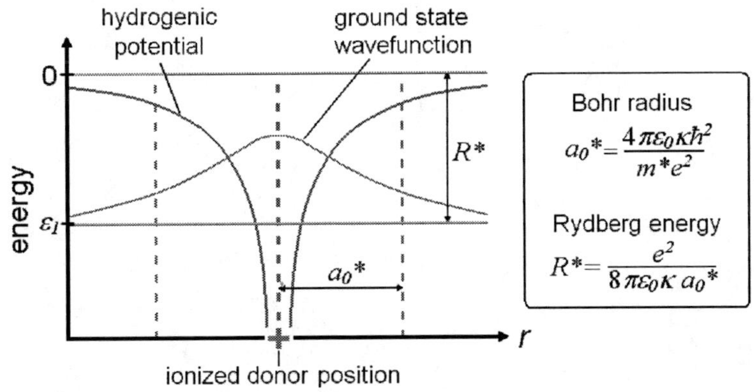

FIGURE 1. Schematic of the effective-mass theory to describe an ionized donor inside a host semiconductor [7]. In this approximation, the donor is regarded as a hydrogen atom with an effective Bohr radius a_0* and effective Rydberg energy R*, as indicated in the inset (in SI units). Here e is the electron charge, m^* is the electron effective mass, and is the dielectric constant; $_1$ is the energy of the one-electron ground state. The potential (green) and ground state wavefunction (red) are plotted as a function of radial distance r.

As an introduction into the physics at play, we can consider the Hamiltonian of a single electron in a fixed hydrogenic potential with a magnetic field B applied in the z-direction (in cgs units):

$$H = \underset{(1)}{\frac{-\hbar^2}{2m}\nabla^2} - \underset{(2)}{\frac{e^2}{r}} + \underset{(3)}{\frac{eB}{2mc}(L_z + gS_z)} + \underset{(4)}{\frac{e^2B^2}{8mc^2}(x^2+y^2)},$$

where m is the electron mass, c is the speed of light, and L_z and S_z are the z-components of the angular momentum and spin, respectively [16]. Of course, the first term (1) gives the kinetic energy, and term (2) is the hydrogenic potential. Term (3) accounts for the Zeeman effect, which lowers the ground state energy as the total angular momentum aligns in response to the applied field. The last term (4) is sometimes referred to as the magnetic parabola; this term accounts for the additional confinement effect of the magnetic field. In systems such as semiconductor quantum dots, term (4) leads to an increase in energy with B, just as particle-in-a-box energy levels increase with decreasing box size [13]. However, for atoms and molecules, the confinement due to the applied field may instead lead to decreased ground state energy due to the fact that the wavefunctions are squeezed closer to the central positive ions. Hence we see that the physics can be subtle, even before considering the complexities introduced by electron-electron interactions and extended arrangements of ions.

In mapping the problem of real atoms in free space onto donor atoms in semiconductors, we can write down a very similar Hamiltonian using the Kohn-Luttinger approach. Moreover, we can compare the relative strength of magnetic and Coulomb interactions by analyzing the ratio of the two characteristic energies $_B*B/R*$, where the effective Bohr magneton scales inversely with effective mass m^*; $_B*=e\hbar/2m^*c$. From a straightforward argument, we see that in order to explore similar regimes, the strength of the magnetic field for the natural atomic system has to be much larger than for the semiconductor system. The correspondance is given by

$$B^{atoms} \leftrightarrow \left(\frac{\kappa}{m^*/m}\right)^2 B^{donors},$$

where is the dielectric constant. For Si donors in GaAs, the scale factor is approximately ~3x10^4. In other words, the application of one tesla to a donor molecule in GaAs allows us to probe the same physics as the application of 30,000 tesla to an equivalent hydrogenic molecule in free space. Clearly, the semiconductor setting, due to the renormalized Coulomb interaction and the fixed position of the ions, can explore regimes that would be difficult to probe in natural molecules.

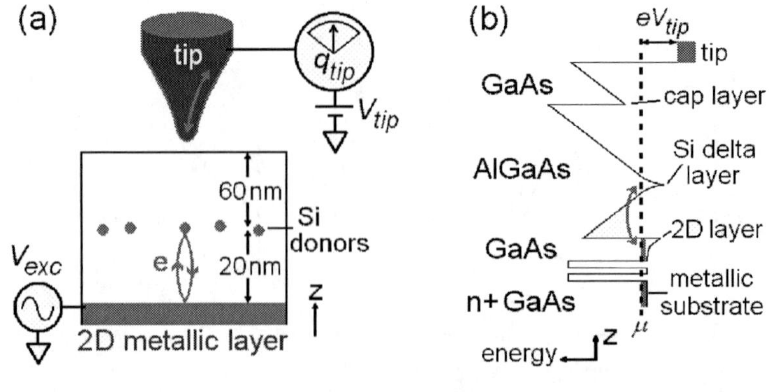

FIGURE 2. (a) Schematic illustration of our measurement. (b) The conduction band profile with respect to the vertical direction of a Si-doped GaAs-AlGaAs heterostructure sample. The delta doped layer is 60 nm from the surface. A two-dimensional electron layer forms the bottom electrode.

EXPERIMENT

Figure 2(a) shows a schematic of our Scanning Charge Accumulation (SCA) method used to probe dopants in a GaAs heterostructure sample. The method is based on the detection of image charge induced at the apex of a sharp metal tip using a cryogenic transistor [17]. The charge moves to the apex in response to electron motion within the sample, essentially providing a measure of the local capacitance [18]. In Fig. 2(a) an electron is shown entering and exiting a Si donor synchronously with an applied AC excitation voltage V_{exc}. The resulting image charge is detected with a noise level of ~0.1 e/√Hz. For the sample employed in reference [12], the donors were incorporated during molecular beam epitaxial (MBE) growth. Figure 2(b) shows the conduction band profile of the sample.

The delta Si doping results in dopants that are well-confined to a plane, but randomly positioned within the layer. For the measurement, the key sample parameters are the sample-surface-to-delta-layer distance of 60 nm and the dopant density within the delta layer of $1.25\text{x}10^{12}$ cm^{-2}; this corresponds to an average spacing between Si donors of 11 nm. The key tip parameters are the radius of curvature of the apex of 50 nm, and its distance from the sample surface of 1 nm. In this experiment the position of the tip was fixed (i.e., not scanned). By comparing the measurements to careful calculations [19], we have determined that the effective area probed is approximately $1.12\text{x}10^{-14}$ m^2. This corresponds to a circle of radius 60 nm which contains about 140 Si donors. In other words, the measurement probed a system consisting of roughly a hundred dopants.

Figure 3(a) shows representative data acquired with the SCA probe immersed in liquid He-3 at a temperature of 0.29 K. We found that the donor contribution to the capacitance consistently showed three broad peaks, labeled A, B and C, over a tip voltage range of 0–1 V; considering a lever-arm factor that scales the voltage (by ~1/12), the voltage range corresponds to an energy range of about 0–8 R^*. Moreover, fine-structure measurements resolved on the sub-millivolt scale consistently showed sharper peaks. As shown in the inset of Fig. 3(a), the shape of the sharp peaks agrees very well with the characteristic capacitance peak expected for single-electrons entering the donor system [20]. In light of the fine-structure measurements, we are able to conclude that each broad peak corresponds to roughly 15 electrons entering the system at nearly the same energy.

Could the broad peaks A, B and C originate from an ensemble of 15 relatively isolated donors each of which accommodates electrons at approximately the same energies? This explanation for the broad peaks is not consistent with Lieb's theorem, which states that the maximum negative ionization for a molecule with K ions is Z+K-1, where Z is the total nuclear charge of the ions [8]. For a single donor this gives one, corresponding to H^- (i.e. the D^- state). That would give only two peaks in a capacitance measurement, which is inconsistent with our observations. However, if we consider closely spaced Si donors effectively forming two-donor molecules (2DMs), then K=2, Z=2, this upper bound is relaxed and the theorem predicts that the system could actually bind up to five electrons (negative ionization of three).

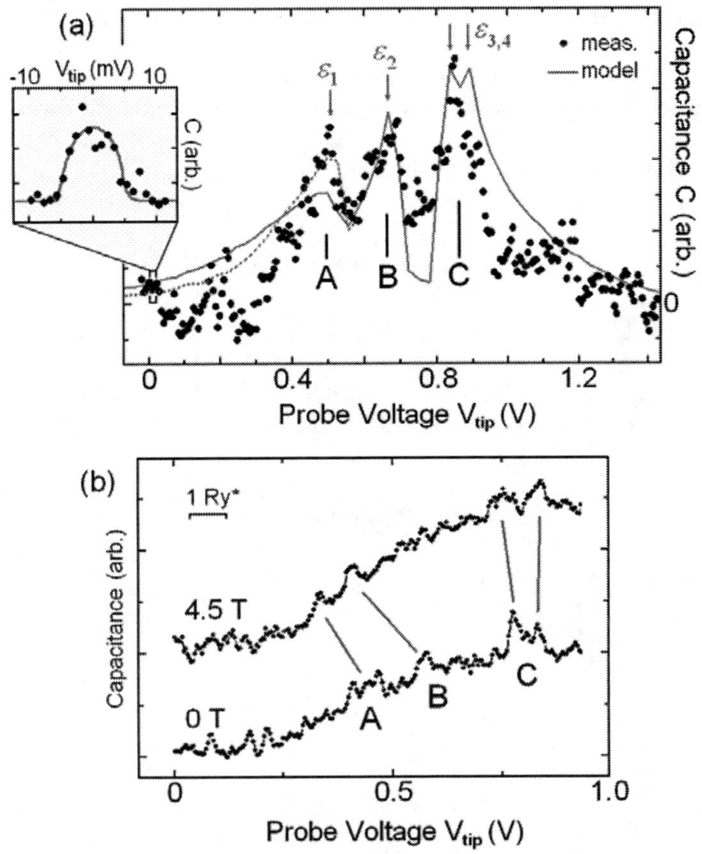

FIGURE 3. (a) Measurement of the donor contribution to the capacitance, averaged over three locations to reduce scatter. The theory curve considers two-atom donor molecules that can each accommodate four electrons. The dashed curve shows the same model with a reduced broadening parameter. The average addition energies for the 1st through 4th electrons are indicated as $_{1-4}$. The inset shows a more sensitive measurement at a single location on an expanded scale, for which we consistently resolve distinct sharp peaks. In this case, the theory curve shows the semi-elliptical peak shape expected for single-electron resonances [20]; the width of the peak is determined mostly by the amplitude of V_{exc}, which was 3.6 mV rms. (b) Capacitance measured in a magnetic field. The two curves were acquired at the same (single) location, with the indicated magnetic field applied perpendicular to the donor plane. They are displayed without subtracting the background capacitance due to charge entering higher layers in the sample; the gray lines serve as a guide to the eye. The observed shift to lower energy of peaks A and B are consistent the expected behavior of strongly localized states, whereas peak C clearly splits into two peaks. This behavior is generally consistent with the donor-molecule interpretation of the resonances.

Motivated by the Lieb's theorem argument, and by the fact that the average spacing between donors of 11 nm is close to the effective Bohr radius of 7 nm, we have developed a theoretical model for the capacitance measurements that considers the addition spectra of pairs of closely spaced silicon atoms – essentially forming two-atom donor molecules. The model uses the configuration-interaction (CI) method in the context of the effective-mass approximation. The CI method solves Schrodinger's equation for the multi-electron system with the use of linear combinations of Slater determinants, employing a variational method to account for electron-electron interactions [21-23]. Our full modeling routine simulates the probed area of the measurement by considering random ensembles of donors, and includes the perturbative influence of the tip. The resulting calculated capacitance curves can be compared directly to the observed capacitance. As shown in Fig. 3(a), we find reasonable agreement between the theory and measurement, with the largest peak C formed from two unresolved resonances.

Figure 3(b) shows data acquired at a single location in both zero magnetic field and with 4.5 T applied in the *z*-direction, i.e., normal to the sample surface. We see that peaks A and B shift to lower energy by about -1 R*, whereas peak C shifts only a small amount – but splits into two peaks. This same behavior has been observed at three sample locations. To the best of our knowledge, no complete theory exists for the addition spectrum of hydrogenic molecules subject to a perpendicular magnetic field, although related studies have been conducted [24,25]. However, we can compare our observation to the well-established behavior of isolated atoms in magnetic field, such as isolated silicon donors in GaAs [26-28]. In the context of the donor-molecule interpretation of our data, the observed -1 R* shift of peaks A and B is consistent with the isolated Si studies in that it is similar to the behavior of the one-electron (D^0) and two-electron (D^-) states. Physically, the dominant effect of the magnetic field is that it squeezes the ground state orbit closer to the positive ion(s), thus lowering the energy. This effect is expected to be reduced for less-localized states such as peak C. While at present no theory exists that predicts the details of the splitting of peak C, it is certainly consistent with our interpretation that it consists of two unresolved states at zero field.

ACKNOWLEDGMENTS

We gratefully acknowledge L. N. Pfeiffer and K. West, who provided the heterostructure sample used in Ref. [12], and R. C. Ashoori, N. Birge, M. I. Dykman, B. Golding, S. D. Mahanti, I. J. Maasilta, S. Rogge and G. A. Steele, for helpful advice and comments. The work was supported by the Michigan State Institute for Quantum Sciences and the National Science Foundation, Grant Nos. DMR-0305461 and DMR-0605801.

REFERENCES

1. F. J. Ruess et al, *Small* **3**, 563-567 (2007).
2. S. E. S. Andresen et al., *Nano Letters* **7**, 2000-2003 (2007).
3. B. E. Kane, *Nature* **393** (6681), 133 (1998).
4. R. Vrijen et al., *Phys. Rev. A* **62**, 012306 (2000).
5. L. C. L. Hollenberg et al., *Phys. Rev. B* **69**, 113301 (2004).
6. L. M. Kettle et al., *Phys. Rev. B* **68**, 075317 (2003).
7. W. Kohn and J. M. Luttinger, *Phys. Rev.* **98**, 915 (1955).
8. E. H. Lieb, *Phys. Rev. A* **29**, 3018 (1984).
9. M. R. Deshpande, J. W. Sleight, M. A. Reed, R. G. Wheeler and R. J. Matyi, *Phys. Rev. Lett.* **76**, 1328 (1996).
10. H. Sellier et al., *Phys. Rev. Lett.* **97**, 206805 (2006).
11. A. K. Geim et al., *Phys. Rev. B* **50**, 8074 (1994).
12. I. Kuljanishvili et al., arXiv:0704.2910v2 [cond-mat.mes-hall]; *Nature Physics*, in-press.
13. R. C. Ashoori, *Nature* **379**, 413-419 (1996).
14. J. H. Davies, *The Physics of Low-dimensional Semiconductors: An Introduction*, Cambridge: Cambridge University Press, Cambridge, 1998.
15. M. Levinstein, S. Rumyantsev and M. Shur, *Handbook Series on Semiconductor Parameters, vol. 2*, London: World Scientific, 1999.
16. S. Gasiorowicz, *Quantum Physics*, New York: Wiley & Sons, 1974.
17. S. Urazhdin, S. H. Tessmer and R. C. Ashoori, *Rev. Sci. Instrum* **73**, 310 (2002).
18. T. A. Kaplan, *J. Stat. Phys.* **122**, 1237 (2006).
19. I. Kuljanishvili, S. Chakraborty, I. J. Maasilta and S. H. Tessmer, *Ultramicroscopy* **102** (1), 7 (2004).
20. R. C. Ashoori et al., *Phys. Rev. Lett.* **68**, 3088 (1992).
21. J. C. Slater, *Quantum Theory of Molecules and Solids, Vol. 1, Electronic Structure of Molecules,* New York: McGraw-Hill, 1963.
22. I. N. Yassievich, A. Blom, A. A. Prokofiev, M. A. Odnoblyudov and K.-A. Chao, *Physica B* **308–310**, 1129 (2001).
23. T. H. Dunning, *J. Chem. Phys.* **90**, 1007-1023 (1989).
24. T. Detmer, P. Schmelcher, F. K. Diakonos, L. S. Cederbaum, *Phys. Rev. A* **56**, 1825-1838 (1997).
25. U. Kappes, *Phys. Rev. A* **51**, 4542 (1995).
26. J. G. S. Lok et al., *Phys. Rev. B* **53**, 9554 (1996).
27. J. M. Shi, F. M. Peeters, G. Q. Hai and J. T. Devreese, *Phys. Rev. B* **44**, 5692-5702 (1991).
28. J. M. Shi and J. T. Devreese, *Phys. Rev. B* **51**, 7714-7724 (1995).

Electron Transport through Protein Fragments

David M. Cardamone and George Kirczenow

Department of Physics, Simon Fraser University, 8888 University Drive, Burnaby, BC, V5A 1S6, Canada

Abstract. By combining *ab initio* and semi-empirical techniques, we construct a theoretical model of electron transport through oligopeptide molecules, i.e. short protein fragments. With no fitting parameters, this model achieves quantitatively accurate agreement with experiment, and so enables the extraction of chemical and physical information, such as bonding geometry and the behavior of the molecules under stretching, from experimental data. Furthermore, the model explains the experimentally observed current rectifying properties of these molecules as a consequence of the hybridization of interfacial states at opposite gold-molecule contacts; under appropriate bias, these poorly conducting localized states mix to form well conducting whole-molecule orbitals. Finally, we predict that oligopeptide molecules can be made to exhibit negative differential resistance by a related mechanism, thus opening the way to many interesting protein-based nanoelectronic devices.

Keywords: molecular electronics, oligopeptide, protein, nanoelectronic, rectification, negative differential resistance
PACS: 81.07Nb, 73.63.-b, 85.65.+h, 87.14.Ee

1. INTRODUCTION

Proteins are nature's nanomachines, assembled by living cells to perform all the multitudinous and varied functions necessary to life. While they are active within an organism, proteins are complex, three-dimensional macromolecules, whose structure is strongly tied to function. In their denatured state, however, they become simple polypeptide chains, linear polymers consisting of a sequence of amino acids (each coded for by one or more three-base-pair DNA codons) held together by peptide bonds.

Such a molecule holds much promise as the main ingredient of a nanoelectronic device. With an alphabet of twenty amino acids, the precisely controllable variability of protein-based nanostructures is unparalleled. Furthermore, their fabrication is chemically straightforward *in vitro*, and can even be envisioned as a highly scalable *in vivo* process via suitable single-celled organisms. Finally, these molecules play central roles in modern biology and medical science; thus, a working knowledge of their nanoelectronic properties implies a plethora of useful physical techniques to aid those important fields of study.

While the potential of protein-based nanoelectronic devices has long been acknowledged [1], only recently have experimentalists succeeded in systematically studying these biomolecules' electron transport characteristics [2-4]. From the point of view of primary protein structure (i.e. particular sequences of amino acids), the recent work of Xiao *et al.* [2, 3] is especially interesting; they used the scanning tunneling microscope break junction (STMBJ) technique [5–7] to measure the single-molecule conductance and current–voltage characteristics of several small thiol-terminated oligopeptide molecules (see Fig. 1), suspended between two gold nanocontacts. These devices were even found to exhibit significant current rectifying behavior [3].

The purpose of this article is to provide an overview of our recent work [8] developing the first theoretical model of electron transport through these systems. By applying Landauer conductance theory [9], we achieve quantitatively accurate agreement with experiment, without the use of any fitting parameter, and are able to understand all the observed phenomena. This allows us to extract important physical and chemical information about the experimental systems, such as the geometry of the metal-molecule interfaces themselves, and the origin of the systems' current rectifying properties. Furthermore, we predict that, at bias voltages only moderately higher than those at which measurements have been carried out to date, these systems will exhibit negative differential resistance (NDR) (i.e. a decrease in the magnitude of current with *increasing* bias), a phenomenon with many important device applications [10–13]. These findings thus open the way to the realization of practical oligopeptide, and ultimately protein-based, nanoelectronic devices.

The paper is organized as follows. In Section 2, we provide a brief overview of the STMBJ technique used by experimentalists to measure the electron transport characteristics of these molecules. In Section 3, we outline the model and formalism we have developed to treat these single-molecule metal–oligopeptide–metal nanoelectronic junctions. Section 4 describes our results, including quantitative agreement with experiment, extraction of physical and chemical

CP995, *Nuclei and Mesoscopic Physics - WNMP 2007*, edited by P. Danielewicz, P. Piecuch, and V. Zelevinsky

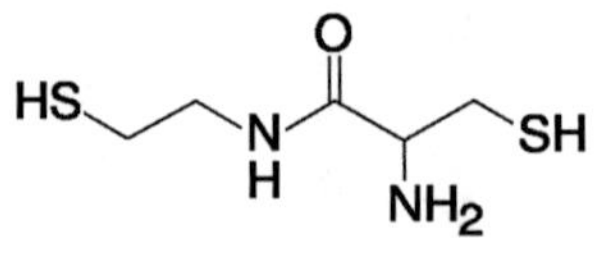

1: cysteamine-Cys

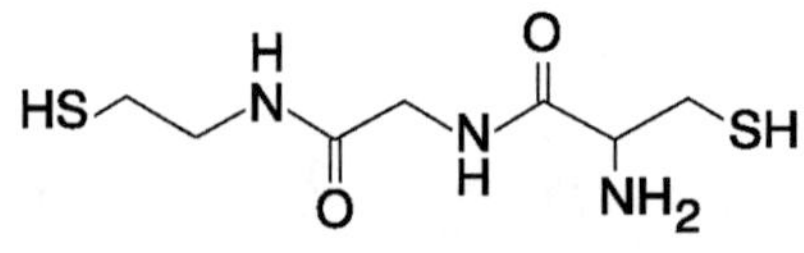

2: cysteamine-Gly-Cys

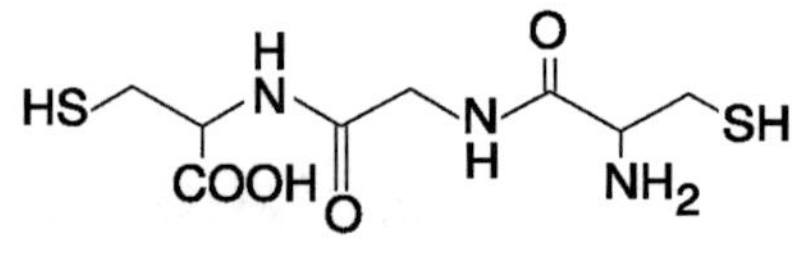

3: Cys-Gly-Cys

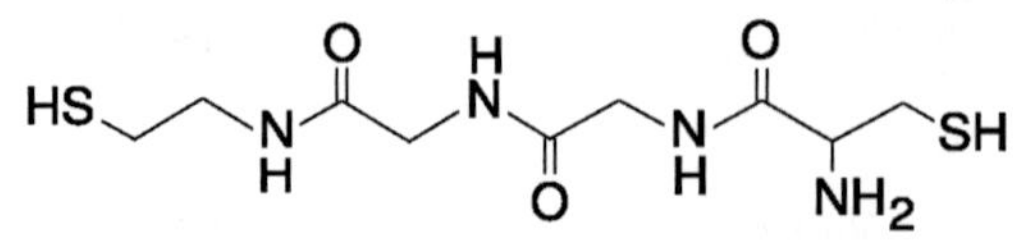

4: cysteamine-Gly-Gly-Cys

FIGURE 1. Structural formulae of the four oligopeptide molecules studied theoretically in this work and Ref. [8], and experimentally in Refs. [2] and [3]. The terminating hydrogens are lost when thiol-bonding to the gold contacts.

information from experiment, an understanding of the observed current rectification, and our prediction of NDR. In Section 5, we offer a summary and conclusions.

2. EXPERIMENTS

We provide here a brief overview of the experimental method [5–7] employed in the STMBJ experiments on oligopeptide molecular wires [2, 3]. A scanning-tunneling microscope (STM) tip and substrate are submerged in an electrolyte solution containing the oligopeptide molecules. Because it is energetically preferred for the thiol (SH) group to lose its hydrogen and bond to gold, a self-assembled monolayer (SAM) of oligopeptides forms on each gold surface. The two SAM-coated gold surfaces are brought into contact, and then pulled apart as conductance between them is monitored.

For small stretching distance, parallel chains of gold atoms can bridge the gap between the two contacts; each such chain contributes approximately one conductance quantum, $g_0 = 2e^2/h$, to the conductance [14–18]. As the tip-to-substrate distance is increased, these nanochains snap one by one, and quantized conductance plateaus can be observed at integer values of g_0, with the total conductance decreasing as the stretching increases.

Once the tip-to-substrate distance becomes too great for gold atomic chains to bridge, however, the much lower conductance of the molecules themselves becomes relevant. As the junction is stretched farther, the falling conductance occasionally arrests at plateaus with conductance much smaller than g_0. After many such experiments, the conductances of such plateaus form histograms which are strongly peaked around small-integer multiples of a single value; like the case of the atomic nanochains, these conductances are interpreted as the conductances of 1, 2, 3, etc. *molecules* in parallel. Thus, the conductance of the lowest peak is identified as the conductance of a single molecule [2,3,5–7]. Once this identification is made, a stretching experiment can be stopped at this conductance, and a full $I-V$ characteristic measured [3].

3. MODEL

We combine *ab initio* and semi-empirical techniques to arrive at a model [8] capable of quantitative agreement with experiments. On one hand, *ab initio* density functional theory (DFT) calculations have for many years provided sufficiently accurate quantum chemical calculations of molecular geometries in similar systems. This is an important advantage because the electronic structure of different configurations can differ, yet the experimental geometry cannot be directly measured due to the small aperture and delicate nature of the gold–oligopeptide–gold junction. On the other hand, arriving at appropriate single-electron states, as required by Landauer conductance theory, within DFT is problematic [19], and furthermore DFT is a rather computationally expensive theory, which can at best treat only a few atoms of the gold contacts, making a quantitative understanding of the rate at which electrons move from lead to molecule and *vice versa* difficult. Thus, we use DFT to calculate molecular geometries, but implement a semi-empirical molecular Hamiltonian before calculating conductance and current. This modular approach has the further advantage that either aspect of the calculation can be updated as methods improve, without the need to change the other.

To simulate the stretching the molecule undergoes in a typical STMBJ experiment, we use a series of DFT optimizations [20]. First, we relax the thiolated oligopeptde's geometry with one gold atom bonded to either sulfur. Then, we increase the gold-gold distance d slightly, and reoptimize the geometry with d held constant. This process is repeated until the DFT geometry optimization no longer results in a molecule which bridges both gold leads (i.e., the junction breaks). For bonding geometries for which one or both sulfur atoms have more than one nearest-neighbor gold atom, we replace the single gold atom with the appropriate fixed nearest-neighbors, and relax the oligopeptide one final time.

A primary motivation to shift to a semi-empirical treatment once the oligopeptide geometry is optimized by *ab initio* means is to include large numbers of gold atoms into the nanocontacts at either side of the junction. We include 50–100 gold atoms in each nanocontact, a number large enough to ensure that spurious geometrical resonances due to the finite size of the cluster are eliminated. The Hamiltonian can then be written as the sum of three terms:

$$H = H_{mol} + H_{leads} + H_c, \tag{1}$$

where H_{mol} is the Hamiltonian of the isolated extended molecule, including both the oligopeptide and the two large gold nanoclusters, while H_{leads} contains the Hamiltonians of the macroscopic electrodes (representing all the gold atoms beyond the nanoclusters), and H_c provides quantum mechanical coupling between these two systems.

For H_{mol}, we use the standard semi-empirical extended Hückel formulation [21, 22] which has been very successful in predicting, reproducing, and explaining experimental results in similar molecular electronics experiments (e.g., Ref. [23], Ref. [24] and the results of Refs. [25] and [7], Ref. [26]). In this approach, H_{mol}'s diagonal elements ε_i are given by experimentally determined ionization energies, and the off-diagonal elements by [22, 27]

$$(H_{mol})_{ij} = \left(1.75 + \Delta_{ij}^2 - 0.75\Delta_{ij}^4\right) S_{ij} \frac{\varepsilon_i + \varepsilon_j}{2}, \tag{2}$$

where $\Delta_{ij} = (\varepsilon_i - \varepsilon_j) / (\varepsilon_i + \varepsilon_j)$, and S_{ij} is the overlap between atomic orbitals i and j, which can be nonzero for $i \neq j$ since atomic orbitals of different atoms are not orthogonal.

While H_{mol} defines a spectrum of discrete molecular orbitals of definite energy, H_{leads} and H_c, possessing an infinite Fock space, provide both decoherent broadening and elastic scattering. We model this with a set of identical, semi-infinite, one dimensional leads:

$$H_{leads} = \sum_{\alpha} \sum_{k=1}^{\infty} \left[\varepsilon_\alpha c_{\alpha,k}^\dagger c_{\alpha,k} + \left(t c_{\alpha,k+1}^\dagger c_{\alpha,k} + \text{H.c.} \right) \right], \tag{3}$$

with ε_α the energy of every site in lead α, and t the hopping matrix element between them. The operator $c_{\alpha,k}^\dagger$ creates an electron in the kth site of lead α, and $c_{\alpha,k}$ destroys one. To best reproduce the well known experimental results for quantized conductance of a gold quantum point contact [14–18] (*i.e.*, our model system *without* any molecule in it), we fix $t = 5\text{eV}$ and take $\varepsilon_\alpha = \varepsilon_a$, where a is the gold orbital to which lead α couples directly. Coupling to H_{mol} is provided by

$$H_c = \sum_{\alpha} \left(t c_{\alpha,1}^\dagger d_a + \text{H.c.} \right), \tag{4}$$

where d_a is the annihilation operator for an electron in the gold orbital a to which lead α is attached.

To every gold orbital not immediately adjacent to the oligopeptide, we attach one such lead α for every nearest neighbor atom in the bulk gold lattice not present in H_{mol}. Thus, gold orbitals interior to the cluster receive no broadening, and the role of the leads is to mimic the effect of the "missing" gold atoms on the spectrum of the nanocluster. According to Dyson's Equation, the full retarded Green's function of energy $G(E)$, including the effects of H_{leads} and H_c to all orders, is

$$G(E) = \left[G_{mol}^{-1}(E) - \sum_{\alpha} \Sigma^{(\alpha)} \right]^{-1}, \tag{5}$$

where

$$G_{mol}(E) = \frac{1}{SE - H_{mol} + i0^+} \tag{6}$$

is the Green's function for the extended molecule in the absence of leads, with the overlap matrix S appearing due to the non-orthogonality of the basis set. The retarded self-energy due to lead α is [28, 29]

$$\Sigma^{(\alpha)}(E) = \begin{cases} t\left[\frac{E-\varepsilon_\alpha}{2t} - i\sqrt{1-\left(\frac{E-\varepsilon_\alpha}{2t}\right)^2}\right], & \left|\frac{E-\varepsilon_\alpha}{2t}\right| < 1 \\ t\left[\frac{E-\varepsilon_\alpha}{2t} + \sqrt{\left(\frac{E-\varepsilon_\alpha}{2t}\right)^2 - 1}\right], & \frac{E-\varepsilon_\alpha}{2t} < -1 \\ t\left[\frac{E-\varepsilon_\alpha}{2t} - \sqrt{\left(\frac{E-\varepsilon_\alpha}{2t}\right)^2 - 1}\right], & \frac{E-\varepsilon_\alpha}{2t} > 1 \end{cases}. \tag{7}$$

When bias voltage is applied to the system, we treat the electrostatic problem as a parallel-plate capacitor with electrolyte dielectric. The two plates are taken to lie at the metal–oligopeptide interface surfaces (separated by the distance d), with the interior of the metal contacts assumed to be at constant potential. The ions of the electrolyte are free to move and screen applied bias. We treat this effect with the linearized Debye-Hückel model of electrolytes [30]: Poisson's equation for the electrostatic potential $\Phi(\vec{r})$ becomes

$$\nabla^2\Phi(\vec{r}) = \frac{2ne^2}{k_B T\varepsilon}\Phi(\vec{r}), \tag{8}$$

where $n = .1M$ is the concentration of the solution, $-e$ is the charge of an electron, ε is the dielectric constant of the solvent (in this case water), k_B is the Boltzmann constant, and T is room temperature [31]. In the electrostatic gauge for which the bias voltage on the two contacts sums to zero, the solution of Eq. (8) is [8]

$$\Phi(z) = \frac{V}{2}\left[\cosh\left(\frac{z}{\lambda}\right) - \frac{\cosh\left(\frac{d}{\lambda}\right)+1}{\sinh\left(\frac{d}{\lambda}\right)}\sinh\left(\frac{z}{\lambda}\right)\right], \tag{9}$$

where $z = 0$ and $z = d$ are the interfaces, V is the applied bias, and the Debye screening length is $\lambda = \sqrt{k_B T\varepsilon/2ne^2}$. For the experimental systems of Refs. [2] and [3], $\lambda = 0.97$nm. In systems, like all those studied here, for which $\lambda \gtrsim d/2$, the electrolyte only poorly screens applied bias; thus, reasonable changes in λ do not radically alter the physics. It should also be noted that, although Eq. (9) is written in a particular gauge, the results of our calculations are gauge invariant.

We use a simple ansatz to include the effect of Φ into H_{mol} [32]:

$$(H_{mol})_{ij} \rightarrow (H_{mol})_{ij} - eS_{ij}\frac{\Phi(\vec{r_i})+\Phi(\vec{r_j})}{2}, \tag{10}$$

where, $\vec{r_i}$ is the center of the ith atomic orbital. That is, each $(H_{mol})_{ii}$ is affected in the usual way by application of the electrostatic potential, while the off-diagonal elements of H_{mol} are modified according to the non-orthogonality of the Hilbert space. Of course, the ε_α are also shifted by $-e\Phi$, as evaluated inside the appropriate gold contact.

In the systems of interest here, Landauer conductance theory provides a good approximation for conductance and current. The probability for an electron of energy E to move from the left set of leads to the right is [9]

$$T(E,V) = \sum_{\substack{\alpha\in\text{left} \\ \beta\in\text{right}}} \text{Tr}\left[\Gamma^{(\beta)}(E,V)G(E,V)\Gamma^{(\alpha)}(E,V)G^\dagger(E,V)\right], \tag{11}$$

where the broadening functions are given by

$$\Gamma^{(\alpha)}(E,V) \equiv -2\text{Im}\left[\Sigma^{(\alpha)}(E,V)\right]. \tag{12}$$

At $V = 0$, the conductance is simply

$$g = g_0 T(E_F,0). \tag{13}$$

Here, E_F is the Fermi energy of the bulk gold electrodes within extended Hückel theory, which we calculate by applying it to a 459-atom spherical gold nanocluster. Finally, the current in Landauer theory is [9]

$$I(V) = \frac{e}{h}\int_{-\infty}^{\infty} dET(E,V)\left[f(E-\mu_R) - f(E-\mu_L)\right], \tag{14}$$

TABLE 1. Comparison of the conductances found in our model [8] (in units of g_0) for various bonding geometries of the fully stretched oligopeptides with those of the STMBJ experiments [2, 3]. In the column headings, the first entry indicates the bonding geometry on the amine side of the molecule, the second the other. Only hollow bonding at both interfaces is compatible with experiment.

molecule	experimental	hollow-hollow	hollow-top	top-hollow	top-top
1: cysteamine-Cys	1.8×10^{-4}	5.4×10^{-5}	1.5×10^{-6}	9.6×10^{-7}	5.8×10^{-9}
2: cysteamine-Gly-Cys	4.2×10^{-6}	4.3×10^{-6}	5.3×10^{-8}	3.7×10^{-8}	1.5×10^{-9}
3: Cys-Gly-Cys	5.3×10^{-6}	4.2×10^{-6}	9.3×10^{-8}	1.0×10^{-7}	1.7×10^{-9}
4: cysteamine-Gly-Gly-Cys	5.0×10^{-7}	4.1×10^{-8}	7.3×10^{-10}	1.9×10^{-10}	2.2×10^{-11}

where $f(E-\mu_{R,L})$ is the Fermi function of the leads on the right (left) side of the extended molecule, and $\mu_{R,L}$ is the corresponding electrochemical potential. For numerical simplicity, we take the zero-temperature limit of Eq. (14); since the lead-induced broadening of H_{mol}'s discrete states is very much greater than k_BT, this is an excellent approximation.

4. RESULTS

4.1. Conductance and Bonding Geometry

The geometry by which the sulfur at each end of a thiol-terminated organic molecule bonds to its gold contact remains an open and important question [33]. We address this for the oligopeptides examined experimentally in Refs. [2] and [3] (see Fig. 1), comparing the data from those works with the results of the model described above and considering the two most likely bonding geometries [33, 34]: the sulfur atom can be either bound to a single gold atom lying on top of the gold surface's first atomic layer, in a linear bonding geometry; or the sulfur atom can lie above a triangular (111) hollow in the first gold layer itself, in a tetrahedral geometry.

Two bonding options at either end of the molecule yields a total of four molecular configurations for each oligopeptide: We present all sixteen equilibrium transmission curves $T(E,0)$ in Fig. 2, and the corresponding conductance results in Table 1. From the table, it is clear that in all cases, only the hollow bonding geometry on both sides of the molecule is compatible with the measured conductance value; the hollow-hollow calculation agrees well with experiment, while configurations with one or both top-bound geometries are at least an order of magnitude too small. Since no fitting to molecular electronics experiments was used in the construction of the model, and similar models have likewise arrived at good agreement with experiment [23, 24, 26], we conclude that these measured conductances correspond to a hollow bonding geometry on both ends of the oligopeptide [8].

The radical difference between the calculated hollow-hollow conductances and those with even one top-bound end can be understood simply: The slower of the two tunneling rates naturally limits the conductance in such a "series" arrangement. Gold-sulfur tunneling is much slower through the top-bound geometry than through the hollow, as the former possesses a single *sp* bond with relatively poor overlap, while the latter has much enhanced tetrahedral sp^3 bonding geometry: The sulfur atom in this case plays a bonding role roughly similar to the carbon in methane. Even if the full extended molecule is considered, the atoms at the interface retain this symmetry, and their orbitals' tendency to hybridize tetrahedrally allows stronger through-tunneling.

Though the distance d between the two gold contacts plays some role in determining conductance, this effect is not as great [8] as might be assumed. As an example, in Fig. 3a we present the conductance of peptide **2** in the top-top geometry as a function of d. The DFT geometry optimizations only allow a total variation of 1Å in d: The rigidity of the molecule prevents configurations for which $d < 14.2$Å, while beyond $d = 15.2$Å, configurations in which the molecule bonds only to one contact are preferred. The oligopeptide molecule itself differs very little from one value of d to another; instead, it reorients itself between the two contacts mainly as a single unit, stretching only by .3Å over the total 1Å range. This is consistent with the persistence lengths of longer polypeptide chains, in general several ångstroms or more. As the molecule is stretched, this small lengthening of bond lengths throughout decreases interatomic tunneling matrix elements slightly, both within the molecule and with the atoms of the lead; thus, the molecular orbitals become less broadened, and the transmission peaks narrower, as Fig. 3b shows.

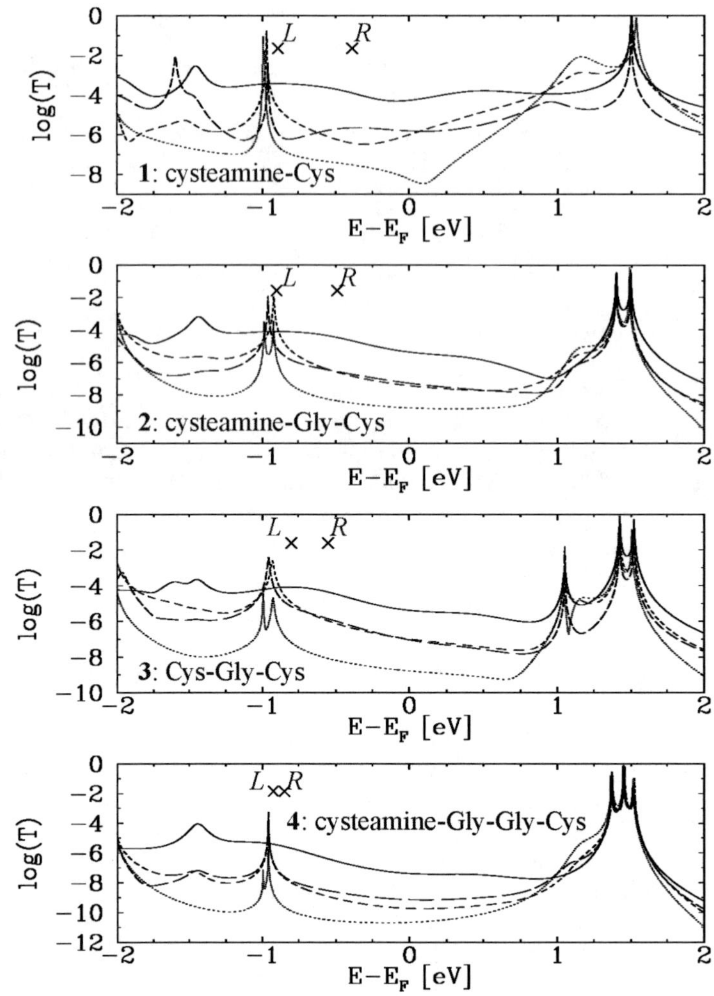

FIGURE 2. Equilibrium transmission functions $T(E,0)$ for various bonding geometries for each of the four fully stretched oligopeptide molecules studied experimentally in Refs. [2] and [3]. Solid lines indicate hollow-site bonding geometry on both sides. Dotted: top-site at both metal-molecule interfaces. Long dashes: hollow-site bonding for the sulfur on the amine-terminated side of the molecule and top-site bonding for the other. Short dashes: *vice versa*. Experimental conductances agree well with the solid curves in all cases. For the hollow-hollow geometry, the energies of the two molecular orbitals responsible for rectification are indicated with crosses. From Ref. [8].

Because the molecule itself distorts very little, the distance an electron must travel along the oligopeptide backbone remains nearly constant as the tip spacing increases. As a consequence, the conductance changes by a measurable amount, but one not at all sufficient to reconcile a top-bound geometry with experiment. Moreover, the DFT geometry optimizations permit even less stretching for hollow-bound geometries: Hollow-hollow configurations stretch by .2Å or less, affecting conductance by 10% or less [8]. The experimental conductance plateaus are nearly unsloped [2, 3]; since the conductance variation indicated by Fig. 3a would be measurable, if it were present, this is further evidence that the oligopeptides are bound in the hollow-hollow configuration.

4.2. Current Rectification and Negative Differential Resistance

One of the most striking results of the experimental studies is the strong current rectifying behavior exhibited by the gold–oligopeptide–gold junctions [3]. Figure 4 compares the current–voltage characteristics [8] calculated in our model for hollow-hollow bonding geometries with the characteristics obtained from experiment. For the three

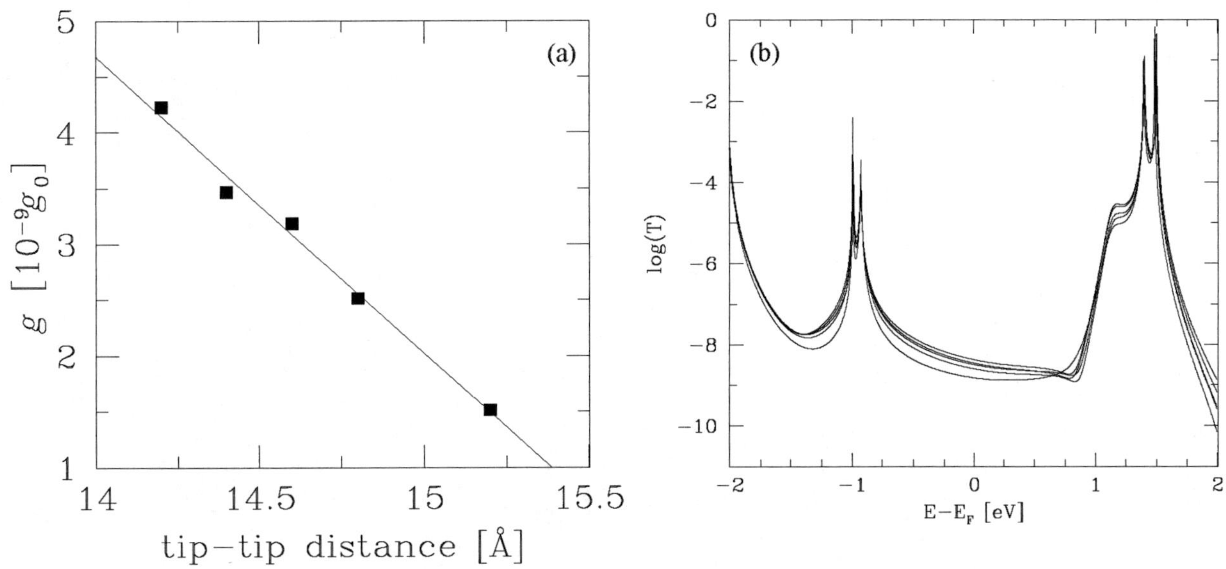

FIGURE 3. (a) Conductance of peptide **2**, cysteamine-Gly-Cys, in the top-top bonding configuration, as a function of the tip-to-tip distance d. Conductance variation is measurable, but not sufficient to reconcile the large difference between the calculated top-bound geometries and the experimental result seen in Table 1. (b) Transmission probabilities for the points in (a). The oligopeptide changes little from one point to another, preferring to reorient itself. The main effect of the slight stretching of bond lengths which does occur as d increases is to narrow the transmission peaks, as the oligopeptide itself couples less strongly to the leads, which provide broadening. From Ref. [8].

oligopeptides for which experimental curves were measured, the shape and magnitude of the model current are in good agreement, without any fitting to the experiments. Furthermore, the current-rectifying ratio is similar: For peptides **2** and **3**, negative bias generates about 150% of the current that positive bias does, while both theory and experiment find a somewhat smaller ratio for **4**.

To understand the current rectification, we turn to a simplified extended molecule [8] consisting of the oligopeptide itself and its six nearest-neighbor gold atoms, three on each side. While use of the full extended molecule is vital to the *quantitative* accuracy we seek throughout this work, such a simplified extended molecule is a useful tool for building a *qualitative* understanding of the numerical results. In particular, Figure 5 shows the movement of the generalized eigenenergies of one such simplified extended molecule under bias voltage.

In general, a strongly transmitting molecular orbital must be delocalized: It must contain significant weight on both metallic nanocontacts. Many states in Fig. 5 appear as nearly straight lines, strongly tracking the electrochemical potential of one or the other electrode; these are interfacial states, strongly localized at only one of the gold–oligopeptide interfaces, and thus they conduct poorly. When two such levels are near to each other in energy, however, they can hybridize into bonding and antibonding combinations. These whole-molecule orbitals can then conduct quite well.

For a symmetric extended molecule, interfacial states always hybridize into bonding and antibonding pairs at equilibrium. Peptide chains, however, are naturally asymmetric: Thus, unhybridized right/left pairs of interfacial states can be present at $V = 0$, and applied bias brings them either closer to or farther from resonance with each other, depending on the sign of V. In this case, applying bias in one direction generates two additional strongly conducting channels, one for the bonding and one for the antibonding state, which are not present if the bias is applied in the opposite direction. Indeed, we find just such a pair in Fig. 5, labeled R and L. In each of the molecules which rectify current, such a pair exists near E_F and is responsible for the behavior [8].

When the bias is increased *beyond* such a transmitting resonance of two interfacial states, the extra channels close, and NDR ($dI/dV < 0$) can result [8]. Such effects have not yet been observed experimentally; they are initially masked by additional transmitting states entering the window between μ_R and μ_L. However, at moderate bias, once an entire neighborhood of transmission states is encompassed within the voltage window, the effect could be observed

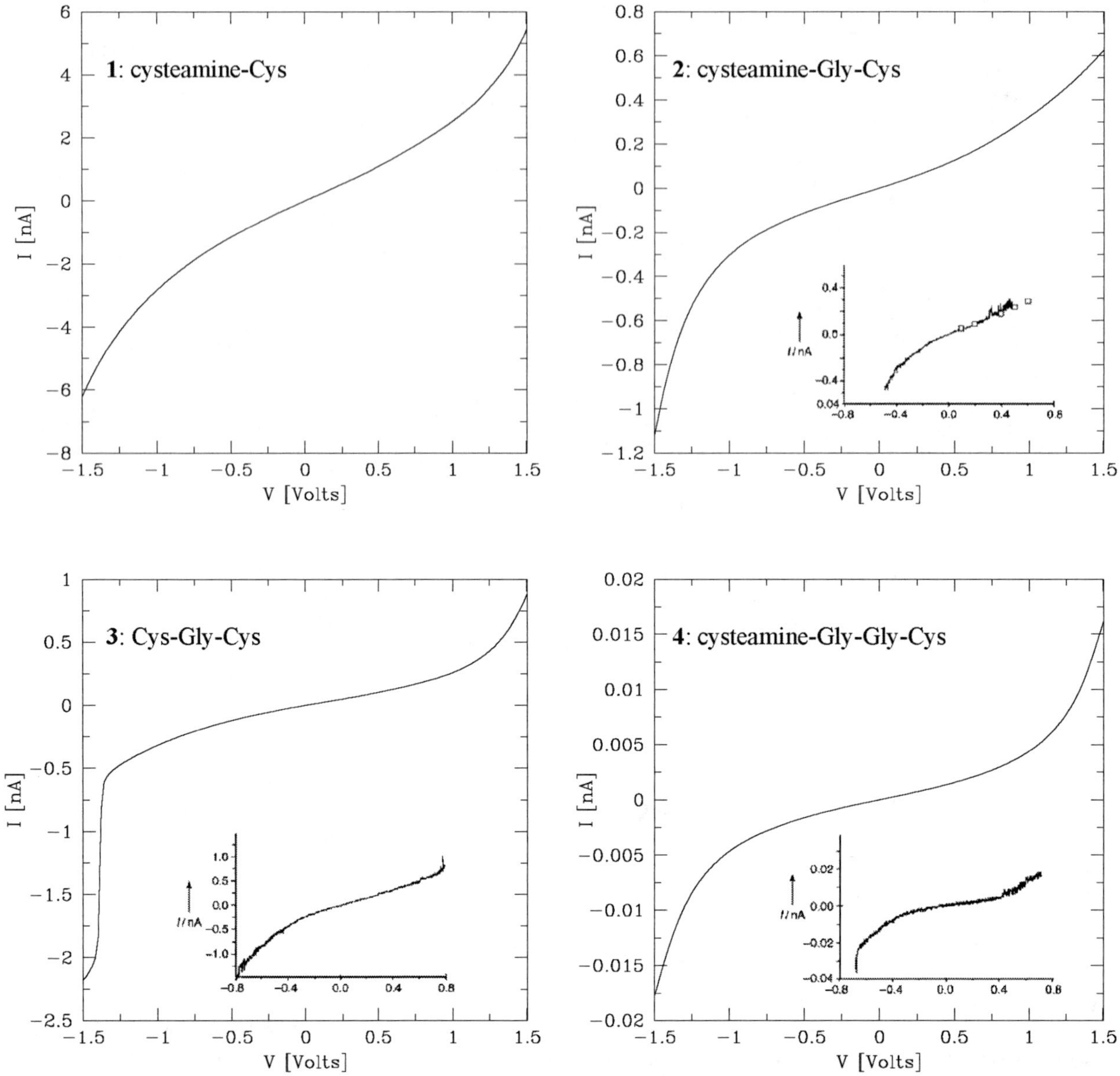

FIGURE 4. Calculated current-voltage characteristics of the four single-molecule oligopeptide devices of Fig. 2 and Table 1. The scale and asymmetry of the current are comparable to those found in the experimental studies of Ref. [3], reproduced here in inset. From Ref. [8].

by increasing the bias voltage past a right/left hybridized interfacial resonance. A calculated example for peptide **2** is shown in Fig. 6.

5. SUMMARY

We have presented the first theoretical model [8] of electron transport through single-molecule metal–oligopeptide–metal nanojunctions. The model achieves very good agreement with STMBJ experiments on these systems, without the use of fitting parameters; this allows us to use it to extract important physical and chemical information from

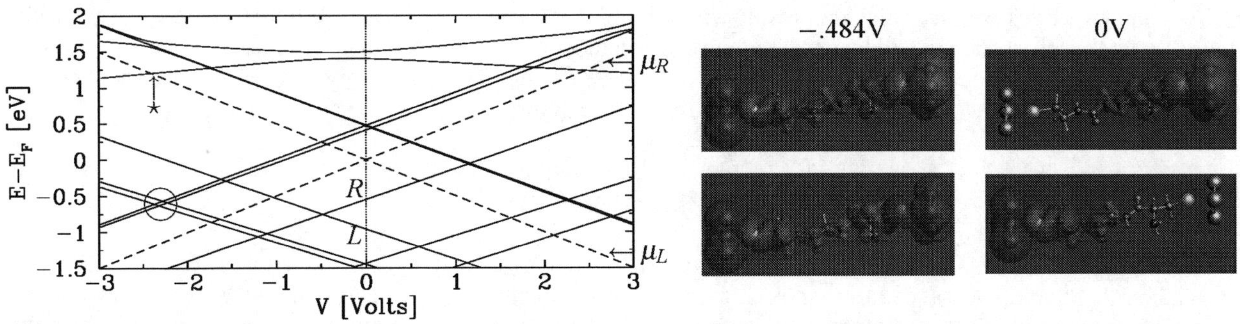

FIGURE 5. (left) Movement with applied bias of the generalized eigenenergies of the fully stretched six-gold-atom simplified extended molecule for peptide **2** in the hollow-hollow configuration. Some lines are doubled by degrees of freedom in the gold contacts; when more gold atoms are included, the molecular orbitals split further, eventually developing into the continuous density of states of an infinite system. Note that the apparent intersections are actually avoided crossings. In equilibrium, states R and L are detuned by .4eV due to molecular asymmetry. They come into resonance only under negative bias, causing current rectification. Without these two states, the location of anticrossings, and thus the I–V characteristic, would be largely symmetric. The dashed lines represent the electrochemical potentials of the right and left electrodes, μ_R and μ_L. NDR can result from increasing bias beyond the last anticrossing in a group; an example is circled. (right) The R and L states at equilibrium, and in resonance with each other due to applied bias. In resonance, they form whole-molecule bonding and antibonding hybrids. Both (left) and (right) are from Ref. [8].

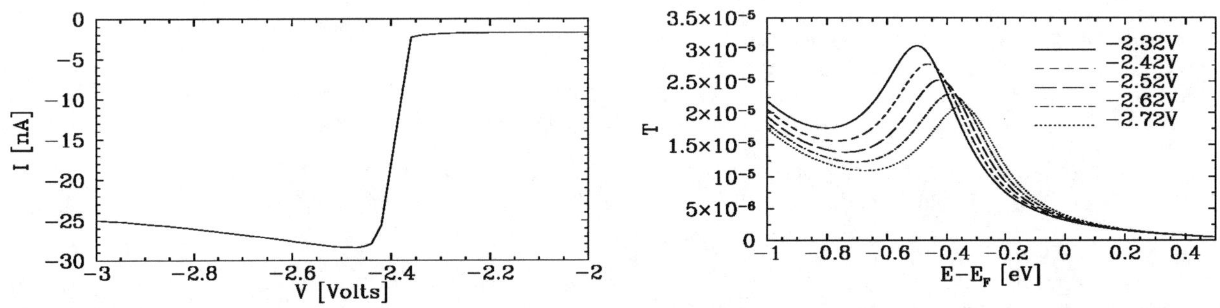

FIGURE 6. (left) Calculated I–V characteristic for the fully stretched hollow-hollow device based on peptide **2** in the medium-bias regime. NDR is visible after the sharp current step due to a state entering the voltage window ($\star$ in Fig. 5). Other oligopeptide molecules have qualitatively similar characteristics. (right) Corresponding $T(E,V)$ curves. NDR corresponds to a gradual decrease in the transmission peak's strength as voltage becomes more negative, due to departure from a resonant anticrossing of interfacial states (the circle in Fig. 5). Both graphs are from Ref. [8].

the experiments. As examples, we have demonstrated that the oligopeptide molecules of Refs. [2] and [3] are bound to hollows in the Au(111) surface at both ends, and investigated the role molecular stretching plays in determining conductance, for various bonding geometries.

We are able to understand the observed current rectification in these systems as a consequence of oligopeptides' asymmetric nature. Bias voltage applied in the appropriate direction brings opposing interfacial states, detuned from each other in equilibrium due to the molecular asymmetry, into resonance with each other, thus opening new current-carrying channels. Furthermore, NDR can result when the bias is increased beyond such an anticrossing of levels, and we predict this phenomenon will be observed with only a moderate increase in bias voltage. Especially because of NDR's many important device applications, this work indicates that oligopeptide nanojunctions comprise an especially promising and interesting avenue for future nanoelectronics research.

ACKNOWLEDGMENTS

We thank Nongjian Tao, Joshua Hihath, and Ross Hill for useful discussions. This work was supported by NSERC and the Canadian Institute for Advanced Research. Some numerical computations presented in this work were performed on WestGrid computing resources, which are funded in part by the Canada Foundation for Innovation, Alberta Innovation and Science, BC Advanced Education, and the participating research institutions. WestGrid equipment is provided by IBM, Hewlett Packard, and SGI.

REFERENCES

1. R. I. Gilmanshin, In K. Sienicki, ed. *Molecular Electronics and Molecular Electronic Devices, Vol. 2*, CRC Press, Boca Raton, 1993, pp. 1–78.
2. X. Xiao, B. Xu, and N. Tao, *J. Am. Chem. Soc.* **126**, 5370–5371 (2004).
3. X. Xiao, B. Xu, and N. Tao, *Angew. Chem. Int. Ed.* **43**, 6148–6152 (2004).
4. J. J. Davis, N. Wang, A. Morgan, T. Zhang, and J. Zhao, *Farad. Discuss.* **131**, 167–179 (2006).
5. B. Xu, and N. J. Tao, *Science* **301**, 1221–1223 (2003).
6. B. Xu, X. Xiao, and N. J. Tao, *J. Am. Chem. Soc.* **125**, 16164–16165 (2003).
7. X. Xiao, B. Xu, and N. J. Tao, *Nano Lett.* **4**, 267–271 (2004).
8. D. M. Cardamone and G. Kirczenow, submittted to *Phys. Rev. B* (2007), arXiv:0708.1041 [cond-mat.mes-hall].
9. S. Datta, *Electron Transport in Mesoscopic Systems*, vol. 3 of *Cambridge Studies in Semiconductor Physics and Microelectronic Engineering*, Cambridge University Press, Cambridge, UK, 1995, ISBN 0 521 59943 1.
10. A. van der Ziel, *Solid-State Electron.* **26**, 333–334 (1983).
11. E. R. Brown, J. R. Söderström, C. D. Parker, L. J. Mahoney, K. M. Molvar, and T. C. McGill, *Appl. Phys. Lett.* **58**, 2291–2293 (1991).
12. T. P. E. Broekaert, B. Brar, J. P. A. van der Wagt, A. C. Seabaugh, F. J. Morris, T. S. Moise, E. A. Beam, III, and G. A. Frazier, *IEEE J. Solid-State Circuits* **33**, 1342–1349 (1998).
13. R. H. Mathews, J. P. Sage, T. C. L. G. Sollner, S. D. Calawa, C.-L. Chen, L. J. Mahoney, P. A. Maki, and K. M. Molvar, *Proc. IEEE* **87**, 596–605 (1999).
14. N. Agraït, J. G. Rodrigo, and S. Vieira, *Phys. Rev. B* **47**, 12345–12348 (1993).
15. H. Ohnishi, Y. Kondo, and K. Takayanagi, *Nature* **395**, 780–783 (1998).
16. A. I. Yanson, G. R. Bollinger, H. E. van der Brom, N. Agraït, and J. M. van Ruitenbeek, *Nature* **395**, 783–784 (1998).
17. K. Hansen, S. K. Nielsen, M. Brandbyge, E. Lægsgaarrd, I. Stensgaard, and F. Besenbacher, *Appl. Phys. Lett.* **77**, 708–710 (2000).
18. H. Mehrez, A. Wlasenko, B. Larade, J. Taylor, P. Grütter, and H. Guo, *Phys. Rev. B* **65**, 195419 (2002).
19. M. Koentopp, C. Chang, K. Burke, and R. Car, arXiv:cond-mat/0703591v2 (2007).
20. Gaussian 98, Revision A.11.3, M. J. Frisch *et al.*.
21. R. Hoffman, *J. Chem. Phys.* **39**, 1397–1412 (1963).
22. J. H. Ammeter, H.-B. Bürgi, J. C. Thibeault, and R. Hoffman, *J. Am. Chem. Soc.* **100**, 3686–3692 (1978).
23. S. Datta, W. Tian, S. Hong, R. Reifenberger, J. I. Henderson, and C. P. Kubiak, *Phys. Rev. Lett.* **79**, 2530–2533 (1997).
24. E. G. Emberly, and G. Kirczenow, *Phys. Rev. B* **64**, 235412 (2001).
25. M. A. Reed, C. Zhou, C. J. Muller, T. P. Burgin, and J. M. Tour, *Science* **278**, 252–254 (1997).
26. J. G. Kushmerick, D. B. Holt, J. C. Yang, J. Naciri, M. H. Moore, and R. Shashidhar, *Phys. Rev. Lett.* **89**, 086802 (2002).
27. The numerical implementation used was the YAeHMOP package (ver. 3.0) by G. A. Landrum and W. V. Glassey.
28. D. M. Newns, *Phys. Rev.* **178**, 1123–1135 (1969).
29. V. Mujica, M. Kemp, and M. A. Ratner, *J. Chem. Phys.* **101**, 6856–6864 (1994).
30. L. D. Landau, and E. M. Lifshitz, *Statistical Physics*, vol. 5 of *Course of Theoretical Physics*, Pergamon, Bristol, Great Britain, 1958.
31. The experimental conditions are described in the supplementary information to Ref. [2], freely available on the Internet at `http://pubs.acs.org/`.
32. E. G. Emberly, and G. Kirczenow, *Chem. Phys.* **281**, 311–324 (2002).
33. X. Li, J. He, J. Hihath, B. Xu, S. M. Lindsay, and N. Tao, *J. Am. Chem. Soc.* **128**, 2135–2141 (2006).
34. D. Krüger, H. Fuchs, R. Rousseau, D. Marx, and M. Parrinello, *Phys. Rev. Lett.* **89**, 186402 (2002).

Photochemistry of Biological Chemosensors, Organic Light-Emitting Diodes, and Inner-shell Electronic Processes

Masahiro Ehara[1] and Hiroshi Nakatsuji[1,2]

[1] *Department of Synthetic Chemistry and Biological Chemistry, Graduate School of Engineering, Kyoto University, Katsura, Nishikyo-ku, Kyoto 615-8510, Japan*
[2] *Quantum Chemistry Research Institute, Kyodai Katsura Venture Plaza, North building 106, 1-36, Goryo Oohara, Nishikyou-ku, Kyoto 615-8245, Japan*

Abstract. The photochemistry of biological chemosensors, organic light-emitting diodes, and inner-shell electronic processes has been investigated by the SAC-CI method. The electronic mechanism of the photo-induced electron transfer has been clarified for the recently developed fluorescent probe. The absorption and emission spectra of the organic light-emitting diodes such as polyphenylenevinylene and fluorene-thiophene have been predicted in high accuracy. Various kinds of phenomena appearing in the core-electronic processes such as shake-up satellites, vibrational excitations, valence-Rydberg coupling, and its thermal effect, have been accurately examined. The present works provides the useful basis for the theoretical investigations of the wide varieties of photochemistry.

Keywords: Biological chemosensor, Organic light-emitting diode, Inner-shell electronic process
PACS: 36.20.Kd Electronic structure and spectra

INTRODUCTION

The photochemistry of biological chemosensors and organic light-emitting diodes is one of the most important subjects. Recently, various kinds of fluorescent probe which work as biological chemosensor have been developed, since it enables the direct and real-time measurements of the enzyme activities. The fluorescence change by the interaction with the enzyme is essential for the measurement, therefore, elucidation of the electronic mechanism of the fluorescence and its color-tuning is important for developing new fluorescent probes. The other photo-functional material, organic light-emitting diodes (OLED), are recognized as one of the promising candidates for the next generation electro-optical devices. For developing the new materials, theoretical prediction and design of the photo-physical properties of OLED molecules are relevant.

Recently, investigations of the core-electronic processes invoked renewal of interest, because significant developments in both high-resolution soft X-ray spectroscopy and accurate state-of-the-art theoretical methods have made us possible to obtain precise knowledge of the core-electronic processes. One can observe vibrational structures in the core-level photoelectron spectrum and thereby discuss the excited-state dynamics. This situation has motivated intensive cooperative researches on the core-electron processes from experimental and theoretical sides.

For these purposes, the SAC-CI method[1-3] is useful; the method has been established as an accurate molecular excited-state theory applicable to the wide varieties of electronic processes in the wide energy region. In the present article, we have reviewed our recent SAC-CI studies on the photochemistry of biological chemosensors, organic light-emitting diodes,[4,5] and inner-shell electronic processes.[6-8]

BIOLOGICAL CHEMOSENSORS

Recently, novel molecular fluorescent probes which can selectively recognize the phosphoprotein have been developed.[9] One of these fluorescent probes, acridine-based probe shows drastic fluorescence change in the presence of Zn^{2+} ion. We studied the photochemistry of this acridine-based fluorescent probe, in particular, the electronic

CP995, *Nuclei and Mesoscopic Physics - WNMP 2007*, edited by P. Danielewicz, P. Piecuch, and V. Zelevinsky

mechanism of the photo-induced electron transfer and the origin of the energy shift in the fluorescence. First, the stable structures of the mono-nuclear and bi-nuclear Zn(II) complexes were determined. The five-fold structure was most stable for the mono-nuclear complex and 3-fold+3-fold structure was stable for the bi-nuclear complex. For the photo-induced electron transfer, relative energy position of the charge transfer state and $\pi\pi^*$ excited state are very important The solvation effect and the geometry relaxation were found to be relevant for these states. These two factors control the photo-induced electron transfer of this fluorescent probe. Characteristic energy shift of the fluorescence was found to be determined by the distance between the acridine unit and Zn(II) ion.

ORGANIC LIGHT-EMITTING DIODES

The optical properties of polyphenylenevinylene (PPV) that is well-known as an efficient light emitter[10-12] have been extensively investigated, however, the accurate theoretical prediction of the transition energies is still challenging. The excited states of the PPV oligomer (Figure 1) were studied by the SAC-CI method. The calculated absorption and emission energies of the lowest singlet (S_1) excited state of PPVn (n=1~4) are compared with the experimental values[11] and other theoretical values[12] in Figure 2. The SAC-CI method predicted the experimental values in high accuracy; the mean deviations are 0.09 eV for absorption and 0.03 eV for emission. The method reproduced the chain-length dependence of the excitation energy much better than the TDDFT/B3LYP calculations; the conventional TDDFT should be carefully used for the transition energies of the molecules with long π-conjugated chains. The excited-state geometry of the PPVn was calculated to be planar by the SAC-CI method.

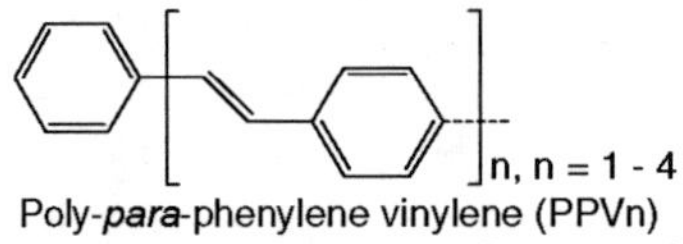

FIGURE 1. Molecular structure of PPVn oligomer.

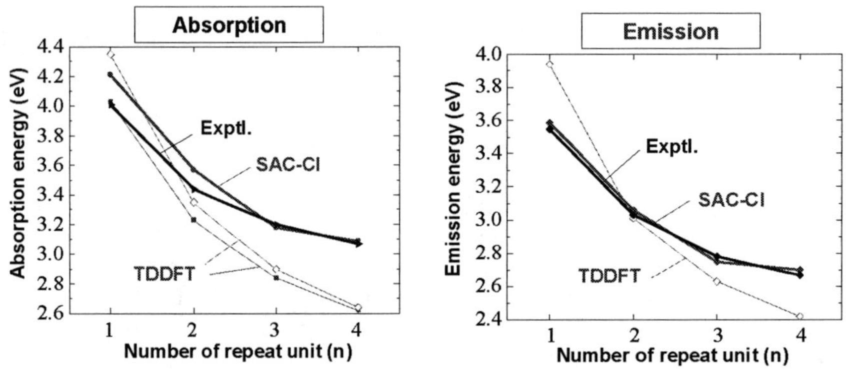

FIGURE 2. Excitation (left) and emission (right) energies of PPVn (n=1-4) plotted to the repeat unit n.

The fluorene co-polymers constitute rigorous candidates of the blue light emitter.[13,14] By introducing the suitable substituents, the optical properties can be controlled because of its flexible structure. Fluorene-thiophene (FT) and fluorene-methyl-thiophene (FMT) in Figure 3 show characteristic electronic spectra.[13]

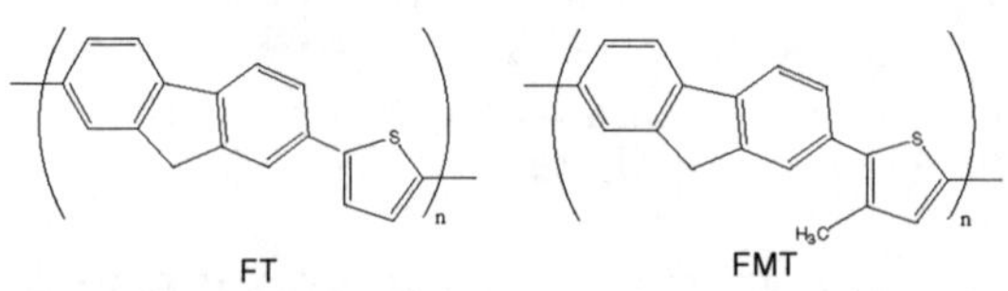

FIGURE 3. Molecular structures of FT and FMT oligomers.

The ground-state potential curves of the FT and FMT monomers along the torsion angle are shown in Figure 4. The dihedral angles of the most stable conformation of FT and FMT are θ=27° and 43°, respectively. The potential energy curves are very flat for the torsion; the relative energy of the conformers for θ=0° ~ 60° are within about 1.0 kcal/mol. This low rotational energy barrier indicates that it allows a wide range of nonplanar conformation of these molecules at room temperature.

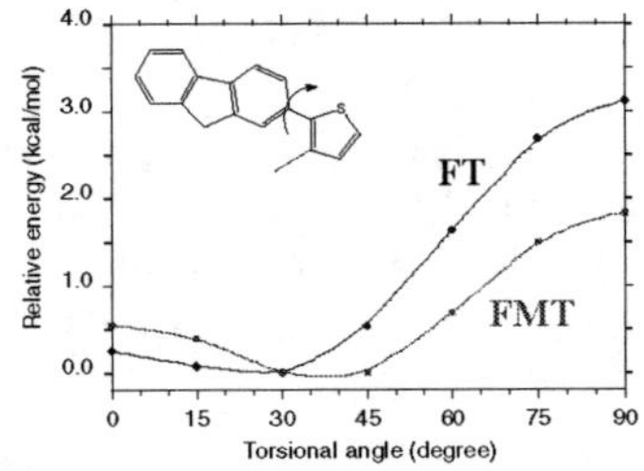

FIGURE 4. The ground-state potential energy curves of FT and FMT monomers.

The optical properties of the low-lying excited states of these molecules are dependent on the torsion angle. The excitation energy and oscillator strength as the function of the torsion angle are displayed in Figure 5. The first to third excited states interact to each other along the torsion. The avoided crossing occurs between the first and third excited states in the region of θ=30°~60°; this interaction is clearly seen in the oscillator strength.

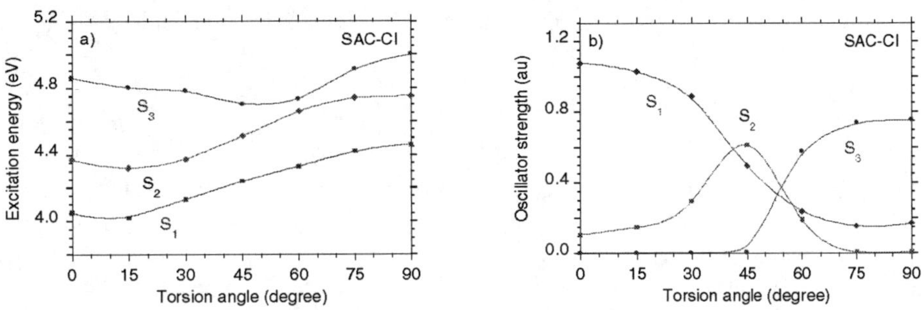

FIGURE 5. SAC-CI (a) excitation energy and (b) oscillator strength of the first three excited states of FT versus torsion angle.

Since the ground-state potential energy surfaces are flat for the torsion and the excitation spectra are dependent on this coordinate, statistical average of the conformers should be considered in order to simulate the absorption spectra. The simulated absorption spectra of FT and FMT, in which thermal distribution at 298 K is taken into account, are compared with the experimental spectra in Figure 6. The simulated spectra with thermal average showed excellent agreement with the experimental spectra. For FT, the contribution of the conformer whose torsion angle is up to θ=~30° is dominant and asymmetric spectrum is obtained. In the case of FMT, the strong shoulder was measured in the higher energy side of the peak. This shoulder is attributed to the second excited state of the conformer of θ=30~45°.

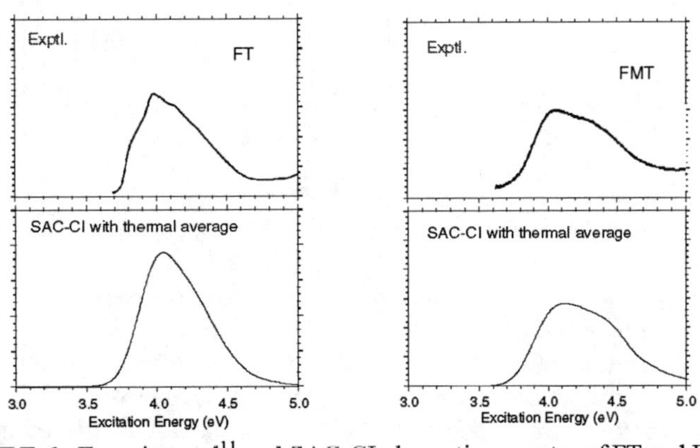

FIGURE 6. Experimental[11] and SAC-CI absorption spectra of FT and FMT.

INNER-SHELL ELECTRONIC PROCESSES

The SAC-CI method can be applicable to the electronic processes in wide energy region. We have investigated wide varieties of the core-electronic processes with the SAC-CI method. There are two approaches in SAC-CI for studying the inner-shell electronic processes. One is based on the general-R method[15-17] in which the neutral ground-state Hartree-Fock is adopted for the reference function. The wave function for core ionization, for example, is given by,

$$\Psi^{SAC-CI}_{general-R} = \left(\sum_i R_i + \sum_{ija} R^a_{ij} + \sum_{ijkab} R^{ab}_{ijk} + \sum_{ijklabc} R^{abc}_{ijkl} + \cdots \right) \exp\left(\sum_I S_I \right) \Phi_{HF} \tag{1}$$

where R and S represent the R- and S-operators in the SAC-CI expansion, respectively, and Φ_{HF} is the closed-shell ground-state Hartree-Fock. In this method, the higher-order R-operators such as triples and quadruples are necessary for describing orbital relaxation and electron correlations of the core-electronic processes. This method is simple and accurate, but, the computational cost is expensive since it includes higher operators. The method is applicable even to the g-u splitting in the molecules with equivalent atoms.

The other method is the open-shell reference SAC-CI method,[18] in which the core-ionized ROHF is used for the reference;

$$\Psi^{OR-SAC-CI}_{SD-R} = \left(\sum_i R_i + \sum_{ija} R^a_{ij} \right) \Psi^{OR-SAC}, \quad \Psi^{OR-SAC} = \exp\left(\sum_I S_I \right) \exp\left(\sum_J S'_J \right) O_K \Phi_{HF} \tag{2}$$

where $O_K \Phi_{HF}$ corresponds to the core-hole ROHF. S and S' represent the excitation and de-excitation S operators, respectively. Since the method start from the core-hole state, it is efficient and accurate; SD-R is enough for describing the core-hole state as given in Eq. (2).

We have applied the SAC-CI method to wide varieties of the core-electronic processes. We present some of the applications in this article. Figure 7 compares the calculated and XPS inner-shell satellite spectra of C1s ionization of CO.[6] There are many shake-up satellites described by the multi-electron processes in the higher energy region of main line. The SAC-CI method reproduced the satellite spectra very accurately and gave the detailed assignments.

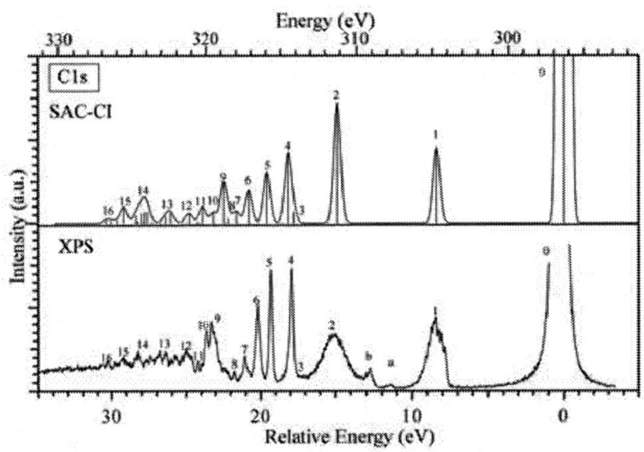

FIGURE 7. SAC-CI and XPS C1s satellite spectra of CO.

For the satellite band at about 8-9 eV, the angle-resolved ion-yield (ARIY) spectroscopy was applied to the vibrational structure.[6] By measuring at the two angles 0° and 90° relative to the polarization vector, symmetry of the states can be resolved; the measurement at 0° mainly observes the direct shake-up in Σ symmetry, while the measurement at 90° observes the conjugate shake-up in Π symmetry. However, the Σ and Π states interfere to each other in the spectra at 0°. Therefore, we decomposed the experimental overlapping spectra using the SAC-CI calculations. Figure 8 shows the theoretically decomposed vibrational spectra; the experimental spectra were successfully decomposed into one Σ state and two Π states. The satellite Σ state shows the excitation to the high vibrational levels reflecting the large geometry change of the state.

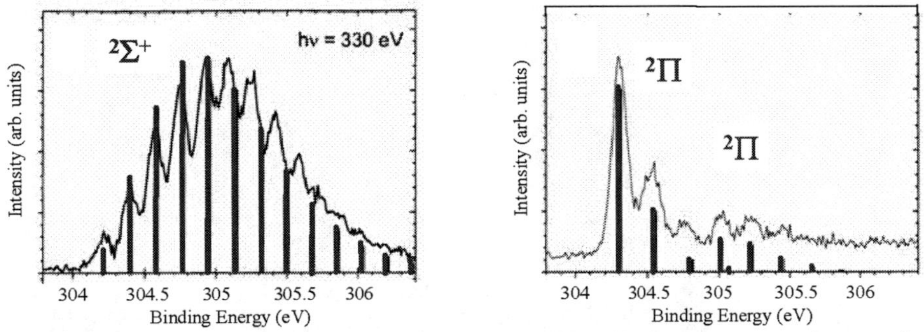

FIGURE 8. XPS and SAC-CI vibrational spectra of C1s satellite bands of $^2\Sigma^+$ and $^2\Pi$ states

ARIY spectra were measured in the O1s excitation region of N_2O.[7] Irregular excitation to the high vibrational levels was observed in the 4sσ state. The potential energy surfaces of the O1s core-excited states were examined to clarify this irregular vibrational progression. Figure 9 shows the potential energy curves and second moments $<r^2>$ of the O1s excited states in Σ and Δ symmetry. This shows the irregular Rydberg behavior in the 4sσ state is caused by the strong valence-Rydberg coupling. Franck-Condon analysis based on the SAC-CI two-dimensional potential energy surfaces of the 4sσ and 3pπ states accurately reproduced the observed vibrational excitations as shown in Figure 10.

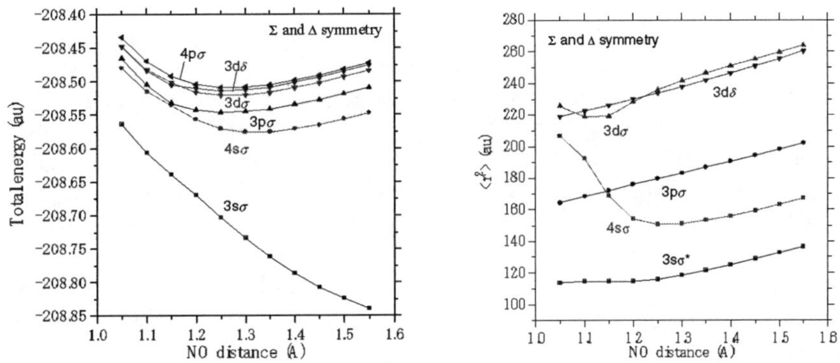

FIGURE 9. Potential energy curves (left) and second moments $<r^2>$ (right) of the O1s excited states of N_2O in Σ and Δ symmetry with the cut of R_{NN}=1.10Å.

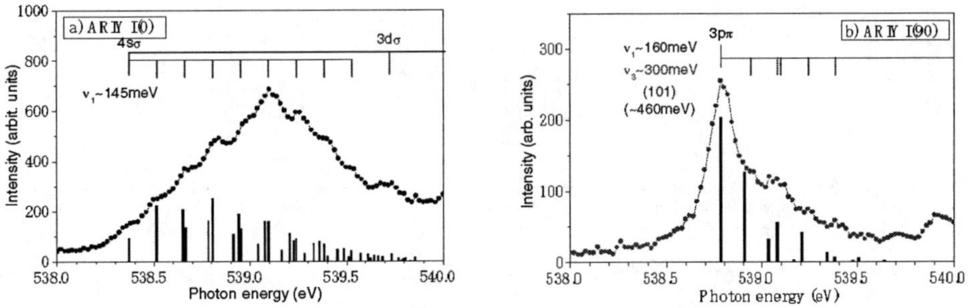

FIGURE 10. Vibrational spectra of a) O1s-4sσ and b) O1s-3pπ excited states. SAC-CI spectra are shown by the vertical solid lines.

Absorption spectra of the vibrationally excited "hot" molecules were observed for the O1s excitations of N_2O.[8] The ARIY spectra are compared in Figure 11. The spectral intensity of the O1s → nsσ Rydberg series, which has σ* character, is significantly reduced by the vibrational excitation. The energy shift was also observed in the spectra from the vibrationally excited states.

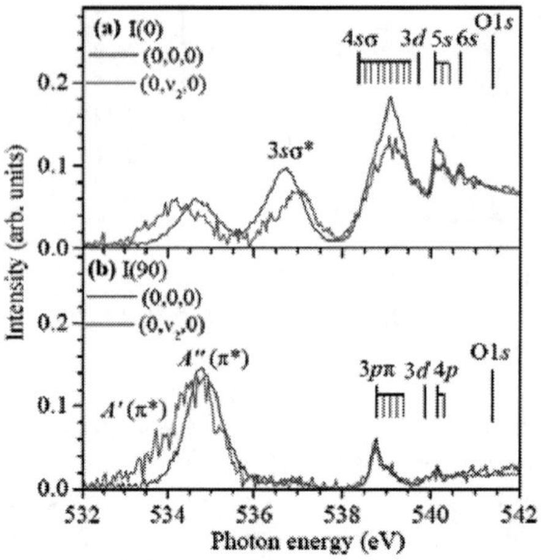

FIGURE 11. ARIY spectra in the N_2O O1s excitation region: blue and red lines show the spectra from vibrationally ground and excited states, respectively.

Assuming a Boltzmann distribution, the vibrational excitations occurs mainly for the bonding coordinate. Therefore, the potential energy curves and the second moments $<r^2>$ of the low-lying excited states along the bending coordinate are examined (Figure 12). The calculated potential energy curves explain the energy shift, that is, the nsσ Rydberg states show blue shift, while π* state shows red shift by vibrational excitation. The second moment is anti-correlated to the amount of valence character and the intensity. The second moments of the nsσ Rydberg states become large as the molecule become bent. This indicates that the mixing of the valence character in these states becomes less as the bond angle decreases. This explains the absorption oscillator strength to the nsσ Rydberg states becomes small in the spectra from the vibrationally excited states. On the other hand, the second moment of π* state is small in the bent structure. This indicates the oscillator strength of this state increases by the vibrational excitation.

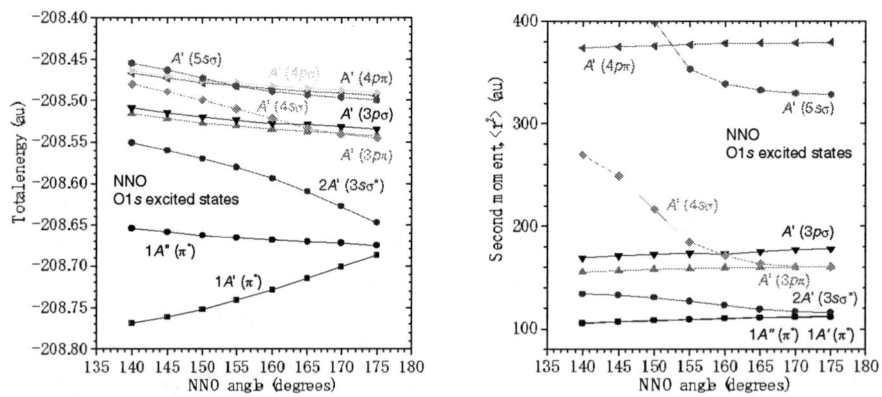

FIGURE 12. Potential energy curves (left) and second moments $<r^2>$ (right) of the low-lying O1s excited states of N_2O along the bending coordinate.

SUMMARY

The SAC-CI method has clarified the electronic mechanism of the photo-induced electron transfer of the acridine-type fluorescent probe. The solvation effect and geometry relaxation in the charge transfer state and ππ* excited state were found to be essential. The SAC-CI method has also predicted the absorption and emission spectra of organic light-emitting diodes in high accuracy. For FT and FMT, the thermal average is shown to be important to simulate the electronic spectra. Various kinds of phenomena appearing in the core-electronic processes such as shake-up satellites, vibrational excitations, and valence-Rydberg coupling have been accurately examined by the SAC-CI calculations.

ACKNOWLEDGMENTS

The authors are grateful to Prof. I. Hamachi and Dr. A. Ojida for the work on biological chemosensor, to Drs. B. Saha, P. Poolmee, and S. Hannongbua for the works of OLED, and to Prof. K. Ueda for the works on inner-shell electronic processes. This study was supported by the Grant for Creative Scientific Research from the Ministry of Education, Science and Sports of Japan.

REFERENCES

1. H. Nakatsuji, *Chem. Phys. Lett.* **59**, 362 (1978); H. Nakatsuji, *Chem. Phys. Lett.* **67**, 329 (1979).
2. H. Nakatsuji, *"SAC-CI method: Theoretical aspects and some recent toipics", in Computational Chemistry, Review of Current Trends*, World Scientific, Singapore, 1997, pp. 62-124.
3. M. Ehara, J. Hasegawa, H. Nakatsuji, *"SAC-CI Method Applied to Molecular Spectroscopy", in Theory and Applications of Computational Chemistry*, edited by C.E. Dykstra, et al., Elsevier, Oxford, 2005, pp. 1099-1141; M.J. Frisch, et al., *GAUSSIAN03*, Gaussian Inc., Pittsburgh, PA, 2003.
4. B. Saha, M. Ehara, H. Nakatsuji, *J. Phys. Chem. A.* **111**, 5473-5481 (2007).
5. P. Poolmee, M. Ehara, S. Hannongbua, H. Nakatsuji, *Polymer,* **46**, 6474-6481 (2005).
6. M. Ehara, K. Kuramoto, H. Nakatsuji, M. Hoshino, T. Tanaka, M. Kitajima, H. Tanaka, Y. Tamenori, A. De Fanis, K. Ueda, *J. Chem. Phys.* **125**, 114304 (2006).
7. T. Tanaka, K. Ueda, R. Feifel, L. Karlsson, H. Tanaka, M. Hoshino, M. Kitajima, M. Ehara, R. Fukuda, R. Tamaki, H. Nakatsuji, *Chem. Phys. Lett.* **435**, 182 (2007).
8. T. Tanaka, M. Hoshino, H. Kato, M. Ehara, N. Yamada, R. Fukuda, H. Nakatsuji, Y. Tamenori, J.R. Harries, G. Pruemper, H. Tanaka, K. Ueda, *Phys. Rev. A*, in press.
9. A. Ojida, I. Hamachi, *Bull. Chem. Soc. Jpn.* **79**, 35 (2006); A. Ojida, Y. Miyahara, J. Wongkongkatep, S. Tamaru, K. Sada, I. Hamachi, *Chemistry an Asian J.* **1**, 555 (2006).
10. J.H. Burroughes, D.D.C. Bradley, A.R. Brown, R.N. Kckay, R.H. Friend, P.L. Burn, A.B. Holmes, *Nature (London)* **347**, 539 (1990).
11. J. Gierschner, H.-G. Mack, L. Luer, D. Oelkrug, *J. Chem. Phys.* **116**, 8596 (2002).
12. A. Pogantsch, G. Heimel, E. Zojer, *J. Chem. Phys.* **117**, 5921 (2002); Y. Han, S.U. Lee, *J. Chem. Phys.* **121**, 609 (2004); Shukla, H. Ghosh, S. Mazumdar, *Phys. Rev. B.* **67**, 245203, 2003; S. Karabunarliev, M. Baumgarten, K. Mullen, *J. Phys. Chem. A* **104**, 8236, 2000.
13. M. Belletete, S. Beaupre, J. Bouchard, P. Blandin, M. Leclerc, G. Durocher, *J. Phys. Chem. B* **104**, 9118 (2000), M. Belletete, M. Bedard, M. Leclerc, G. Durocher, *J. Mol. Struct.(THEOCHEM)*, **679**, 9 (2004).
14. V. Lukes, D. Wegh, P. Hrdlovic, M. Stefko, K. Matsuzna, V. Laurinc, *Syn. Met.* **148**, 179 (2005).
15. H. Nakatsuji, *Chem. Phys. Lett.* **177**, 331 (1991).
16. M. Ehara, H. Nakatsuji, Chem. Phys. Lett. **282**, 347 (1998).
17. M. Ehara, M. Ishida, K. Toyota, H. Nakatsuji, *"SAC-CI general-R method: Theory and Applications to the Multi-Electron Processes", in Reviews in Modern Chemistry*, edited by K. D. Sen, World Scientific, Singapore, 2002, pp. 293-319.
18. Y. Ohtsuka, H. Nakatsuji, *J. Chem. Phys.* **124**, 054110 (2006).

Interplay of Orbital Paramagnetism and Diamagnetism in Nanometer Scale 2DEG Strips

Michael J. Harrison

Department of Physics and Astronomy, Michigan State University, East Lansing, MI 48824-2320, USA

Abstract. The role of boundary corrections to Landau diamagnetism for macroscopic free electronic systems has been discussed [1-8], and orbital paramagnetic effects have been experimentally observed in small systems [9]. The correction terms can theoretically be either paramagnetic or diamagnetic depending on the increase in potential energy at the surface which defines the perimeter of the electron system. In this work calculations of orbital magnetism have been performed for nanometer scale 2DEG strips laterally confined by an attractive transverse parabolic potential. The emergence of orbital paramagnetism depends sensitively on areal electron density, strip width, and parabolic potential strength such as may be produced by a split gate with negative voltage applied. Selected graphical results display considerable variation in the orbital paramagnetic contribution compared to the Landau diamagnetic susceptibility.

Keywords: Two dimensional orbital paramagnetism, Landau diamagnetism, nanometer width 2DEG strips
PACS: 73.63.Nn, 73.90.+f, 75.20.-g, 75.75.+a

INTRODUCTION

Surface corrections to Landau diamagnetism in macroscopic size free electron systems have been discussed by a number of workers [1-8] over an extended period of time. The corrections are very sensitive to the exact boundary conditions at the surface or system perimeter, depending on the increase in potential at the surface relative to the Fermi energy [1]. The correction term can be either paramagnetic or diamagnetic depending on the surface boundary conditions which define the perimeter of the electron system. Measurements of the orbital magnetism of much smaller isolated Ga-As heterostructures in the shape of squares several times 10^3 nm on a side have revealed a paramagnetic orbital contribution 100 times larger than the magnitude of the Landau diamagnetic susceptibility [9]. In this paper, I calculate the orbital magnetism of narrow 2DEG strips formed between the arms of a split-gate electrode [10] which have a negative potential V. The electronic systems considered range in width from 10 nm up to 100 nm, and extend indefinitely along their length. The lateral electron confinement which defines the strip width is represented in our calculation by a parabolic potential energy that is a maximum on the split-gate electrode and vanishes on the center line bisecting the 2DEG strips along their lengths. We further assume temperatures sufficiently low to adopt a zero temperature approximation, and areal electron densities n_0 commonly achieved in two-dimensional heterostructures, i.e. $n_0 \approx$ several times 10^{11} or 10^{12} cm^{-2} .

MODEL 2DEG STRIP IN PERPENDICULAR MAGNETIC FIELD

Consider a 2DEG strip in the x-y plane running parallel to the x-axis, which bisects the strip. A magnetic field $\mathcal{B}$ points in the positive z direction. A quadratic confining potential energy U(y) varies in the y direction within the region $-L_y \leq y \leq L_y$, where $2L_y$ represents the width of the 2DEG strip, as indicated in Fig. 1.

CP995, *Nuclei and Mesoscopic Physics - WNMP 2007*, edited by P. Danielewicz, P. Piecuch, and V. Zelevinsky

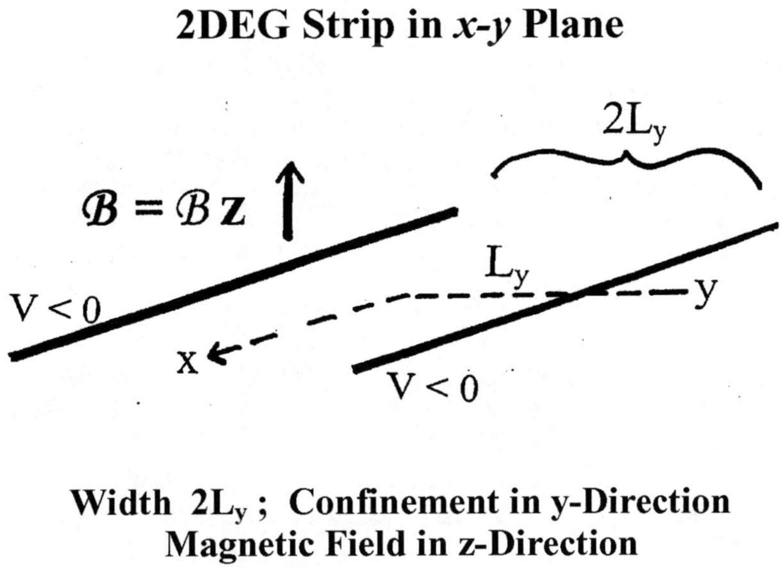

FIGURE 1. Schematic diagram of 2DEG strip of width $2L_y$ in a perpendicular magnetic field, bounded laterally by two arms of a split gate at voltage V< 0.

With the Landau gauge choice of $\mathbf{A} = (-\mathcal{B}\,y, 0, 0)$ for the vector potential and $U(y) = \gamma^2 m y^2/2$ as the confining parabolic potential we have the one-electron Hamiltonian

$$H = [(p_x - \mathcal{B} y\, |e|\,/c)^2/2m] \;+\; [p_y^2/2m] \;+ [\gamma^2 m y^2/2] \;, \tag{1}$$

where e = -|e| is the electronic charge, m is the electron mass, and γ^2 measures the strength of the attractive transverse parabolic potential from the confining split-gate electrode. For periodic boundary conditions over a length L_x in the x direction, the eigenfunctions of this Hamiltonian are known to be [11]

$$\Psi_{n,k}(x,y) \;=\; (e^{ikx} / \sqrt{L_x})\, u_n(y - Y) \;, \tag{2}$$

where $u_n(y)$ is a normalized harmonic oscillator wave function for level number n, $Y = Y(k) = \hbar k \Omega/\omega^2 m$, $\Omega = |e|\mathcal{B}/mc$ is the cyclotron frequency, and $\omega^2 = \Omega^2 + \gamma^2$. Because $- L_y \leq Y \leq L_y$ within the strip, the quantum number k is restricted to the range $-k_{max} \leq k \leq k_{max} = \omega^2 m L_y / \hbar\, \Omega$, with $2N_0$ electrons in the area $2L_xL_y$.

The energy eigenvalues to which the eigenfunctions given by Eq. (2) correspond are [11]

$$\varepsilon_{n\,k} \;=\; \hbar\omega\, (n + ½) + (\hbar^2 k^2 / 2m)\, (1 - \Omega^2 / \omega^2). \tag{3}$$

The emergence of paramagnetism with the presence of an attractive transverse parabolic potential may be suggested by considering the way energy levels change with magnetic field. When $\gamma^2 = 0$ and $\mathcal{B} > 0$, the usual Landau levels are obtained, and for the simplest case with n = 0 one has for these lowest levels $\varepsilon_{0k} = \hbar\Omega / 2$, so that $\partial\varepsilon_{0\,k} / \partial\mathcal{B} > 0$ results in diamagnetism, since the energy levels rise above their zero field values, so that the magnetization $\mathcal{M}$ is negative: $\mathcal{M} \sim -\partial\varepsilon_{0k}/\partial\mathcal{B} < 0$. But for $\gamma^2 > 0$ and $\mathcal{B} > 0$, one finds from Eq. (3) that whenever k is bounded according to $(m/2\hbar\gamma^2)\omega^3 < k^2 < k_{max}^2$ we have $\partial\varepsilon_{0k}/\partial\mathcal{B} < 0$, so that $\mathcal{M} \sim -\partial\varepsilon_{0k}/\partial\mathcal{B} > 0$, and contributions towards paramagnetism occur. By summing up the contributions of all occupied states by integration the net magnetism of the system is then obtained. Since the system has symmetry about the y=0 center line we shall consider the magnetization only within the region $0 \le y \le L_y$ with area L_xL_y containing N_0 electrons. The allowed separation of k values is $\delta k = 2\pi/L_x$, so that the total number of states including spin in any Landau band n associated with the area L_xL_y is $2k_{max}/\delta k = m\omega^2 (L_xL_y)/\pi\hbar\Omega \equiv D$. It is instructive to write D as

$$D = 2\,[\,L_xL_y\,\mathcal{B}/\varphi_0][\,1 + (\hbar\gamma/2\mu\mathcal{B})^2]\,, \tag{4}$$

where the magnetic flux quantum is $\varphi_0 \equiv hc/|e|$ and $\mu \equiv |e|\hbar/2mc$ is the effective magneton. We note that D is greater than the total number of flux lines threading the area L_xL_y by a factor which depends on the strength of the confining parabolic potential relative to the electron spin Zeeman energy in the magnetic field.

The spread of energy eigenvalues, Q, constituting the Landau band width for a given quantum number n is given by

$$(\varepsilon_{n,kmax} - \varepsilon_{n,0}) = m\omega^2\gamma^2L_y^2/2\Omega^2 \equiv Q. \tag{5}$$

The density of states within a Landau band n, $\rho_n(\varepsilon) = [2/\delta k]/[\partial\varepsilon_{nk}/\partial k]$, is then calculated to be

$$\rho(\varepsilon) = [D/2Q^{1/2}]\,[\,\varepsilon - \hbar\omega/2\,]^{-1/2} \tag{6}$$

for n = 0, as illustrated in Fig. 2.

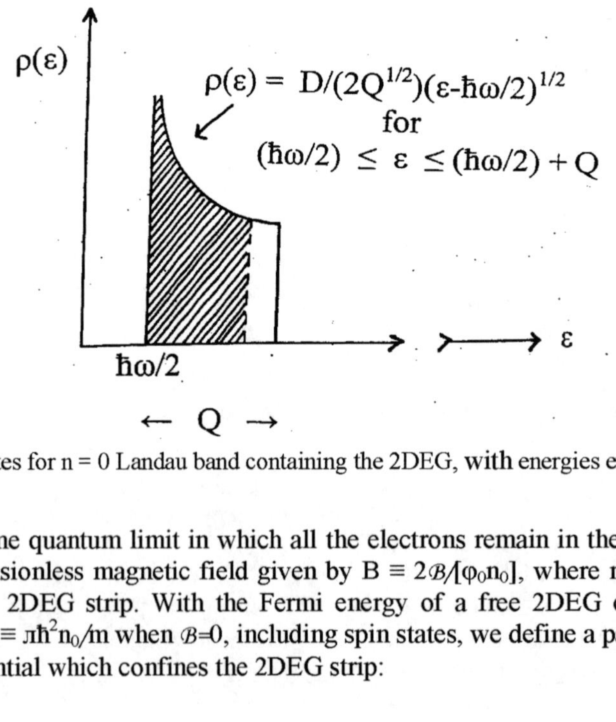

FIGURE 2. Density of states for n = 0 Landau band containing the 2DEG, with energies extending up to [ħω/2] + Q.

We shall explore the extreme quantum limit in which all the electrons remain in the n = 0 band at T = 0 K. It is convenient to define a dimensionless magnetic field given by $B \equiv 2\mathcal{B}/[\varphi_0 n_0]$, where $n_0 \equiv N_0/[L_xL_y]$ is the average electron areal density in the 2DEG strip. With the Fermi energy of a free 2DEG of areal density n_0 and with effective mass m given by $E_F \equiv \pi\hbar^2 n_0/m$ when $\mathcal{B}$=0, including spin states, we define a parameter K that expresses the strength of the parabolic potential which confines the 2DEG strip:

$$K \equiv (\,\hbar\gamma/E_F)^2\,. \tag{7}$$

With these definitions of dimensionless B and K, the number of states D in a Landau band may be written as

$$D = B\, N_0 \,[\, 1 + K/B^2] \;, \tag{8}$$

which illustrates directly how low values of dimensionless field B may nevertheless result in large numbers of states in a Landau band provided a confining parabolic potential energy is present.

We can readily establish an approximate relationship that connects the dimensionless parameter K with the negative statvoltage $V_{ST} < 0$ applied to the arms of the split gate electrode that confines the 2DEG strip. If we regard the two arms of the split gate as sufficiently long, then their potentials separately satisfy LaPlace's equation in cylindrical coordiates in the region $|y| < L_y$ between them. Accordingly, by superposition, the net potential from the two arms, for points at $|y| < L_y$ between them, may be written in the form

$$\xi(y) \;=\; A\,[\,\log\,(L_y - |y|\,) + \log\,(L_y + |y|\,)\,]\,, \tag{9}$$

where A is a constant to be determined. To order y^2, we may write $\xi(y) \approx A\log L_y^{\,2} - Ay^2/L_y^{\,2}$, and, dropping the constant term while taking $A = -V_{ST}$, we have $\xi(y) \approx V_{ST}\,(y^2/L_y^{\,2})$ for $|y| < L_y$. Note that this approximation for the potential $\xi(y)$ takes on the split gate negative statvoltage $V_{ST} < 0$ applied at $y=\pm L_y$. Comparison of this approximate potential $\xi(y)$ with the parabolic potential energy in Eq. (1) shows that $\gamma^2 \approx -2|e|V/[300mL_y^{\,2}]$, where $V<0$ is the negative mks voltage on the split gate electrode. From the expression for E_F and that of K in Eq. (7) we also obtain $\gamma^2 = \pi^2\hbar^2 n_0 K/m$. Setting these two expressions for γ^2 equal to each other we obtain the following approximate relation between V and K:

$$-V = [300\,\pi^2\hbar^2 n_0^{\,2} L_y^{\,2}]\;K/\,[2|e|m]\ \text{volts.} \tag{10}$$

The Fermi level of the electronic system at T = 0 K, $\eta(0)$, when all electrons are in the lowest Landau band may be calculated from

$$\int_{\hbar\omega/2}^{\eta(0)} \rho(\varepsilon)\, d\varepsilon \;= N_0 \;, \tag{11}$$

where $\rho(\varepsilon)$ is given by Eq. (6) incorporating the definitions of D and Q. The integration in Eq. (11) is readily performed and the result is solved for the Fermi level in the upper limit:

$$\eta(0) = [\hbar\omega/2] \;+ 4Q[N_0/D]^2 \,. \tag{12}$$

At T=0 K, the Free Energy of the system is identical to the Energy E(0) given by

$$E(0) \;=\; \int_{\hbar\omega/2}^{\eta(0)} \varepsilon\, \rho(\varepsilon)\, d\varepsilon \;\;. \tag{13}$$

Again using Eq.(6) for $\rho(\varepsilon)$ the integration in Eq.(13) is performed to obtain the following result:

$$E(0) = [D/6Q^{1/2}]\,[\eta(0) - \hbar\omega/2]^{1/2}\,[\eta(0) + \hbar\omega] \;, \tag{14}$$

where $\eta(0)$ is given by Eq.(12) and all other quantities have been previously defined.

The magnetization $\mathcal{M}$ of the system is given by the derivative of E(0) with respect to magnetic field $\mathcal{B}$, times (-1):

$$-\mathcal{M} = \partial E(0)/\partial \mathcal{B} \ . \tag{15}$$

By carrying out the differentiation in Eq. (15), noting the dependence on $\mathcal{B}$ of all quantities entering E(0) in Eq. (14) we find that subsequent algebraic manipulation gives the result:

$$\mathcal{M}(\mathcal{B}) = \mu N_0 M(B, K) \ , \tag{16}$$

where M(B,K) is a function of dimensionless field B given by:

$$M(B,K) = \{ [16/3] [\pi L_y^2(n_0/2)KB^{-3}[1+K/B^2]^{-2} - [1+K/B^2]^{-1/2} \} , \tag{17}$$

which is the average orbital magnetization per particle in units of effective Bohr magnetons. From Eq. (17) it is apparent that in the absence of transverse parabolic potential confinement, when K=0, the value of M(B,0) reduces to (-1) which is the known value of the diamagnetic moment per electron in an unbounded 2DEG at zero temperature.

The function M(B,K) depends sensitively on L_y, n_0, and K. We have plotted graphs of M(B,K) versus both dimensionless field B and actual field $\mathcal{B}$ in Tesla, using the relationship $B = [2\mathcal{B}/\varphi_0 n_0] \times 10^4$. In Fig. 3, M(B,K) is plotted vs. B for a strip width of 100 nm and specified values of areal density and confining split gate voltage. In Fig. 4, M(B,K) is plotted vs. $\mathcal{B}$ in Tesla for the same specified values used in Fig. 3. In Fig. 5, M(B,K) is plotted vs. B for a strip width of 10 nm and different specified values of areal density and confining split gate voltage. In Fig. 6, M(B,K) is plotted vs. $\mathcal{B}$ in Tesla for the same parameter values employed for Fig. 5. In all four examples the values of the average magnetization per electron can vastly exceed the magnitude of the Landau diamagnetic moment for an unbounded 2DEG, by almost two orders of magnitude. And the magnetic fields required are well within achievable laboratory values.

POSSIBLE NEW MATERIAL

It may be possible to construct an heterostructure which is strongly paramagnetic based on the orbital paramagnetism of confined 2DEG strips discussed above. Illustration in Fig. 7 suggests how this may be done. A set of confined 2DEG strips arranged in a plane parallel to each other and uniformly separated by a distance a is stacked together with similar parallel planes containing confined 2DEG strips that are located periodically with separation l. The volume magnetization density is then given by $|\mathbf{m}| = (f/l)\, n_o \mu M(B,K)$, where $f=(2L_y/a)$ is the fraction of a plane covered by 2DEG strips.

ACKNOWLEDGMENTS

I would like to thank Professor Walter Kohn for calling my attention to the 1953 paper of Frank S.Ham , and Dr. George J. Perkins for invaluable help with graphical software.

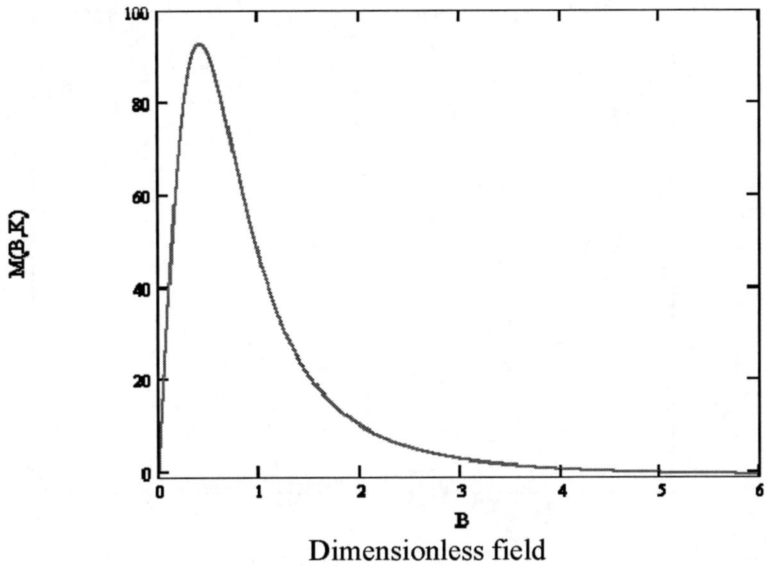

FIGURE 3. Magnetization per particle in units of effective Bohr magnetons as a function of dimensionless field for a strip of width 100 nm, areal density 10^{12} cm^{-2}, and K = 0.531, corresponding to –V = 50 mV.

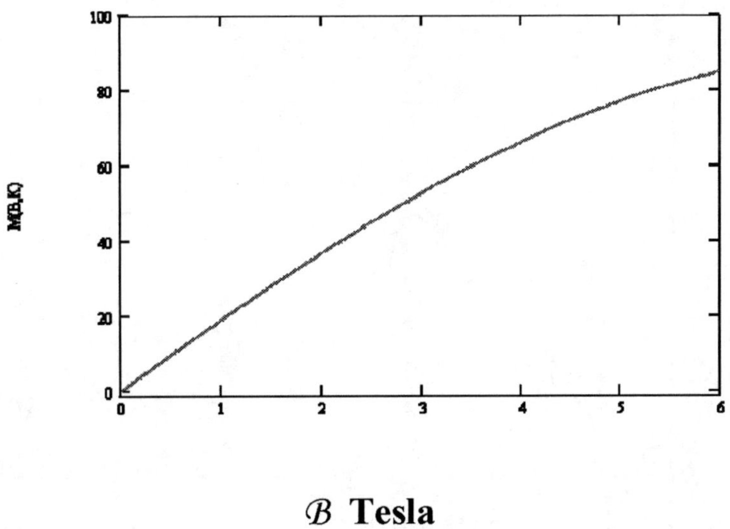

FIGURE 4. Magnetization per particle in units of effective Bohr magnetons as a function of field in Tesla, for a strip of width 100 nm, areal density 10^{12} cm^{-2}, and K = 0.531, corresponding to –V = 50 mV.

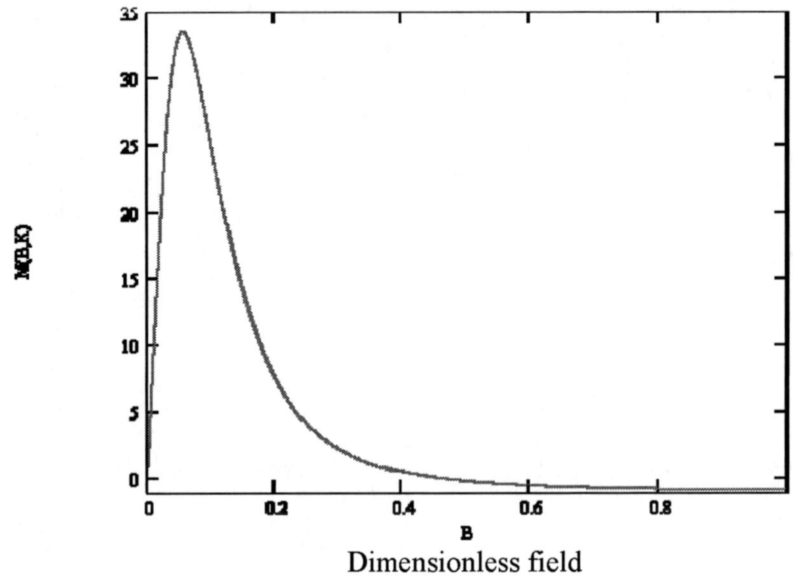

FIGURE 5. Magnetization per particle in units of effective Bohr magnetons as a function of dimensionless field for a strip of width 10 nm, areal density 5.0×10^{12} cm^{-2}, and K = 0.01, corresponding to –V=0.24 mV.

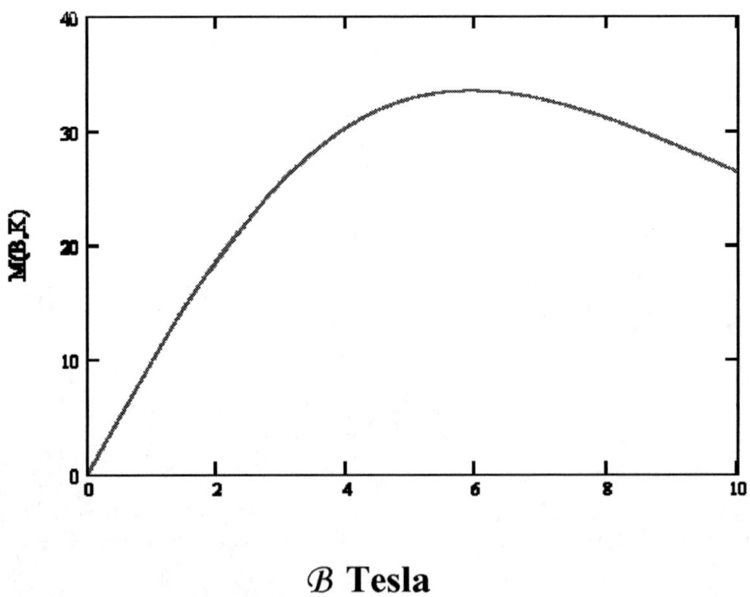

FIGURE 6. Magnetization per particle in units of effective Bohr magnetons as a function of field in Tesla, for a strip of width 10 nm, areal density 5.0×10^{12} cm^{-2}, and K = 0.01, corresponding to –V=0.24 mV.

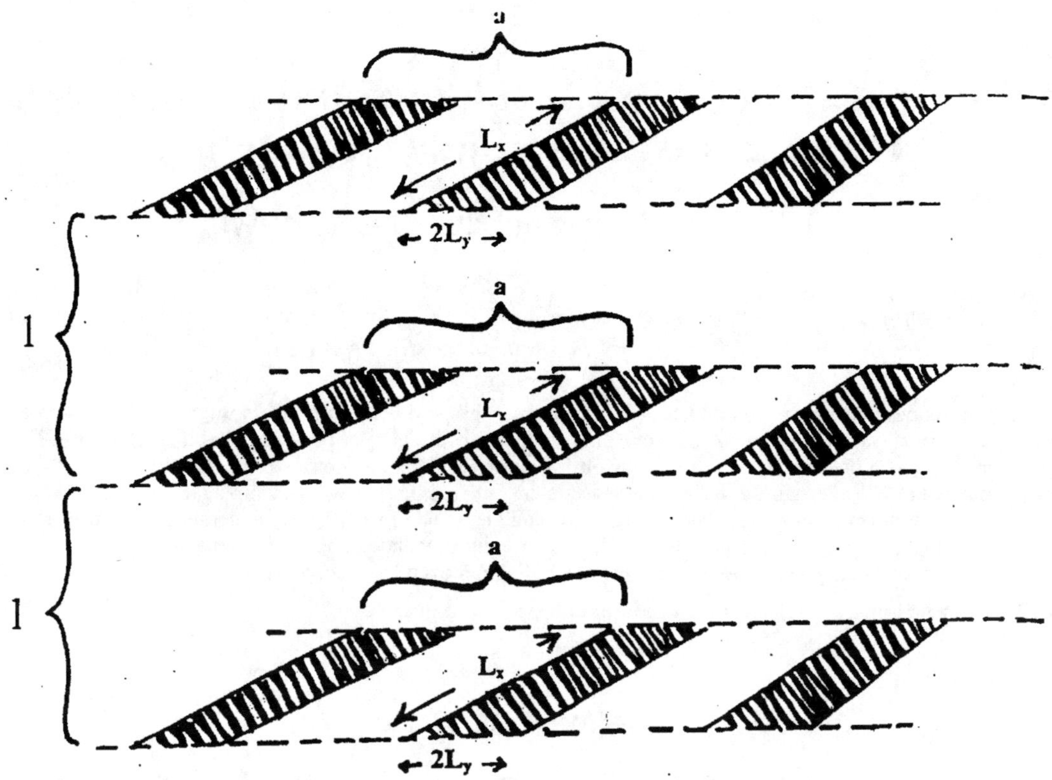

FIGURE 7. Possible magnetic material consisting of regular plane arrays of parallel confined 2DEG strips stacked parallel to each other with periodic separation between planes.

REFERENCES

1. F. S. Ham, Phys. Rev. **92** (5), 1113 (1953).
2. R. V. Denton, Z. Physik **265**, 119 (1973).
3. F. Meier and P. Wyder, Phys. Rev. Lett. **30** (5), 181 (1973).
4. M. Robnik, J. Phys. A: Math Gen. **19**, 3619 (1986).
5. J. M. van Ruitenbeek, Z. Phys. D **19**, 247 (1991).
6. F. von Oppen, Phys. Rev. B **50** (23), 17151 (1994).
7. D. Ullmo, K. Richter, and R. A. Jalabert, Phys. Rev. Lett **74** (3), 383 (1995).
8. J-X Zhu and Z. D. Wang, Phys. Lett. A **203**, 144 (1995).
9. L. P. Levy, D. H. Reich, and L. Pfeiffer & K. West, Physica B **189**, 204 (1993).
10. T. J. Thornton, M. Pepper, H. Ahmed, and D. Andrews, & G.J. Davies, Phys. Rev. Lett. **56** (11), 1198 (1986).
11. M. J. Harrison, Phys. Rev. B, **45** (7), 3815 (1992).

Instabilities of Doped Mott Insulators and the Properties of High-Temperature Superconductors

Mike Guidry*, Yang Sun† and Cheng-Li Wu**

*Department of Physics and Astronomy, University of Tennessee, Knoxville, TN 37996-1200 USA
†Department of Physics, Shanghai Jiao Tong University, Shanghai 200240, People's Republic of China
**Physics Department, Chung Yuan Christian University, Chung-Li, Taiwan 320, ROC

Abstract. Copper oxides become superconductors rapidly upon doping with electron holes, suggesting a fundamental pair-condensation instability. The Cooper mechanism explains normal superconductivity as an instability of a fermi-liquid state, but high-temperature superconductors derive from a Mott-insulator normal state, not a fermi liquid. We show that competing antiferromagnetism and *d*-wave pairing on a singly-occupied lattice implies two fundamental instabilities having large implications for superconducting and pseudogap states in the cuprates. The first instability generalizes the Cooper mechanism to doped Mott insulators; the second instability implies that for intermediate ranges of doping the ground states become exquisitely sensitive to external perturbations, permitting rich disorder even for a universal phase diagram.

Keywords: Superconductivity, antiferromagnetism, Cooper pairs, doped Mott insulators
PACS: 71.10.-w, 74.20.-z

1. INTRODUCTION

High-temperature superconductivity (HTSC) was discovered two decades ago [1], but its theoretical interpretation remains controversial [2]. We believe that this is primarily because of two fundamental, and related, issues: (1) Most models emphasize a limited aspect of the (complex) problem, so few solvable models capture a sufficient range of essential physics. (2) The complex behavior of these compounds obscures the superconducting mechanism because the absence of solvable models incorporating a wide enough range of physics makes it difficult to separate essential features from secondary ones in interpreting data. Despite the lack of general agreement on the mechanism for high-temperature superconductivity, we believe that there *is* broad agreement on six issues:

1. High-temperature superconductivity is related intimately to an interplay among *d*-wave singlet pairing, antiferromagnetism, and charge degrees of freedom for strongly-correlated electrons.
2. The superconducting state is a condensate of (*d*-wave) Cooper pairs having many of the properties of an ordinary BCS (Bardeen–Cooper–Schrieffer) superconductor.
3. The precursor to the superconducting state is an antiferromagnetic Mott insulator (insulating antiferromagnetic state in which the insulator properties derive specifically from strong on-site Coulomb repulsion), unlike for ordinary superconductors where the precursor is a normal fermi liquid (an interacting system of fermions having excitations that can be put into one-to-one correspondence with those of a non-interacting fermion system).
4. The pseudogap state, corresponding to an observed partial gapping of states in the underdoped region above the superconducting transition temperature, is not a "normal" normal state, and is not well described by traditional fermi liquid and BCS concepts. Understanding the pseudogap state is necessary to understanding HTSC.
5. The rapid onset of HTSC with hole doping suggests that the Mott insulator already contains within it a hidden propensity to superconductivity, though no adequate theoretical explanation has been given for this.
6. There is evidence for a rather universal cuprate phase diagram (at least for hole-doped compounds), but there is at the same time strong evidence for a broad variety of disorder in these same compounds.

In this paper we use dynamical symmetries implemented in terms of Lie algebras and generalized coherent state to address all of these issues simultaneously. Specifically, we demonstrate in a solvable model motivated by the phenomenology that the ground state of a minimal implementation of competing *d*-wave pairing, antiferromagnetism, charge, and spin is an antiferromagnetic Mott insulator that is fundamentally unstable with respect to condensing electron hole pairs (and thus becoming a superconductor) at any finite hole doping. Furthermore, the same solution

CP995, *Nuclei and Mesoscopic Physics - WNMP 2007*, edited by P. Danielewicz, P. Piecuch, and V. Zelevinsky

implies pseudogap states having many properties that are in quantitative agreement with data. Finally, we shall show that this classical ground state is unique at half filling and at larger doping, but becomes highly degenerate in the underdoped region, leading naturally to a system that is extremely sensitive to lattice perturbations and thus to a variety of induced inhomogeneity and other emergent behavior.

1.1. Parent States for Superconductors

Band theory suggests that cuprates at half lattice filling should be metals, but they are in fact insulators with antiferromagnetic (AF) properties. This behavior is thought to result from a Mott-insulator normal state, where the insulator properties follow specifically from strong on-site Coulomb repulsion. Doping the normal states with electron holes produces a rapid transition to a superconducting (SC) state, with a pairing gap typically appearing for about 5% hole density per copper site in the copper–oxygen plane. Furthermore, at lower doping a partial energy gap appears at temperatures above the SC transition temperature T_c that is termed a pseudogap (PG), with the SC gap and PG having opposite doping dependence at low doping [3].

Parent states of normal superconductors are fermi liquids and normal superconductors are described by Bardeen–Cooper–Schrieffer (BCS) theory [4], which assumes the condensation of zero-spin, zero-momentum fermion pairs into a new collective state with long-range coherence of the wavefunction. The key to understanding normal superconductivity was the demonstration by Cooper [5] that normal fermi liquids possess a fundamental instability: an electron pair above a filled fermi sea can form a bound state for *vanishingly small attractive interaction.* In normal superconductors the attraction is provided by interactions with lattice phonons, which bind weakly over a limited frequency range because electrons and the lattice have different response times. However, it is the Cooper instability, not the source of the attractive interaction, that is most fundamental: a weak electron–electron interaction alone cannot produce a superconducting state, but the Cooper instability can (in principle) produce a superconducting state for *any* weakly-attractive interaction.

1.2. The Rapid Onset of Superconductivity

The rapid onset of superconductivity with hole doping in high-T_c compounds suggests a fundamental instability with respect to pair condensation, but this (and the origin of PG states) is difficult to interpret within the BCS framework since the superconductor appears to derive from a Mott insulator, not a fermi liquid. As for normal superconductivity, we believe that the key to understanding HTSC is not the attractive interaction leading to pair binding (as important as that is), but rather the nature of the instability that produces the superconducting state. Since at larger doping the high-temperature superconducting state exhibits many properties of a normal (d-wave) BCS superconductor, this instability must reduce to the Cooper instability at larger doping, but evolve into something more complex at lower doping where the normal state approaches a Mott insulator and a pseudogap exists above the SC transition temperature.

To account for rapid onset of superconductivity with hole-doping, Laughlin [6] (see also [7, 8]) proposed a modified Hamiltonian with an attractive term that partially overcomes the on-site repulsion. Then the insulator at half filling is actually a "thin, ghostly superconductor", which fails to superconduct only because its long-range order is disrupted at very low doping, ostensibly by fluctuations due to low superfluid density. This proposed new state is termed a *gossamer superconductor.* It might provide a justification for the resonating valence bond (RVB) state [9], which assumes implicitly that quantum antiferromagnets should exhibit superconductivity, contrary to the observation that cuprate ground states at half-filling appear to be best described as a spin-density-wave vacuum with long-range AF order and no superconductivity [10]. In the RVB model it is usually assumed that the long-range Néel order of the ground state at exactly half filling is replaced quickly by the RVB spin-liquid ground state upon hole-doping, with details lacking. Laughlin [6] contends that the real issue for validity of the RVB picture is not whether all quantum antiferromagnets are secretly superconductors, but whether some are. The gossamer superconductor is then proposed as a second kind of antiferromagnetism—distinguished by a small background superfluid density—that is the true normal state in the cuprates, and is the harbinger of a spin-liquid RVB ground state for low hole-doping.

The gossamer state is created by hand: a strong attractive term is added to the Hamiltonian, which justifies modifying Gutzwiller projectors such that they only partially suppress double occupancy [7]. We shall show that a model implementing competition of d-wave pairing with antiferromagnetic correlations on a lattice with no double occupancy has Mott insulator properties at half filling, but is unstable toward developing a singlet pairing gap under infinitesimal hole-doping. Thus, many features motivating the idea of gossamer superconductivity are natural consequences of AF and SC competition on a lattice having strict no double occupancy at half filling. We shall argue further that a pseudogap with correct properties is a natural consequence of the same theory, thus accounting for both the precocious onset of a pairing gap and the appearance of pseudogaps at low doping in the cuprates. Finally, we shall discuss the implications of these results for gossamer superconductivity and for the RVB model.

1.3. Spatial Inhomogeneity

High-temperature superconductors exhibit a variety of spatial inhomogeneities such as stripes or checkerboards, particularly in the hole-underdoped region and near magnetic vortex cores [11, 12, 13, 14, 15]. The relationship of this inhomogeneity to the unusual properties of these systems is unsettled. Does it favor superconductivity, does it oppose superconductivity, or is it a sideshow? Particularly vexing is that this rich variety of inhomogeneity seems superficially to be inconsistent with evidence for a universal cuprate phase diagram (at least for hole-doped compounds). How does one reconcile a seemingly universal phase diagram with a bewildering array of inhomogeneity for individual compounds?

Dopant atoms support SC globally by enhancing charge carrier density, but may suppress SC locally through atomic-scale disorder. McElroy et al [16] found strong disorder in atomically-resolved scanning tunneling microscope images of the SC gap for Bi-2212. They concluded that this disorder derives primarily from dopant impurities and that charge variation between nanoregions is small, implying inhomogeneity that is tied to impurities but that need not couple strongly to charge. We shall show that such inhomogeneities are a generic consequence of perturbations on the AF and SC correlations, largely independent of specifics and not necessarily coupled to charge variation. Further, we show that these properties are consistent with a global cuprate phase diagram, are tied intimately to the nature of pseudogap states, and imply a linkage among pseudogaps, inhomogeneity, and emergent behavior. Thus we shall propose a testable hypothesis for separating primary from derivative features in the rich high-T_c data set.

1.4. Competing Order and Preformed Pairs

Considerable attention has been paid to the issue of pseudogap (PG) states for high-temperature superconductors and an explanation of the PG states may be central to understanding the origin of the superconductivity. Two general ideas have dominated discussions of the pseudogap. The first is *preformed pairs*, which assumes that the PG is associated with the formation of (singlet, d-wave) Cooper pairs at a pseudogap temperature T^*, but they only condense into a state of long-range order (a superconductor) at a lower temperature T_c. The second contends that the properties of the pseudogap originate in *competing order*, where some other form of order suppresses superconductivity for $T_c < T < T^*$. The order competing with superconductivity is often assumed to be antiferromagnetism or charge degrees of freedom, and competing-order approaches often assume a quantum phase transition between states dominated by the two forms of order, with quantum-critical scaling behavior that is assumed to account in some way for many HTSC features.

Preformed pairs and competing order have generally been viewed as mutually exclusive alternatives, and a large number of experiments supporting one or the other view have been published. The contradictory nature of these results raises the issue of whether this either/or characterization may be too restrictive and whether the preformed pair and competing order pictures need be mutually exclusive. This question is related to another one of fundamental importance. The competing order picture, at least at a mean-field level, suggests that the pseudogap is a phase, but a concerted effort has failed to establish conclusively any order parameters characterizing the PG states. Thus the prevailing view is that the transition to the pseudogap state is not a true phase transition. Yet there is much evidence for competing order in the cuprates and various mean-field pictures account for many cuprate properties by treating the PG state as a phase with a finite value for some order parameter. Thus, another question of vital importance for understanding high-temperature superconductivity is whether competing order is consistent with the absence of a well-defined phase associated with the pseudogap. In this paper we shall claim that the instability responsible for rich susceptibility to inhomogeneity in the PG states suggests that these states are uniquely susceptible to quantum fluctuations in the pairing and particularly antiferromagnetic degrees of freedom, and that this property can reconcile the successes of a mean-field approach to the pseudogap with the absence of well-defined order parameters associated with the pseudogap.

2. GENERALIZED SU(4) COHERENT STATES

Let us now introduce a formalism capable of dealing with the foregoing issues, by virtue of being complex enough to incorporate the essential physics, and yet amenable to systematic solutions that may be compared with observations.

2.1. A Complete Set of Operators

We propose to solve for the doping and temperature dependence of observables in a theory that incorporates on an equal footing d-wave superconductivity and antiferromagnetism. To do so, we shall employ the tools of Lie algebras, Lie groups, and generalized coherent states [17, 18, 19, 20, 21, 22]. To construct a Hamiltonian embodying these degrees of freedom and expected conservation laws for charge and spin in the many-body wavefunction, we require at a minimum three staggered magnetization operators $\vec{\mathscr{Q}}$ to describe AF, creation and annihilation operators $D^\dagger$ and D for d-wave singlet pairs and a charge operator M to describe superconductivity, and three spin operators $\vec{S}$ to impose spin conservation. However, this set of 9 operators is physically incomplete since scattering of singlet pairs (antiparallel spins on adjacent sites) from the AF particle–hole degrees of freedom can produce triplet pairs (parallel spins on adjacent sites), which are not part of the operator set. The mathematical statement of this incompleteness is that the operator set $\{\vec{\mathscr{Q}}, D^\dagger, D, M, \vec{S}\}$ does not close a Lie algebra under commutation. As demonstrated in Refs. [17, 18, 20], a complete operator set results if we add to these operators the six triplet pair operators $\vec{\pi}^\dagger$ and $\vec{\pi}$. This set of 15 operators $\{\vec{\mathscr{Q}}, D^\dagger, D, \vec{\pi}^\dagger, \vec{\pi}, M, \vec{S}\}$ closes the Lie algebra SU(4). The explicit forms for these operators in both momentum and coordinate space, and the corresponding SU(4) commutation algebra, may be found in Refs. [17, 18, 20].

2.2. SU(4) Symmetry and No Double Occupancy

A critical feature of this symmetry structure is that the SU(4) algebra closes only if the 2-dimensional lattice on which the generators are defined has no doubly-occupied sites [19]. Thus, the SU(4) algebra embodies the minimal theory that describes AF and d-wave SC competition through a many-body wavefunction that conserves charge and spin, and that has no components corresponding to double site occupancy on the lattice. The Hamiltonian restricted to one-body and two-body terms is unique, with the general form

$$H = H_0 - \tilde{G}_0\left[(1-\sigma)D^\dagger D + \sigma \vec{Q}\cdot\vec{Q}\right] + g'\vec{S}\cdot\vec{S}, \tag{1}$$

where H_0, $\tilde{G}_0$ and g' are constants, $D^\dagger$ creates d-wave singlet pairs, $\vec{Q}$ is the staggered magnetization, $\vec{S}$ is spin, and $\sigma = \sigma(x) = \chi(x)/(\chi(x) + G_0(x))$, where $G_0(x)$ and $\chi(x)$ are the effective SC and AF coupling strengths. The doping is characterized by a parameter $x = 1 - n/\Omega$ for an n-electron system, with Ω the maximum number of doped holes (or doped electrons for electron-doped compounds) that can form coherent pairs, taking the normal state at half filling as the vacuum. Data and SU(4) symmetry suggest that $P \simeq 0.25x$, where P is the standard hole-doping parameter, normalized to the number of copper sites [19]. The zero-temperature ground state corresponds to a superposition of singlet and triplet fermion pairs.

2.3. Symmetry Limits and Coherent States

Three dynamical symmetry limits have exact solutions [17]. The SO(4) limit ($\sigma = 1$) is a collective AF state, the SU(2) limit ($\sigma = 0$) is a collective SC state, and the SO(5) limit ($\sigma = 1/2$) is a critical dynamical symmetry interpolating between the SC and AF limits [17]. For other values of σ an approximate solution can be obtained using generalized coherent states. We characterize AF in these solutions by the staggered magnetization $Q \equiv \langle Q_z \rangle$ or alternatively by β defined through

$$Q = 2\Omega\beta(n/(2\Omega) - \beta^2)^{1/2}, \tag{2}$$

and singlet d-wave pairing through $\Delta = \langle D^\dagger D \rangle^{1/2}$. Figure 1 illustrates that the coherent state solution is capable of giving a quantitative description of both the superconducting and pseudogap transition temperature.

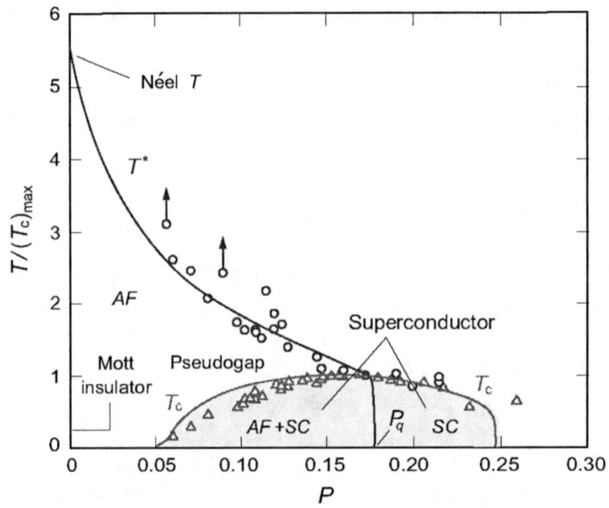

FIGURE 1. SU(4) cuprate phase diagram compared with data. Strengths of the AF and singlet pairing correlations were determined by global fits to cuprate data [20]. The PG temperature is T^* and the SC transition temperature is T_c. The AF correlations vanish, leaving a pure singlet d-wave condensate, above the critical doping P_q. Dominant correlations in each region are indicted by italic labels. Data from Refs. [23, 24, 25, 26, 27, 28, 29] (arrows indicate lower limits).

3. GROUND STATE ENERGY SURFACES

We shall solve for the ground-state properties of this theory using generalized coherent states [22]. Our immediate interest is the ground-state total energy surface, which is the expectation value of the Hamiltonian in the ground coherent state. The formalism for constructing the SU(4) coherent state and the ground state energy surface has been developed extensively in Refs. [17, 18, 20], to which we refer for details. The resulting energy surface is a function of order parameters for singlet pairing and antiferromagnetism, with the hole-doping as a control parameter.

Figure 2 illustrates the SU(4) total energy surface in coherent state approximation as a function of an AF order parameter $Q = \langle \vec{\mathscr{Q}}.\vec{\mathscr{Q}} \rangle^{1/2}/\Omega$ and an SC order parameter $\Delta = \langle D^\dagger D \rangle^{1/2}/\Omega$ as doping $x \simeq 4P$ (for P holes per copper lattice site) is varied. The explicit expression for the energy surface is [18]

$$E = -\chi\Omega^2[(1-x_q^2)\Delta^2 + Q^2] \qquad \Delta \equiv \tfrac{1}{2}[\tfrac{1}{4} - (Q - x/2)^2]^{1/2} + \tfrac{1}{2}[\tfrac{1}{4} - (Q + x/2)^2]^{1/2}, \tag{3}$$

where x_q is the critical doping (see Fig. 2 and Ref. [20]). The vertical lines bounding the curves for different doping in Fig. 2 represent the constraints

$$|Q| \leq \tfrac{1}{2}(1-x) \qquad (x(1-x))^{1/2} \leq 2\Delta \leq (1-x^2)^{1/2}$$

that result from SU(4) symmetry within a finite valence space. Specifically, $|Q|$ must lie between 0 and $n/2$ (where n is electron number) because of the number of spins available, and SU(4) symmetry then relates this constraint on Q to the one on Δ.

3.1. Pairing Instability with Hole Doping

From Fig. 2, the energy surface at half filling ($x = P = 0$) implies a $T = 0$ ground state with AF order but no pairing order ($Q_0 \neq 0$ and $\Delta_0 = 0$, where the subscript zero denotes the value at the minimum of the energy surface). The ground state for $x = 0$ may be interpreted as an antiferromagnetic Mott insulator [17, 18]. From Fig. 2(a), the energy surface retains strong AF character for small x with $Q_0 \neq 0$, but from Fig. 2(b) the ground state differs qualitatively from that at half filling even for infinitesimal hole-doping. Specifically, for any non-zero attractive pairing strength a finite singlet d-wave pairing gap develops spontaneously for *any non-zero x*, and Δ_0 has increased to half its value at optimal doping by $P \simeq 0.03$ ($x \simeq 0.12$). We term this *precocious pairing*.

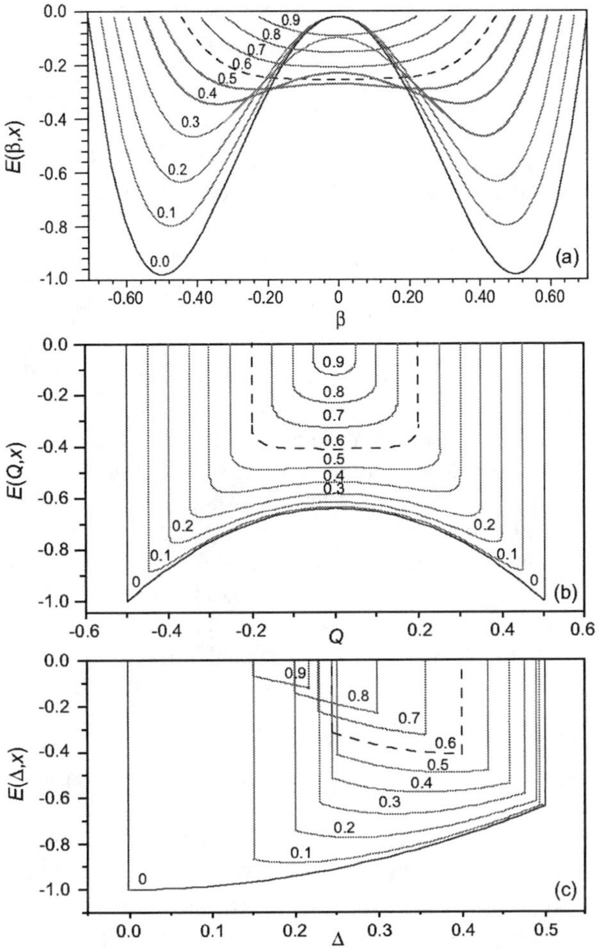

FIGURE 2. Total energy vs. AF order parameters β (top) and Q (middle), and SC order Δ (bottom); curves labeled by hole-doping $x \simeq 4P$, where P is the number of holes per copper site. Energy is in units of $\chi\Omega^2/4$, with χ the AF coupling strength. The black dashed line indicates the critical doping $x = x_q$ (see Ref. [20]); color coding indicates energy surfaces favoring SC (red), AF (blue), and AF + SC (green). The finite, doping-dependent ranges of the energy contours in the order parameters reflect the finite valence space (single band) of the microscopic model.

The instability against condensing pairs displayed graphically in Fig. 2 may also be understood analytically. From the $T = 0$ solution for Δ given in Eq. (24b) of Ref. [20]a, we find

$$\left.\frac{\partial \Delta}{\partial x}\right|_{x=0} = \left.\frac{1}{4}\frac{x_q^{-1} - 2x}{(x(x_q^{-1} - x))^{1/2}}\right|_{x=0} = \infty, \tag{4}$$

displaying explicitly the pairing instability at $x = 0$.

Thus, at half filling the lattice is a Mott insulator with long-range AF order and no pairing gap, but upon infinitesimal hole-doping a finite singlet pairing gap and a ground state representing strong competition between AF and SC order appears. This spontaneous development of a finite singlet pairing gap for infinitesimal hole-doping has been obtained in the coherent-state approximation subject to SU(4) symmetry. Since the SU(4) algebra closes only if the lattice has no double occupation, this precocious pairing has occurred without invoking double occupancy. Weak SU(4) breaking or doping-dependent effective interactions could delay the onset of the instability from $P \sim 0$ to a small doping fraction, as observed. However, we propose that the pair-condensation instability of the SU(4) symmetry limit represents the essential physics governing the emergence of superconductors from doped Mott insulators.

The source of precocious pairing is SU(4) invariance (which ensures no double occupancy for a minimal model of antiferromagnetism, d-wave pairing, charge, and spin). SU(4) symmetry requires that

$$Q^2 + \Delta^2 + \Pi^2 = \tfrac{1}{4}(1 - x^2), \tag{5}$$

where $\Pi = \langle \vec{\pi}^\dagger \vec{\pi} \rangle^{1/2}/\Omega$ is the triplet pair correlation [20]. But for the pure AF SU(4) solution, $Q^2 = \frac{1}{4}(1-x)^2$. Thus, even the AF limit has finite pair gaps unless the hole doping fraction x vanishes identically.

These results have several important implications. We shall discuss three: (1) the interpretation of cuprate data at low hole-doping, (2) the implications for gossamer superconductivity, and (3) the implications for models of the RVB type.

3.1.1. Cuprates at Low Hole Doping

Cuprate data for low doping suggest that normal compounds at half filling are AF Mott insulators, that a finite pairing gap develops by $P \simeq 0.05$, and that a pseudogap develops in the underdoped region having opposite doping dependence than the singlet pairing gap. These results are consistent with a Mott insulator state at half filling that evolves rapidly into a state with a finite singlet pairing gap at very low hole-doping. However, since at low doping both the singlet pairing and AF correlation energies are substantial, the AF fluctuations prevent development of strong superconductivity until near optimal doping, where the zero-temperature AF correlations are completely suppressed at a quantum phase transition. We have demonstrated separately that this same coherent-state SU(4) theory gives a pseudogap having the observed doping behavior, and that the pseudogap may be interpreted either as arising from competing AF and SC degrees of freedom, or fluctuations of pairing subject to SU(4) constraints [20].

3.1.2. The Generalized Cooper Instability and Gossamer Superconductivity

The results presented here suggest that an inherent instability toward condensation of Cooper pairs with hole-doping occurs naturally in a minimal model of d-wave pairing interacting with AF correlations on a lattice with no double occupancy. Thus, the rapid onset of superconductivity with hole-doping in the cuprates results from a Cooper-like instability (instability against condensing pairs for non-zero attractive pairing interaction), but for d-wave pairs in an AF Mott insulator. This solution reduces formally to ordinary d-wave BCS theory if the antiferromagnetic interaction vanishes, and to an antiferromagnetic Mott insulator if the pairing vanishes [20]; thus it represents a minimal self-consistent generalization of the Cooper instability to doped Mott insulators, and demonstrates that the rapid onset of superconductivity upon hole doping a Mott insulator results from the traditional Cooper instability modified by strong on-site Coulomb repulsion and antiferromagnetic correlations. Strongly-attractive electron interactions, as in the gossamer hypothesis, are not precluded; however, they don't seem to be necessary for the pairing instability.

3.1.3. SU(4) and Resonating Valence Bond States

Resonating valence bond models assume implicitly that the half-filled state is in some sense practically a spin liquid, though it looks like an AF state. Our results give independent support for a picture rather similar to this, but without explicit RVB assumptions: the SU(4) ground state at half-filling is an antiferromagnetic Mott insulator, but its wavefunction can reorganize spontaneously into a superconductor when perturbed by a vanishingly-small hole doping in the presence of a vanishingly small pairing interaction. For low hole doping this superconductor is strongly modified by AF correlations. Below T_c this gives a d-wave superconducting state weakened by AF correlations; for a range of temperatures above T_c the pairing gap vanishes but strong AF correlations in a basis of fermion pairs leads to a pseudogap that may be interpreted either in terms of preformed pairs or as competing AF and SC order. Finally, the AF competition weakens with hole doping until the pure superconductor emerges near optimal doping.

The properties of the corresponding SU(4) coherent state at low hole-doping presumably share many features with RVB states. Triplet pairs are essential for a complete set of operators in the minimal SU(4) model (for example, no double occupancy is enforced by the SU(4) Lie algebra, which fails to close without triplet pairing operators), and a mixture of singlet and triplet pairs is essential to describe the AF states at half filling in the highly-truncated SU(4) fermion basis. But the significance of triplet pairs relative to singlet pairs decreases rapidly with doping [20]. Therefore,

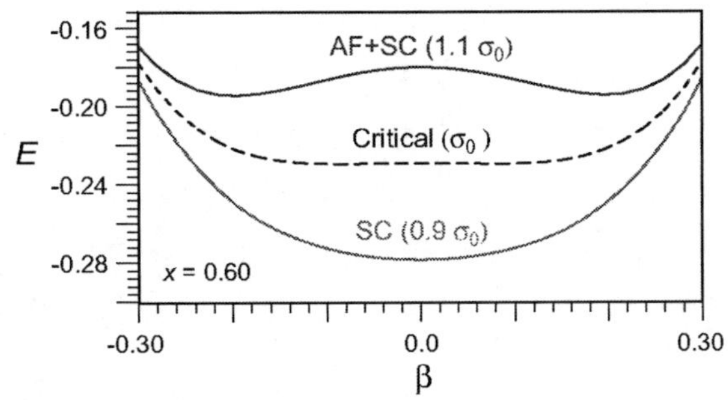

FIGURE 3. Energy surfaces as a function of the AF order parameter β for fixed doping $P = x/4 = 0.15$. The dashed line corresponds to $\sigma = \sigma_0 = 0.6$, the upper curve to a 10% increase in σ, and the lower curve to a 10% reduction in σ.

the underdoped SU(4) ground states could have significant overlap with a singlet spin liquid. Furthermore, as already noted, SU(4) states at low doping lead naturally to a pseudogap that decreases in size with increased hole-doping and exhibits fermi arcs [21], in quantitative accord with data.

The SU(4) coherent state justifies many features of RVB models, but it is richer than typical RVB applications because it accounts even-handedly for both AF and SC on a lattice with no double occupancy. As a consequence, the SU(4) variational wavefunction is more complex than a mere singlet spin liquid. Conversely, the SU(4) coherent-state model is simpler in many respects than RVB models because superconductivity and antiferromagnetism are accounted for quantitatively in a minimal theory having only (dressed) electron degrees of freedom: there are no pair bosons, no gauge fields, and no spinons or holons (which have formal justification in one dimension, but are less obviously justified in higher dimensions, and for which there is little direct evidence in cuprate superconductors). The SU(4) coherent state represents a minimal extension of the BCS formalism to incorporate d-wave pairing in the presence of strong AF correlations and large effective on-site electron repulsion. It requires no Gutzwiller projection because the symmetry enforces no double occupancy on the lattice. It exhibits a type of spin–charge separation (see Ref. [30]), but not through topological spinons and holons: in the fermion basis, charge is carried both by singlet fermion hole pairs having a spin of 0 and charge -2, and triplet fermion hole pairs having a spin of 1 and charge -2, but spin is carried solely by the triplet hole pairs. As we have discussed in References [17, 18, 19, 20, 21], these features permit a model that describes many observed features of the cuprates from half-filling to the overdoped region in a unified manner.

4. CRITICAL DYNAMICAL SYMMETRY AND INHOMOGENEITY

The energy surfaces at constant doping in Fig. 2 fall into three general classes: AF+SC (e.g., $x = 0.1$), SC (e.g., $x = 0.9$), and critical (e.g., $x = 0.6$, which marks a quantum phase transition). Curves in the AF+SC class have minima at finite and large β_0, and small but finite Δ_0, where the subscript zero denotes the value of the order parameter at the minimum of the energy surface. Curves in the class SC are characterized by $\beta_0 = 0$ and finite Δ_0. Of most interest in the present context are the surfaces that are near critical in Fig. 2, which correspond to broken SU(4) $\supset$ SO(5) dynamical symmetry [17] and are rather *flat* over large regions of parameter space. This implies that there are many states lying near the ground state with very different values for β and Δ. Thus the surface is critically balanced between AF and SC order, and small perturbations can drive it strongly from one to the other. This defines a critical dynamical symmetry of the SU(4) algebra [17]; we shall term this situation *dynamical criticality*.

4.1. Dynamical Criticality and Extreme Sensitivity to Perturbations

The extreme sensitivity of critical surfaces to perturbations is illustrated in Fig. 3. Each set of curves is associated with a fixed value of doping $x = 0.6$ (equivalently, $P = 0.15$), with the solid line corresponding to $\sigma = 0.6$, the dashed line to a 10% increase in σ (AF perturbation), and the dotted line to a 10% reduction in σ (SC perturbation). The effect on the energy surface versus Δ (not shown) is significant but less dramatic: Δ_0 is shifted, but remains finite in all three cases. We see that this small fluctuation in σ can alter the energy surface between AF+SC (finite β_0 and Δ_0) and SC ($\beta_0 = 0$ and finite Δ_0). This sensitivity is specific to the critical (broken SO(5)) dynamical symmetry. The AF

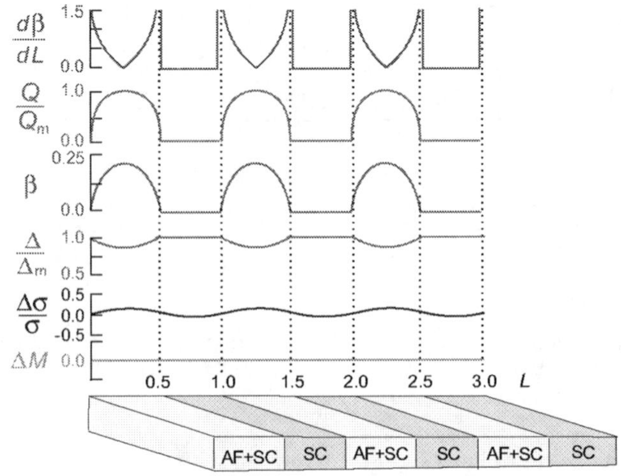

FIGURE 4. A small, spatially-periodic variation in the AF and SC coupling can produce inhomogeneity without significant charge variation. The charge is M and a subscript m denotes maximum values.

region near $x = 0$ and the d-wave superconducting region at larger hole doping (see Fig. 2) are very stable against such perturbations.

The AF instability displayed graphically in Fig. 3 may also be understood analytically. From the $T = 0$ solution for Q given by Eqs. (24b) and (14) of Ref. [20]a, we find

$$\left.\frac{\partial Q}{\partial x}\right|_{x=x_q} = -\frac{1}{4}\left.\frac{x_q + x_q^{-1} - 2x}{[(x_q - x)(x_q^{-1} - x)]^{1/2}}\right|_{x=x_q} = -\infty, \tag{6}$$

and a small change in doping will cause a large change in antiferromagnetic correlations near $x = x_q$. This is a consequence of SU(4) symmetry, which requires that Q vanish for $x \geq x_q$, and be finite for $0 < x < x_q$.

It is not hard to conjecture mechanisms altering the ratio of AF to SC coupling locally. For example, Ref. [16] found that nanoscale disorder is tied to influence of dopant impurities. Nunner et al [31] (see also [32]) compared these results with Boboliubov–de Gennes calculations and proposed that out-of-plane dopant atoms can modulate pairing on a scale comparable to the lattice spacing, through lattice-distortion modification of electron–phonon coupling or superexchange.

4.2. Spatial Inhomogeneity Induced by Background Perturbations

Figure 4 is constructed from the expectation value of (1) in coherent state approximation, assuming a 1-D spatial perturbation, $\sin(2\pi L)$, with $\sigma = 0.6$. It illustrates schematically how a small (10%) periodic fluctuation in the AF and SC coupling for a critically symmetric underdoped compound can lead to inhomogeneity. In this example, 1-dimensional spatial variations of the coupling ratio σ give fluctuations in order parameters leading to stripes in which AF+SC ($\sigma > 0.6$) and SC ($\sigma < 0.6$) are favored alternately. Also shown are the responses of AF fluctuations $d\beta/dL$ to this variation in σ. (We do not intend this as a realistic model of a stripe phase, but as a cartoon indicating how such a model could be built.) Figure 4 indicates that a small spatial variation in the coupling ratio σ can produce regions having large AF and weaker SC order, interspersed with regions having significant SC correlations but no AF order. Although not plotted, the AF+SC regions also exhibit small triplet pair densities, which vanish in the SC regions. The nature of the energy surface dictates that the primary fluctuation between stripes is in the AF order, which can jump between zero and the maximum allowed by the SU(4) model between adjacent stripes; the pairing order changes are much smaller and pairing is finite in both the SC and SC+AF regions. The variation of $d\beta/dL$ indicates that appreciable softness in AF and SC may occur on the boundaries between regions.[1]

4.3. The Role of Charge

The expectation value of the charge M is not a function of the coupling ratio σ in the SU(4) coherent state, so the critical energy surface fluctuations responsible for alternating AF+SC and SC stripes in Fig. 4 cause no charge variation ($\Delta M = 0$). Data indicate that the relative charge variation for Bi-2212 surface nanoscale patches is less than 10%, implying heterogeneity not strongly coupled to charge [16]. This view is supported by analyses of heat capacity and NMR data on Bi-2212 and YBCO that find a universal phase behavior for cuprates, with little static charge modulation [33, 34].

Of course, the variation in σ could itself be due to a charge modulation. From the preceding discussion, we may expect that if a charge modulation occurs in either the AF region near half filling or the SC region at larger doping, its effect will be small because the energy surfaces are not critical there. However, if a charge modulation occurs in the underdoped region where the energy surfaces are near critical its effect could be amplified by dynamical criticality even if σ is not altered significantly, since this is equivalent to a doping modulation. Thus, we suggest a mechanism for producing strong inhomogeneity without necessarily invoking charge fluctuations as the cause, but that can in particular cases for underdoped compounds be produced by a charge modulation. Such a mechanism could make understandable strong inhomogeneity in the face of a universal phase diagram, and resolve competing experimental claims regarding the role of charge variation in producing inhomogeneity.[2]

4.4. Self-Organization Versus Dopant Impurities

Inhomogeneity caused by electronic self-organization is often contrasted with that caused by dopant impurities. Our discussion implicates both as sources of nanoscale structure. The proximate cause may be impurities that perturb the SU(4) energy surface but the criticality of that surface, which greatly amplifies the influence of impurities, results from the self-organizing, doped Mott insulator encoded in the SU(4) algebra. Note a recent analysis [32] suggesting, from a different perspective, that intrinsic amplification of impurity effects is required to explain nanoscale structure in Bi-2212.

4.5. Amplification of Proximity Effects

Giant proximity effects are observed in the cuprates where non-SC copper oxide material sandwiched between superconducting material can carry a supercurrent, even for a thickness much larger than the coherence length [36]. Phenomenology indicates that pre-existing nanoscale SC patches can precipitate such effects [37]. We find similar possibilities on fundamental grounds, but also suggest that the inhomogeneity need not pre-exist. Dynamical criticality renders even a homogeneous pseudogap phase unstable against large fluctuations in the AF and SC order. Thus, proximity of superconducting material to pseudogap material, coupled with perturbations from background impurity fields, can trigger dynamical nucleation of nanoscale structure and giant proximity effects, even if no static inhomogeneity exists beforehand.

Because electronic disorder in cuprate superconductors can exist on a scale comparable to the coherence length (~ 15 Å), a proximity effect may be expected once static nanoscale disorder forms whereby superconducting character leaks between patches. This effect can be greatly enhanced if patches have energy surfaces that are critical since a

---

[1] The minimal patch size that can support SU(4) coherent states is crucial to the present argument. Experience (e.g., Ref. [35]) suggests that dynamical symmetry can be realized in fermion valence spaces having as few as several particles. This would be consistent with inhomogeneities on scales comparable to atomic dimensions, as required by data [16].

[2] As an aside, we note that inhomogeneity without charge modulation has a suggestive analog in astronomy. We might assume, erroneously, that density in the bright arms of a spiral galaxy (a form of "stripe order") is much larger than that between the arms. Actually, spiral arms are prominent because the mass (gravitational charge) there is more visible: galactic dynamics preferentially generates hot, luminous stars in the spiral arms. In analogy, the structure in Fig. 4 is not due to charge variation but rather to modulation of the charge "visibility" by (quantum) dynamics.

small perturbation that flips the energy surface of a patch has a non-perturbative influence far out of proportion to its perturbative size (a small tail wagging a very large dog).

In experiments with one unit cell thickness La_2CuO_4 AF barrier layers between superconducting $La_{1.85}Sr_{0.15}CuO_4$ samples, Bozovic et al [38] found that the two phases did not mix, with the barrier layer completely blocking a supercurrent. These results were interpreted to rule out many models of high-temperature superconductors, in particular models in which SC and AF phases are nearly degenerate like the SO(5) model [39]. The absence of a proximity effect between the AF and SC phases (but presence of a strong proximity effect between pseudogap and SC material) is plausibly consistent with the SU(4) model: the AF phase is not rotated directly into the SC phase but rather evolves with increased doping into a competing SC and AF phase, which then is transformed into a pure SC state at a critical doping point near optimal doping, where the AF correlations vanish identically [20]. (For $T = 0$, this critical doping point marks a quantum phase transition.)

4.6. Dynamical Criticality, Emergence, and Complexity

The spontaneous appearance of properties in a many-body system that do not pre-exist in its elementary components is termed emergence; systems with emergent behavior are said to exhibit complexity (see Dagotto [40] for a review). Complexity results when multiple potential ground states exist and the choice between them is extremely sensitive to weak external perturbations.

The amplification effect implied by SU(4) dynamical criticality can facilitate emergent behavior and associated complexity, as exemplified by giant proximity effects and the perturbatively-induced structure of Fig. 4. More generally, critical dynamical symmetry may represent a fundamental organizing principle for complexity in strongly-correlated fermi systems. For example, critical dynamical symmetries have long been known in nuclear structure physics [41].

4.7. Varied Inhomogeneity but a Universal Phase Diagram

The SU(4) coherent state method that yields the variational energy surfaces discussed here admits quasiparticle solutions that generalize the BCS equations, giving a rich, highly universal phase diagram in agreement with much available data [20], and a quantitative description of fermi arcs [21]. Therefore, the propensity of the pseudogap state to a broad variety of induced inhomogeneity, and a quantitative model of the cuprate phase diagram (including pseudogaps) that exhibits highly universal character, both follow directly from a model that implements AF and d-wave SC competition in a doped Mott insulator. This natural coexistence of a universal phase diagram with a rich susceptibility to disorder could reconcile many seemingly disparate observations in the cuprate superconductors. For example, since the SU(4) pseudogap state has both pairing fluctuations and critical dynamical symmetry, it could support Nernst vortex states as perturbations on a homogeneous phase having incipient nanoscale heterogeneity.

5. THE PSEUDOGAP, MEAN FIELDS, AND FLUCTUATIONS

Finally, let us address whether the pseudogap is explicable within a mean-field context. In particular, can the pseudogap be described in terms of competing order when little experimental evidence supports a distinct phase associated with the PG? The coherent-state approximation is mean-field, in a similar sense as BCS. However, the corresponding mean field is quite rich because it is defined in terms of singlet and triplet fermion (not boson) pairs that are strongly perturbed by antiferromagnetic and pairing correlations. This leads to unusual properties in the resulting solutions. In addition, although the general coherent state solution is mean field, the SU(4) model has exact solutions in three (physically relevant) dynamical symmetry limits that are not mean field. These observations have the following implications:

1. The SU(4) model embodies competing order. However, because of its basis of singlet and triplet fermion pairs the pseudogap state may be interpreted also in terms of preformed pairs [20]: The Hamiltonian implements a competition between antiferromagnetism and pairing interactions, but the eigenstates in the pseudogap region correspond to rich superpositions of fermion pairs.
2. Because the SU(4) coherent state solution is mean field, it does not include the effect of quantum fluctuations on the ground-state wavefunction. However, comparison of the coherent state solution with the three exact SU(4) solutions indicates that in all three cases the coherent state solution gives a ground-state energy that coincides

with the exact solution, but the wavefunction of the coherent state solution provides only the expectation value of the energy, not the quantum fluctuations.[3] This explains why the SU(4) coherent state solution is successful in describing quantities such as energy gaps and the critical temperatures T_c and T^* that are determined primarily by the ground-state energy.

3. The SU(4) energy surfaces provide a guide to the expected quantum fluctuations. The coherent state solutions imply that the ground-state in much of the pseudogap region is inherently unstable against both antiferromagnetic and pairing fluctuations, because many solutions having different combinations of antiferromagnetism and superconductivity become nearly degenerate. This instability (coinciding with broad, flat regions of the energy surface) occurs predominantly in the pseudogap region, implying unique features for the pseudogap state. For our Néel state solution at half filling and the superconducting solution in optimally and overdoped regions the energy surfaces have sharp minima in both the AF and pairing directions, suggesting that these define stable phases with small influence from quantum fluctuations. For the underdoped region below T_c our solutions indicate an energy surface that is soft in the AF direction but localized with respect to pairing, giving a solution that is superconducting, but with the superconductivity modified significantly by AF fluctuations.
4. Thus, although the SU(4) coherent state is a mean field solution, the energy surfaces evaluated in this approximation (and compared with exact many-body solutions in the three symmetry limits) already give a strong guide to the expected role of quantum fluctuations around the mean-field solution. Inclusion of the effect of fluctuations in the PG region should lead to a richer PG state that should not compromise previous successes such as the description of the doping dependence of T^* or the behavior of fermi arcs.

Therefore, an extended theory including fluctuations associated with the second instability described above may be expected to produce a PG state that is more complex than in the mean-field limit. This mean field plus quantum fluctuations pseudogap state would be a superposition of SU(4)states having expectation values near zero for pairing and AF order parameters, but with large fluctuations in both, and a strong susceptibility to external perturbations from the lattice. The transition to this pseudogap state would be expected to exhibit a rapid evolution in properties instead of a clear phase transition. The resulting theory should retain the correct predictions of Fig. 1. However, quantum fluctuations of the pseudogap state, coupled with the predicted extreme sensitivity to perturbations, would set the stage for a complex set of phenomena (Nernst vortex fluctuations, spatial inhomogeneity, giant proximity effects, and other emergent properties), all describable within a framework unifying the competing order and preformed pairs pictures.

6. SUMMARY AND CONCLUSIONS

In summary, we have examined a minimal theory of d-wave superconductivity and antiferromagnetism on a lattice with no double occupancy. The resulting total energy surface implies ground states that are antiferromagnetic Mott insulators at half filling, but are unstable against developing singlet d-wave pairing gaps upon hole-doping, thereby generalizing the Cooper instability to doped Mott insulators. Many properties motivating the idea of gossamer superconductivity are explained, without invoking the gossamer hypothesis. Our results support the assumption of resonating valence bond models that the state with AF order at half filling would really like to be a spin-singlet liquid. However, the SU(4) variational coherent state is simpler to implement, and yet contains a broader range of physics, than a spin-singlet liquid, and accounts for many cuprate properties across the entire physical hole-doping range.

In addition, we have shown that competing antiferromagnetism and d-wave superconductivity, constrained by no double lattice occupancy, leads on general grounds to energy surfaces in hole underdoped cuprates that may be critically balanced between antiferromagnetic and superconducting order. These surfaces can be flipped between dominance of one order or the other by small fluctuations in the ratio of the antiferromagnetic to pairing strength. Therefore weak perturbations in the underdoped region, and near vortex cores or magnetic impurities, can produce amplified inhomogeneity having the spatial dependence of the perturbation but the intrinsic character of an SU(4) symmetry. (The symmetry defines the possible states; the perturbation selects among them.) Our results show that

[3] We may illustrate with two specific examples: (a) The solutions of the coherent state gap equations for a Néel state and for the PG state above T_c are formally equivalent, but (as noted above) they lead to fundamentally different dependence of the corresponding energy surfaces on antiferromagnetism and pairing, implying that the full wavefunction of the two states (including quantum fluctuations) are very different, but that the exact solution still corresponds to a superposition of states described by mean fields having approximate SU(4) symmetry. (b) The solutions for the coherent-state gap equations for the SU(2) superconducting limit and the SO(5) limit are formally equivalent, but again their energy surfaces are qualitatively different, implying very different roles for quantum fluctuations.

such effects could, but need not, imply spatial modulation of charge. More generally, we have suggested that critical dynamical symmetry may be a fundamental organizing principle for emergent behavior in correlated fermion systems.

Thus, by using techniques that are well-known in general many-body physics, we conclude that the properties of high-temperature superconductors, including the rapid development of a superconductor from a Mott insulator, the properties of pseudogap states, and rich disorder within a universal phase diagram, are understandable in terms of a minimal generalization of traditional BCS theory and the Cooper instability to include self-consistently the role of antiferromagnetism and on-site Coulomb repulsion. The generality of our solution implies that any realistic theory deriving superconductivity from a doped Mott insulator should contain features similar to those discussed here.

REFERENCES

1. J. G. Bednorz and K. A. Müller, *Z. Phys.* **B64**, 189 (1986).
2. See D. A. Bonn, Nature Phys. **2**, 159 (2006); M. R. Norman and C. Pépin, *Rep. Prog. Phys.* **66**, 1547 (2003), and Refs. therein.
3. T. Timusk, and B. Statt, *Rep. Prog. Phys.* **62**, 61 (1999); M. Norman, D. Pines, and C. Kollin, Adv. Phys. **54**, 715 (2005).
4. J. Bardeen, L. N. Cooper, and J. R. Schrieffer, *Phys. Rev.* **108**, 1175 (1957).
5. L. N. Cooper, *Phys. Rev.* **104**, 1189 (1956).
6. R. B. Laughlin, arXiv:cond-mat/0209269 (2002).
7. F. C. Zhang, *Phys. Rev. Lett.* **90**, 207002 (2003).
8. P. Coleman, Nature **424**, 625 (2003).
9. P. W. Anderson, *Science* **235**, 1196 (1987).
10. D. Vaknin, et al, *Phys. Rev. Lett.* **58**, 2802 (1987); J. M. Tranquada, et al, *Phys. Rev. Lett.* **60**, 156 (1988); S. Chakravarty, B. I. Halperin, and D. R Nelson, *Phys. Rev. Lett.* **60**,1057 (1988); J. D. Reger and A. P. Young, *Phys. Rev.* **B37**,5978 (1988); E. Dagotto, *Rev. Mod. Phys.* **66**, 763 (1994).
11. S. M. Hayden et al, *Nature* **429**, 531 (2004).
12. J. M. Tranquada et al, *Nature* **429**, 534 (2004).
13. J. E. Hoffman, et al, *Science* **295**, 466 (2002).
14. M. Vershinin, et al, *Science* **303**, 1995 (2004).
15. T. Hanaguri et al, *Nature* **430**, 1001 (2004).
16. K. McElroy, et al, *Science* **309**, 1048 (2005).
17. M. W. Guidry, Rev. Mex. de Fisica **45 Suplemento 2**, 132 (1999); M. W. Guidry, L.-A. Wu, Y. Sun, and C.-L. Wu, *Phys. Rev* **B 63**, 134516 (2001).
18. L.-A. Wu, M. W. Guidry, Y. Sun, and C.-L. Wu, *Phys. Rev* **B 67**, 014515 (2003).
19. M. W. Guidry, Y. Sun, and C.-L. Wu, *Phys. Rev.* **B70**, 184501 (2004).
20. Y. Sun, M. W. Guidry, and C.-L. Wu, *Phys. Rev.* **B73**, 134519 (2006); *Phys. Rev.* **B 75**, 134511 (2007).
21. M. W. Guidry, Y. Sun, and C.-L. Wu, arXiv:0705.0822.
22. W.-M. Zhang, D. H. Feng, and R. Gilmore, *Rev. Mod. Phys.* **62**, 867 (1990).
23. P. Dai, et al, Science **284**, 1344 (1999).
24. M. Oda, et al, Physica **C 281**, 135 (1997).
25. T. Watanabe, et al, Phys. Rev. Lett. **84**, 5848 (2000).
26. J. C. Campuzano, et al, Phys. Rev. Lett. **83**, 3709 (1999).
27. J. Wuyts, V. V. Moshchalkov, and Y. Bruynseraede, Phys. Rev. **B53**, 9418 (1996).
28. M. Takigawa, et al, Phys. Rev. **B43**, 247 (1991).
29. T. Ito, K. Takenaka, and S. Uchida, Phys. Rev. Lett. **70**, 3995 (1993).
30. P. W. Anderson, *Physics Today*, p. 42 (October, 1997).
31. T. S. Nunner, B. M. Anderson, A. Melikyan, and P. J. Hirschfeld, *Phys. Rev. Lett.* **95**, 177003 (2005).
32. A. C. Fang, et al, *Phys. Rev. Lett.* **96**, 017007 (2006).
33. J. W. Loram, J. L. Tallon, and W. Y. Liang, *Phys. Rev.* **B69**, 060502(R) (2004).
34. J. Bobroff, et al, *Phys. Rev. Lett.* **89**, 157002 (2002).
35. C.-L. Wu, D. H. Feng and M. W. Guidry, *Adv. in Nucl. Phys* **21**, 227 (1994).
36. I. Bozovic, et al, *Phys. Rev. Lett.* **93**, 157002 (2004).
37. G. Alvarez, M. Mayr, A. Moreo, and E. Dagotto, *Phys. Rev.* **B71**, 014514 (2005).
38. I. Bozovic, G. Logvenov, M. A. J. Verhoeven, P. Caputo, E. Goldobin, and T. H. Geballe, *Nature* **422**, 873 (2003).
39. E. Demler, A. J. Berlinsky, C. Kallin, G. B. Arnold, and M. R. Beasley, *Phys. Rev. Lett.* **80**, 2917 (1998).
40. E. Dagotto, *Science* **309**, 257 (2005).
41. W.-M. Zhang, D. H. Feng, and J. N. Ginocchio, *Phys. Rev. Lett.* **59**, 2032 (1987); *Phys. Rev.* **C37**, 1281 (1988).

Mesoscopic Fluctuations of the Pairing Gap

S. Åberg*, H. Olofsson* and P. Leboeuf†

*Mathematical Physics, LTH, Lund University, P.O. Box 118, S-221 00 Lund, Sweden
†Laboratoire de Physique Théorique et Modèles Statistiques, CNRS, Bât. 100, Université de Paris-Sud, 91405 Orsay Cedex, France

Abstract. A description of mesoscopic fluctuations of the pairing gap in finite-sized quantum systems based on periodic orbit theory is presented. The size of the fluctuations are found to depend on quite general properties. We distinguish between systems where corresponding classical motion is regular or chaotic, and describe in detail fluctuations of the BCS gap as a function of the size of the system. The theory is applied to different mesoscopic systems: atomic nuclei, metallic grains, and ultracold fermionic gases. We also present a detailed description of pairing gap variation with particle number for nuclei based on a deformed cavity potential.

Keywords: Semiclassical Methods, BCS theory, Order/Chaos, Pairing Gap, Fluctuations.
PACS: 03.65.Sq, 05.45.Mt, 21.10.Dr, 74.20.Fg

INTRODUCTION

Bohr, Mottelson and Pines were first to apply the Bardeen, Cooper, Schrieffer (BCS) theory of superconductivity to finite size systems, namely to describe pairing in atomic nuclei [1]. A consequence of the finite size of the system is the appearance of shell structure. This implies fluctuations of the pairing gap as a parameter, like the particle number, is varied. In this contribution we shall discuss how these fluctuations can be described in a semiclassical theory.

We first discuss pairing in nuclei obtained from the odd-even mass difference. Pairing gaps calculated from different mass models are compared, both with respect to average gaps and to fluctuations. In the next section a semiclassical theory of fluctuations of the BCS pairing gap [2] is presented. This provides analytic expressions of fluctuations of the pairing gap, where the dynamics of the underlying classical system (chaos/order) is an important parameter. The theory is first applied to pairing fluctuations in nuclei, where a detailed comparison with data is performed. We also utilize explicit periodic orbits, taken from a deformed cavity model, to describe the detailed variation of the pairing gap with particle number. Nano-sized metallic grains are also studied, where, due to a lack of experimental data, we compare our theoretical results to other existing numerical results. Interaction strength, external potential, as well as the number of particles can be experimentally tuned in ultracold fermionic gases, and we discuss the size of pairing fluctuations for such systems, as obtained from our theory.

ODD-EVEN MASS DIFFERENCE IN NUCLEI

The systematic difference between the ground-state mass of odd and even nuclei constitutes an important indicator of pairing in nuclei. The pairing gap can be calculated from binding energies, B, utilizing the three–point measure

$$\Delta_3(M) = B(M) - \frac{1}{2}[B(M+1) + B(M-1)], \tag{1}$$

where M is the neutron N or proton Z number. In the presence of other possible interactions, this quantity has been shown to be a good measure of pairing correlations [3] provided M is taken as an odd number. In that case, it is easy to see that there is no contribution from the mean field in Δ_3, while if M is taken as an even number, an extra contribution to Δ_3 from the single-particle levels appears in the extreme single-particle model as $\frac{1}{2}(e_{i+1} - e_i)$, where e_i is the last occupied single-particle level. In Fig. 1 we see the systematic difference in the average of Δ_3 when M is an odd and even number. Restricting to Δ_3-values obtained from odd numbers of M, a fit of the pairing gap gives

$$\bar{\Delta} = \frac{2.7}{A^{1/4}} \text{ MeV}, \tag{2}$$

CP995, *Nuclei and Mesoscopic Physics - WNMP 2007*, edited by P. Danielewicz, P. Piecuch, and V. Zelevinsky

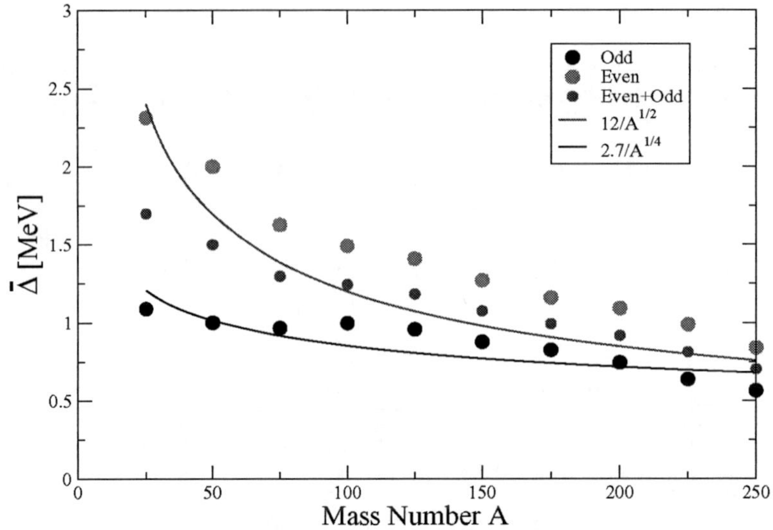

FIGURE 1. Average pairing gap, Δ, versus particle number A, as obtained from Eq. (1) using measured masses [4]. The averaging is performed in a region of fixed mass number, A, over all available isotopes. The particle number M (neutron or proton number) is either odd (lower black dots), even (upper red dots), or both even and odd (middle blue dots). The curves, $2.7/A^{1/4}$ MeV (lower line) and $12/A^{1/2}$ MeV (upper line), are obtained from fits to cases with odd M, and both odd and even M-values, respectively.

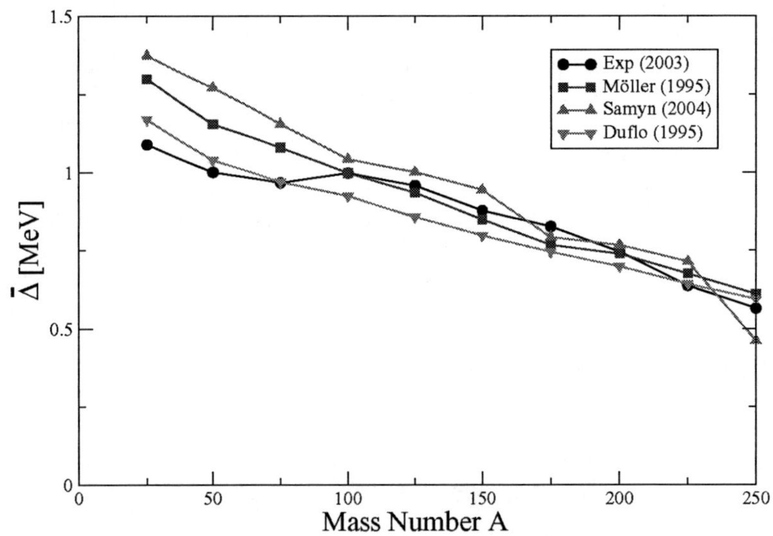

FIGURE 2. Average pairing gap versus mass number A using the three-point measure, Eq. (1), with odd M. Binding energies are obtained from different theoretical mass models, Möller et al [5] (black dots), Samyn et al (red triangles pointing up) [6] and Duflo et al (magenta triangles pointing down) [7]. The theoretical curves are compared to experimental results (black dots), cf Fig. 1.

where $A = N + Z$ is the total number of nucleons. If, however, also cases with M is an even number are included, the usually employed pairing gap value,

$$\bar{\Delta} = \frac{12}{A^{1/2}}\ \text{MeV}\,, \tag{3}$$

gives a better fit. The fitted difference Δ_3(even M)-Δ_3(odd M) = $\frac{1}{2}(e_{i+1} - e_i)$ is found to be inversely proportional to the mass number, A, and vary as $50/A$ MeV. The pairing gap as defined by Eq. (1) can also be extracted from theoretical mass models, and in Fig. 2 we compare the average value of Δ_3 versus particle number for three different mass models. Two of them are based on mean field theory. The first one is a non-self-consistent macroscopic-microscopic model [5], the second one is a self-consistent calculation based on Skyrme-Hartree-Fock-Bogoliubov [6] while the third one

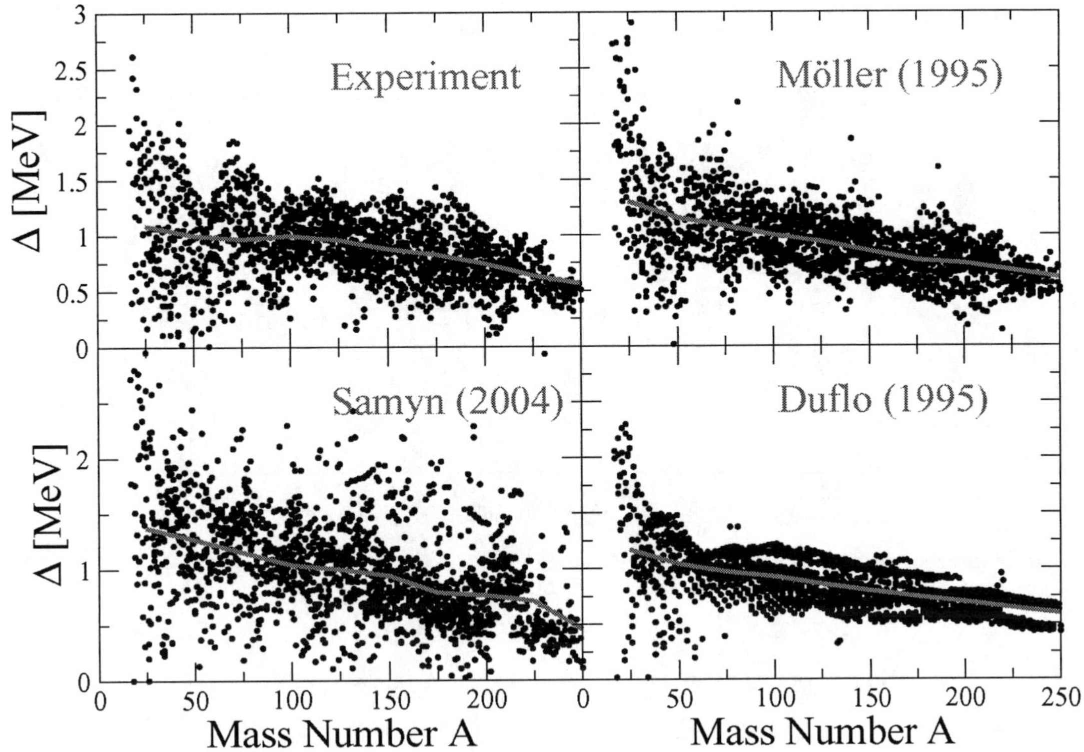

FIGURE 3. Pairing gaps extracted from Eq. (1) with odd M versus mass number A from experimental data, and three different theoretical mass models. Each black point corresponds to a nucleus. The solid red lines show the average pairing gap as shown in Fig. 2.

[7] is a shell–model based calculation with parameterized monopole and multipole terms. Also the experimental mean value of the pairing gap is shown in Fig. 2, and it is seen that all three mass models give similar results in good agreement with experimental numbers. If, however, not average values, but pairing gaps obtained from all nuclei are shown (Fig. 3), it is clear that the average values give a poor description of the result. The fluctuations (RMS value) of the pairing gaps are indeed very large, and exhibit different variation with particle number for the different mass models, see Fig. 4. All mass formula show the same tendency of decreasing fluctuations with increasing mass number, as is also seen from experimental masses. However, considerably larger pairing fluctuations are seen in the mass model by Samyn et al [6], where, particularly for large mass numbers, almost three times larger fluctuations are obtained, as compared to pairing gap fluctuations obtained from measured masses. The mass model based on the shell–model [7] gives systematically too small fluctuations, while the mass model by Möller et al [5] gives pairing fluctuations closest to experimental data.

It is thus clear that the fluctuations of the pairing gap is a most important property, and in the next section we shall present a semiclassical theory for pairing fluctuations based on periodic orbits [2].

PERIODIC ORBIT DESCRIPTION OF PAIRING FLUCTUATIONS

The many-body Hamiltonian,

$$H = \hat{H}_1 + \hat{V}_{\text{pair}} = \sum_k e_k a_k^\dagger a_k - G \sum_{k\ell} a_k^\dagger a_{\bar{k}}^\dagger a_{\bar{\ell}} a_\ell, \tag{4}$$

incorporates a one-body part, $\hat{H}_1$ (typically obtained from a deformed mean-field), and a two-body pairing interaction between time-reversed states, $\hat{V}_{\text{pair}}$. All two-body matrix elements are assumed to take the same value, G (seniority

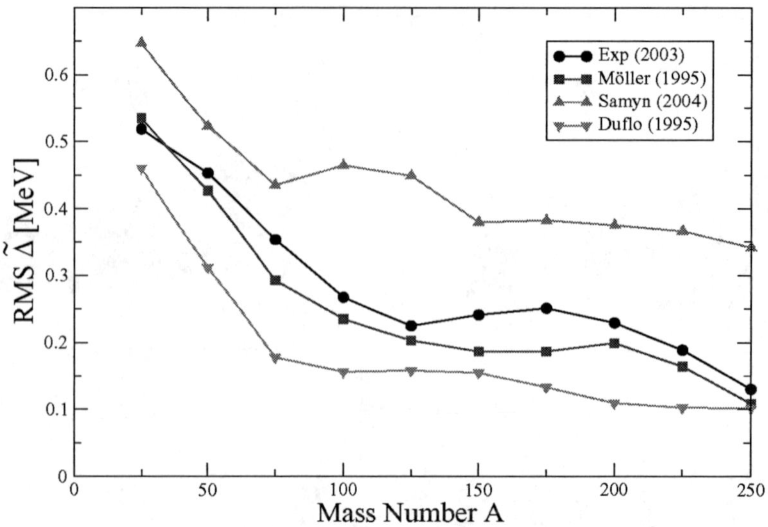

FIGURE 4. Root mean square (RMS) values of pairing gaps Δ obtained from Eq. (1) using odd M-values for three theoretical mass models, and experimental data. See caption of Fig. 2 for notations.

interaction). In the mean-field approximation in pairing space a pairing gap, or pairing "deformation",

$$\Delta = \left\langle G \sum_k a_k^\dagger a_{\bar{k}}^\dagger \right\rangle, \tag{5}$$

is determined by the BCS gap equation [8]

$$\frac{2}{G} = \sum_k \frac{1}{\sqrt{(e_k - \lambda)^2 - \Delta^2}}, \tag{6}$$

that can be written as

$$\frac{2}{G} = \int_{-L}^{L} \frac{\rho(\varepsilon) d\varepsilon}{\sqrt{\varepsilon^2 + \Delta^2}}, \tag{7}$$

where $\rho(\varepsilon)$ is the single–particle level density, and we have put the Fermi energy, λ, to zero. The energy cut off is at $\pm L$.

Following semiclassical approaches, the pairing gap as well as the single-particle density of states are divided in a smooth part and a fluctuating part, $\Delta = \bar{\Delta} + \widetilde{\Delta}$ and $\rho = \bar{\rho} + \widetilde{\rho}$, respectively. In the weak coupling limit $\bar{\Delta} \ll L$, the smooth part of the gap is given by the well known solution $\bar{\Delta} = 2L\exp(-1/\bar{\rho}G)$ (see e.g. Ref. [8]). In semiclassical theory the fluctuating part of the density $\widetilde{\rho}$ can be calculated from purely classical properties [9],

$$\widetilde{\rho}(\varepsilon) = 2\sum_p \sum_{r=1}^{\infty} A_{p,r} \cos(rS_p/\hbar + \nu_{p,r}), \tag{8}$$

where the sum is over all primitive periodic orbits p (and their repetitions r) of the classical underlying effective single-particle Hamiltonian, H_1. Each orbit is characterized by its action S_p, stability amplitude $A_{p,r}$, period $\tau_p = \partial S_p/\partial \varepsilon$ and Maslov index $\nu_{p,r}$ (all evaluated at energy ε). Assuming $\widetilde{\Delta} \ll \bar{\Delta}$ and $\bar{\Delta} \ll L$ gives after some algebra [2]

$$\widetilde{\Delta} = 2\frac{\bar{\Delta}}{\bar{\rho}} \sum_p \sum_{r=1}^{\infty} A_{p,r} K_0(r\tau_p/\tau_\Delta) \cos\left(\frac{rS_p}{\hbar} + \nu_{p,r}\right), \tag{9}$$

where all classical quantities involved are evaluated at the Fermi energy. $K_0(x)$ is the modified Bessel function of second kind, and

$$\tau_\Delta = \frac{h}{2\pi\bar{\Delta}} \tag{10}$$

is a time corresponding to the pairing gap, that we may call the pairing time. Since $K_0(x) \propto \exp(-x)/\sqrt{x}$ for $x \gg 1$, the Bessel function exponentially suppresses all contributions for times $\tau \gg \tau_\Delta$ (making the sum in Eq. (9) convergent).

Since the value of the actions depend on the shape of the mean–field potential, Eq. (9) predicts generically fluctuations of the pairing gap as one varies, for instance, the particle number, or the shape of the system at fixed particle number. The fluctuations result from the interference between the different oscillatory terms that contribute to $\widetilde{\Delta}$. When the motion is regular (integrable), continuous families of periodic orbits having the same action, amplitude, etc, exist. The coherent contribution to the sum (9) of these families of periodic orbits produces large fluctuations. In contrast, in the absence of regularity or symmetries, incoherent contributions of smaller amplitude coming from isolated unstable orbits are expected for chaotic dynamics.

The second moment of the fluctuations may be obtained from Eq. (9) as [2]

$$\left\langle \widetilde{\Delta}^2 \right\rangle = 2\frac{\bar{\Delta}^2}{\tau_H^2} \int_0^\infty d\tau K_0^2(\tau/\tau_\Delta) K(\tau), \tag{11}$$

where $\tau_H = h/\delta$ is the Heisenberg time ($\delta = \bar{\rho}^{-1}$ is the single–particle mean level spacing at Fermi energy), and $K(\tau)$ is the spectral form factor, i.e. the Fourier transform of the two-point density–density correlation function [10].

The structure of the form factor $K(\tau)$ is characterized by two different time scales. The first one, the smallest of the system, is the period $\tau_{\min}$ of the shortest periodic orbit. The form factor is zero for $\tau \leq \tau_{\min}$, and displays non-universal (system dependent) features at times $\tau_{\min} \lesssim \tau \ll \tau_H$. As τ further increases, the function becomes universal, depending only on the regular or chaotic nature of the dynamics, and finally tends to τ_H when $\tau \gg \tau_H$. The result of the integral (11) thus depends on the nature of the dynamics, and on the relative value of τ_Δ with respect to $\tau_{\min}$ and τ_H. In the simplest approximation, all the non–universal system–specific features are taken into account only through $\tau_{\min}$ [11], and one can write $K(\tau) = 0$ for $\tau < \tau_{\min}$ and, for $\tau \geq \tau_{\min}$, $K(\tau) = \tau_H$ for integrable systems and $K(\tau) = 2\tau$ for chaotic systems with time reversal symmetry. We assume a generic regular system; the analysis does not apply to the harmonic oscillator, whose form factor is pathological.

This finally gives the expressions for fluctuations of the pairing gap (normalized to the single–particle mean level spacing), $\sigma = \sqrt{\left\langle \widetilde{\Delta}^2 \right\rangle}/\delta$, assuming regular dynamics

$$\sigma_{\text{reg}}^2 = \frac{\pi}{4}\frac{\bar{\Delta}}{\delta} F_0(D) \ , \tag{12}$$

and assuming chaotic dynamics,

$$\sigma_{\text{ch}}^2 = \frac{1}{2\pi^2} F_1(D) \ , \tag{13}$$

where we have introduced the function [2]

$$F_n(D) = 1 - \frac{\int_0^D x^n K_0^2(x) dx}{\int_0^\infty x^n K_0^2(x) dx}. \tag{14}$$

The argument D is defined as

$$D = \frac{\tau_{\min}}{\tau_\Delta} = \frac{2\pi}{g}\frac{\bar{\Delta}}{\delta}, \tag{15}$$

where the parameter g is the ratio between the Heisenberg time and the time of the shortest periodic orbit,

$$g = \frac{\tau_H}{\tau_{\min}}. \tag{16}$$

This parameter is often called "dimensionless conductance". It expresses the energy range, $L_{\max}(= g)$ (in dimensionless units; energies are divided by the mean energy spacing, δ), in the single-particle spectrum where spectrum fluctuations show universal properties [10]. For a system where the corresponding classical dynamics is regular/chaotic (with time reversal symmetry), the statistical properties of the one-body energies, e_k of Eq. (4) are described by Poisson/GOE (Gaussian Orthogonal Ensemble) statistics, respectively. The limit $g \to \infty$ (i.e. D=0), corresponds to a universal situation when the full spectrum corresponds to pure GOE (if chaotic) or Poisson (if regular) statistics.

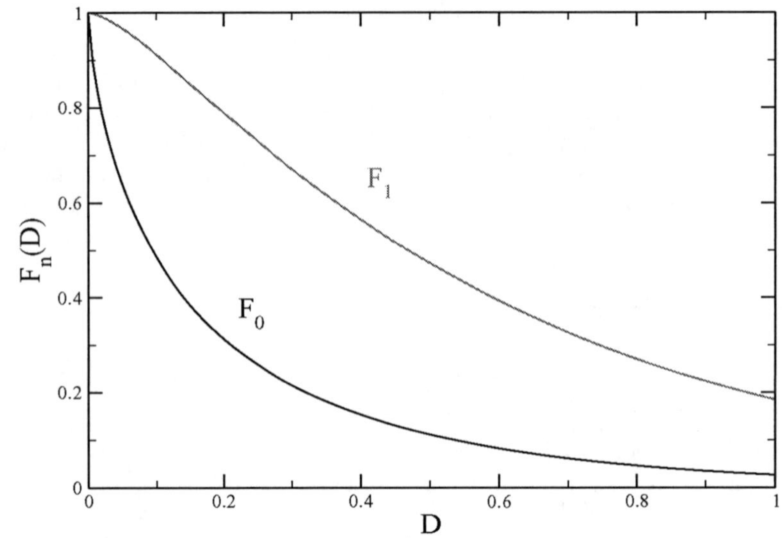

FIGURE 5. The two functions F_0 and F_1, see Eq. (14), versus the argument D.

The dimensionless parameter D can also be expressed as the system size, $2R$, divided by the coherence length of the Cooper pair, $\xi_0 = \hbar v_F/(2\bar{\Delta})$, where v_F is Fermi velocity,

$$D = 2R/\xi_0. \tag{17}$$

The Cooper pairs can thus be considered as restricted by the system size if $D < 1$.

In Fig. 5 we show the two functions F_0 and F_1 of Eq. (14) versus the argument D. For small values of D, $F_0 \approx 1$ and $F_1 \approx 1$, and a universal behavior appears, i.e. the fluctuations do not depend on the system properties, only the dynamics, implying Poisson and GOE statistics for the full spectrum in the regular and chaotic cases, respectively. In the other limit when D becomes large, F_0 and F_1 go to zero and all pairing fluctuations disappear. This is, for example, the situation in bulk systems with a large number of particles.

In Fig. 6 pairing gap fluctuations are shown versus the pairing gap for different values of the dimensionless conductance, g, for regular as well as for chaotic dynamics. The plot covers a large range of parameter values and is shown in log-log scale. Pairing gaps of the order of the mean level spacing or smaller, $\bar{\Delta} < \delta$, (ultrasmall regime), are not treated by the present theory and corresponding region is shaded in the figure. In this region the Anderson condition [12] implies no BCS pairing. As mentioned above, non-universal behavior (deviating from GOE or Poisson) appears when $D > 0$, corresponding to finite values of the dimensionless conductance, g.

FLUCTUATIONS OF PAIRING GAP IN NUCLEI

In a previous section we studied pairing gap fluctuations in nuclei as obtained from the odd-even mass difference, where binding energies were taken from different mass models as well as from data, see Fig. 4. We now would like to use the semiclassical theory developed in previous section to compare pairing gap fluctuations, and compare to these results. Notice that the theory does not contain any parameters. Once the system is defined, and the variable D has been determined, the fluctuations only depend on the corresponding classical dynamics, being regular or chaotic. In nuclei ground states the dominating dynamics is expected to be regular, although elements from chaos may be present, see Ref.[13].

To evaluate the root mean square value (RMS) of the pairing fluctuations from the theoretical expressions, Eqs. (12) and (13), we only have to determine the parameter D. The size of the nucleus, $2R = 2 \times 1.2A^{-1/3}$ fm, and size of the pairing correlation length, $\xi_0 = \hbar v_F/(2\bar{\Delta}) = 11.3A^{-1/4}$ fm, give (using Eqs. (2) and (17)) $D = 2R/\xi_0 = 0.22A^{1/12} = 0.27 - 0.33$ for mass numbers in the interval, A=25-250. The dependence of the pairing fluctuations, assuming regular dynamics, with the (average) pairing gap is shown in Fig. 7 for the three cases, D=0, 0.3 and 1. If the pairing correlation length is smaller than the system size ($D > 1$) pairing fluctuations are quite small. Largest fluctuations appear in the

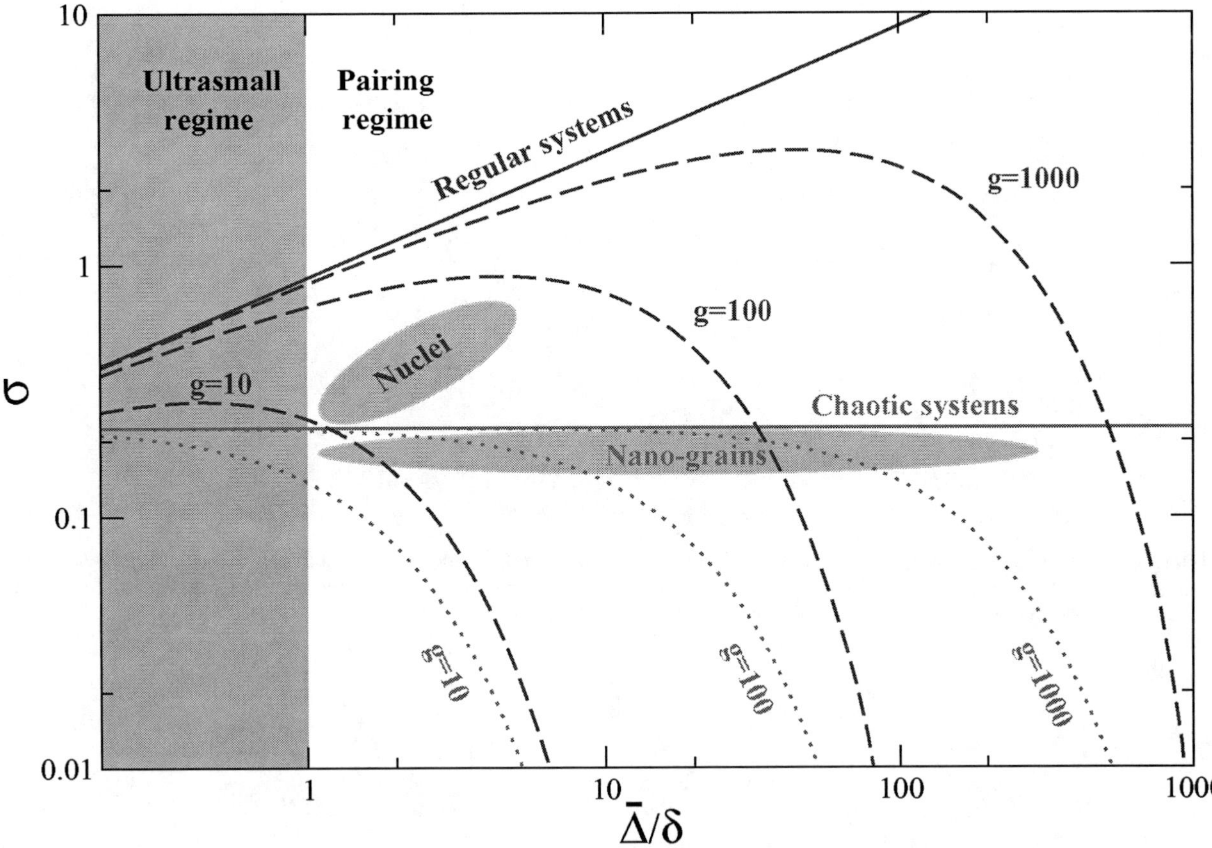

FIGURE 6. Fluctuations of the pairing gap as a function of the mean value for mesoscopic systems (log-log scale; all quantities normalized with the single-particle mean spacing). Regular and chaotic dynamics are shown by blue and red lines, respectively. The dashed curves correspond to different values of the dimensionless conductance, g, and the limiting case of $g \to \infty$ is shown by solid lines. The results are valid in the pairing regime $\bar{\Delta}/\delta > 1$. Applications to Nuclei and Nano-grains are marked out. Ultracold atomic gases may be controlled to appear in major parts of the figure. From Ref. [2].

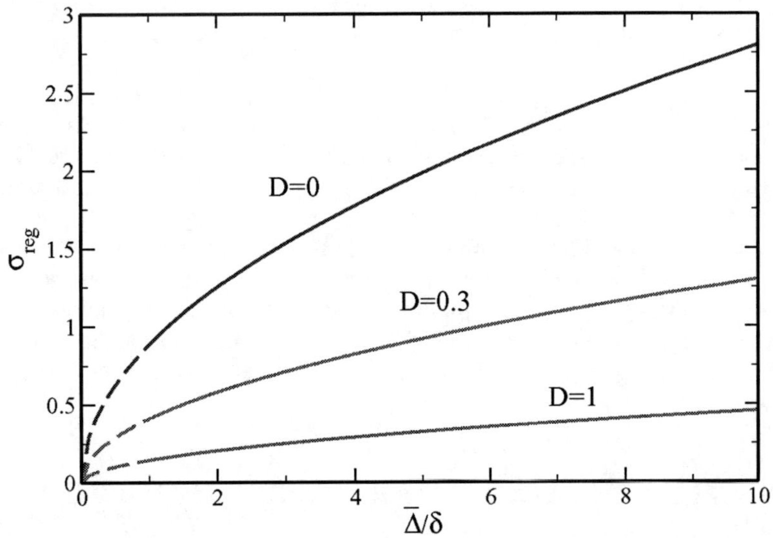

FIGURE 7. Fluctuations of pairing gap versus average pairing gap for different values of D. Regular dynamics is assumed. $D = 0.3$ approximately corresponds to nuclei. Fluctuations and pairing gaps are expressed in units of the mean level spacing, δ.

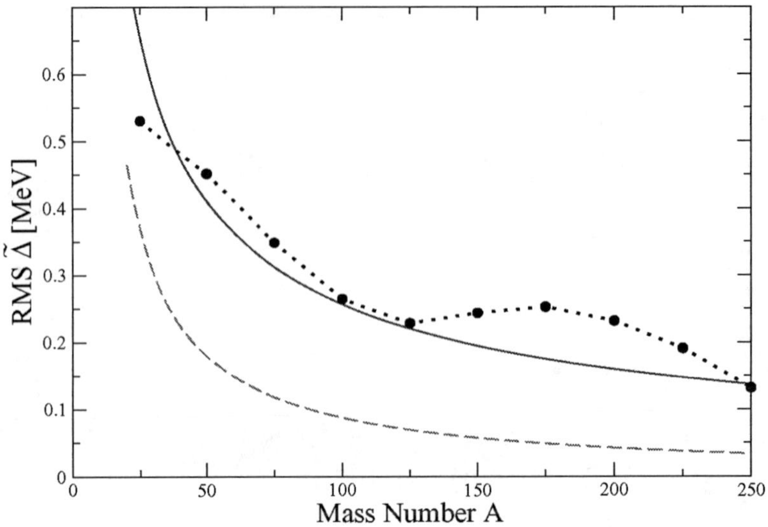

FIGURE 8. RMS values of pairing gap fluctuations in nuclei versus mass number A, obtained from measured masses (dashed line connected by filled circles). Semiclassical calculations of the fluctuations assuming chaotic (Eq. (13)) and regular (Eq. (12)) dynamics are shown by red dashed and blue solid lines, respectively. From Ref. [2]

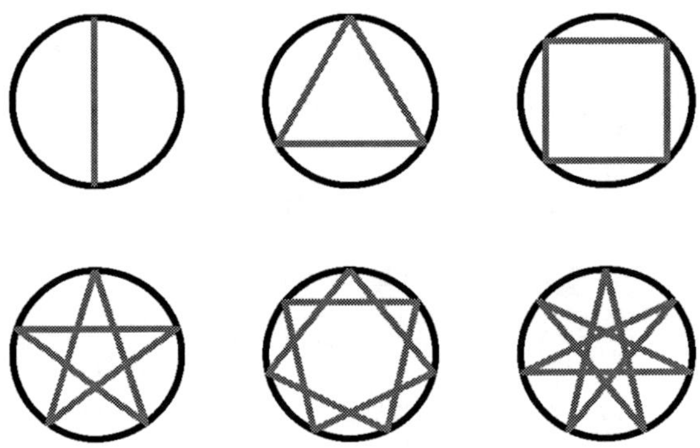

FIGURE 9. A sample of classical periodic orbits of the cavity potential. The orbits shown correspond to the index (v,w)=(2,1), (3,1), (4,1), (5,2), (7,2) and (7,3), counted from upper left to lower right figure.

universal limit when D=0. The small values of D for nuclei ($D \approx 0.3$) implies that the Cooper pairs are non-localized. The pairing gap fluctuations are thus substantial, and are about half the value at the universal limit.

We may compare the fluctuations to the experimental pairing gap fluctuations by inserting the above values for D, and the average values of Δ from Eq. (2), in Eqs. (12) and (13), assuming regular or chaotic dynamics. The resulting curves are compared to the experimental one in Fig. 8. Note the good agreement between the theoretical pairing gap fluctuations assuming regular dynamics, and the experimental curve, both in the overall amplitude and in the A–dependence. In Ref.[13] it was discussed the possibility that the dynamics of the nuclear ground state is mixed regular and chaotic. Making this assumption in the calculation of fluctuations of the pairing gap results in a curve that is very close to the purely regular curve in Fig. 8. That is, fluctuations of the pairing gap cannot distinguish if there is a chaotic component in the nuclear ground state.

SHELL STRUCTURE IN PAIRING GAP FROM PERIODIC ORBIT THEORY

Shell effects in nuclei are also seen in the pairing gap. One may go beyond a statistical description, and use Eq. (9) to obtain a detailed description of the variation with neutron or proton numbers. For that purpose, we assume for the nuclear mean field a simple hard-wall cavity potential. The shape of the cavity at a given number of nucleons is fixed

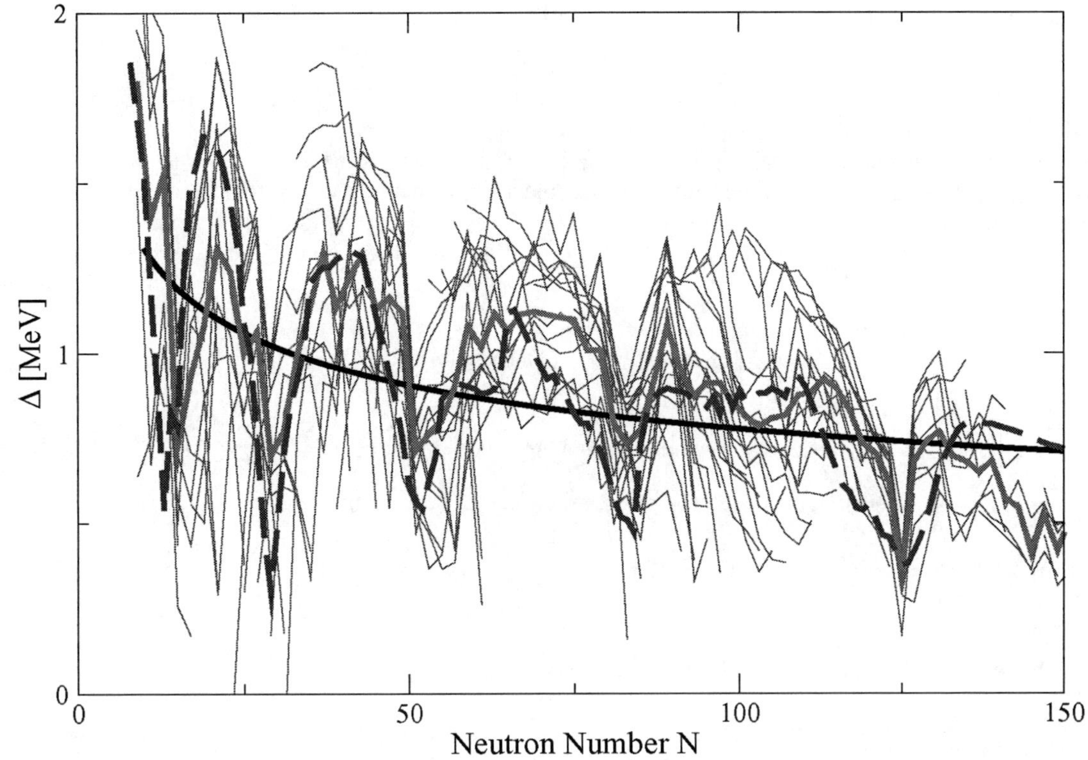

FIGURE 10. Nuclear pairing gaps for neutrons. Experimental isotope sequences are connected by thin solid lines. The blue dashed line shows average pairing gaps. Calculations from the cavity model are shown by the solid red line. Average behavior (Eq. (2)) is shown by solid black line. Data from Ref. [4].

by minimization of the energy against quadrupole, octupole and hexadecapole deformations [14]. To simplify, we take $N = Z$. The periodic orbits of the spherical cavity (a few short orbits are shown in Fig. 9) are used in Eq. (9), with modulations factors that take into account deformations and inelastic scattering [15]. This gives

$$\tilde{\Delta} = \frac{\bar{\Delta}}{\bar{\rho} E_0} \sum_{v,w} A_{vw} M_{vw}(x) \kappa_\chi(\ell_{vw}) K_0\left(\frac{\ell_{vw}\bar{\Delta}}{2\bar{k}_F R E_0}\right) sin(\bar{k}_F R \ell_{vw} + \nu_{vw}\frac{\pi}{2}), \tag{18}$$

where $M_{vw}(x)$ is a modulation factor for perturbative deformations, $\kappa_\chi(\ell_{v\omega})$ a modulation factor for inelastic scattering, and x stands for the three considered deformation degrees of freedom, quadrupole (ε_2), octupole (ε_3) and hexadecapole (ε_4). The summation is carried out over the two indices (v, w) including the periodic orbits shown in Fig. 9. We set the average of $\tilde{\Delta}$ to zero, as was done with the experimental data. In Fig. 10 we compare the theoretical result $\tilde{\Delta}(N)$ to the experimental value averaged over the different isotopes at a given N. The agreement is excellent; the theory describes all the main features observed in the experimental curve.

PAIRING FLUCTUATIONS IN NANO-SIZED METALLIC GRAINS

Experiments in the 90's have explored the superconducting properties of nanometer scale aluminum grains [16]. Irregular shape of the grains implies chaotic dynamics, and energy levels are found to follow GOE (there are no further symmetries than time-reversal), see Ref. [17]. The existence of a superconducting gap was demonstrated in the regime $\bar{\Delta} > \delta$, whereas no gap was observed when $\bar{\Delta} < \delta$. The transition occurs around $N \sim 5000$, where N is the number of conduction electrons in the grain. The N dependence of the average gap $\bar{\Delta}$ is poorly understood. We will adopt for grains the thin-film value $\bar{\Delta} \approx 0.38 \times 10^{-3}$ eV [17]. The mean level spacing is $\delta = (2E_F)/(3N) \approx 2.1/N$ eV, whereas the dimensionless conductance, $g \approx 2.6N^{2/3}$. Eq. (15) gives $D \approx 4.4 \times 10^{-4} N^{1/3}$, which ranges from 0.05 to 0.02 when N varies between 10^3 and 10^5. This means that the variance will be close to its "universal" value (GOE limit) obtained by setting $F_1(D) = 1$ in Eq. (13), namely [18]

$$\sigma_{\text{ch}}^2 = 1/2\pi^2. \tag{19}$$

The typical range of variation of pairing gap fluctuations for nano-grains is marked out by a grey area in Fig. 6.

There are no explicitly measured pairing gap fluctuations for nanosized metallic grains to compare the present theory. We may, however, compare to other independent calculations, namely the condensation energy calculated of chaotic grains [20], and variation of the pairing gap with particle number in a regular cubic shaped grain [21].

The BCS condensation energy is defined as the total energy difference between the paired and the unpaired system. With our choice of gap equation (7) it is written as

$$E_{\text{C}} = E_{\text{tot}}(\Delta) - E_{\text{tot}}(\Delta = 0) = \int_{-L}^{L} \rho(e) e v^2(e) de - \frac{\Delta^2}{4G} - \int_{-L}^{0} \rho(e) e dx \tag{20}$$

where $v^2(e) = \frac{1}{2}\left(1 - \frac{e}{\sqrt{e^2+\Delta^2}}\right)$. Inserting the semiclassical approximations of $\rho = \bar{\rho} + \widetilde{\rho}$ and $\Delta = \bar{\Delta} + \widetilde{\Delta}$ and expanding $E_{\text{C}} = \bar{E}_{\text{C}} + \widetilde{E}_{\text{C}}$ to lowest order in fluctuating properties, assuming $\widetilde{E}_{\text{C}} \ll \bar{E}_{\text{C}}$, gives

$$\widetilde{E}_{\text{C}} = -2\sum_{p}\sum_{r=1}^{\infty} A_{p,r}\left[Q_{p,r} + qK_0(r\tau_p/\tau_\Delta)\right]\cos\left(\frac{rS_p}{\hbar} + \nu_{p,r}\right) \tag{21}$$

where K_0 is the modified Bessel function of second kind, $q = \bar{\Delta}^2\left(1 - \frac{1}{2\bar{\rho}G}\right)$ and

$$Q_{p,r} = \int_{-L}^{0} \cos\left(\frac{r\tau_p}{\hbar}\right) e \left(1 + \frac{e}{\sqrt{e^2 + \bar{\Delta}^2}}\right) de \tag{22}$$

The second moment, $\sigma_{\text{C}}^2 = \left\langle \widetilde{E}_{\text{C}}^2 \right\rangle$, is thus obtained as

$$\sigma_{\text{C}}^2 = \frac{2}{\hbar^2}\int_0^{\infty} \left[Q(\tau) + qK_0(\tau)\right]^2 K(\tau) d\tau, \tag{23}$$

where $K(\tau)$ is the spectral form factor. Using in addition the estimates $G = \frac{0.224}{\bar{\rho}}$ and $L = \frac{\bar{\Delta}}{2} e^{1/0.224}$ [20] we show in Fig. 11 the fluctuations of the condensation energy, σ_{C}, versus the level distance in the system, assuming regular and chaotic motion. As mentioned, there is no experimental data to compare to, but we may compare to a numerical calculation of the σ_{C} by Sierra *et. al.* [20] where they use the Richardson's solution of the pairing problem and random matrix theory (GOE) for generating the spectrum. We see that our semiclassical theory, assuming chaotic dynamics, agrees well with the random matrix calculation of Ref. [20]. The random matrix model applied in [20] happens to be a reasonable approximation for the considered nano-grains, since $D \approx 0$, i.e. the nano-grains have properties which are close to the universal limit.

Garcia-Garcia et al [21] studied shell effects on the pairing gap in a cubic geometry as a function of the mean pairing gap, $\bar{\Delta}/\delta$. The cubic geometry implies classically regular motion, and we may thus use Eq. (12) above to calculate the fluctuation of the paring gap, normalized to the mean pairing gap $\bar{\Delta}$ instead of the mean level spacing δ, $\sigma_{\text{reg}}\delta/\bar{\Delta} = \sqrt{\frac{\pi}{4}\frac{\delta}{\bar{\Delta}}}$, where we have set $F_0 = 1$ since for small grains $D \approx 0$. This gives a very good agreement to the pairing fluctuations calculated in Ref. [21]. It is interesting to note that by making the system chaotic, the pairing fluctuations decreases substantially and become (see Eq. (13)), $\sigma_{\text{ch}}\delta/\bar{\Delta} = \frac{1}{\sqrt{2}\pi}\frac{\delta}{\bar{\Delta}}$, i.e. $\sigma_{\text{ch}}/\sigma_{\text{reg}} = \sqrt{\frac{2}{\pi^3}\frac{\delta}{\bar{\Delta}}} \approx 0.05$ if $\bar{\Delta}/\delta = 20$.

PAIRING FLUCTUATIONS IN ULTRACOLD FERMIONIC GASES

Recently, a large interest has emerged in studying trapped atomic gases of bosons and fermions. The gases are ultracold and dilute, and provide the possibility to study new quantum phenomena in the physics of finite many-body systems.

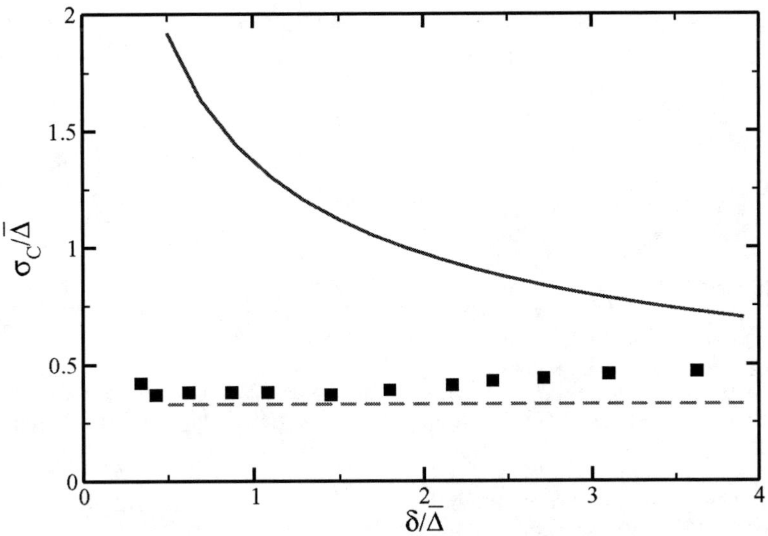

FIGURE 11. Fluctuations of the condensation energy as a function of the mean level spacing of the system. The blue solid and red dashed lines show the numerical calculations of regular and chaotic fluctuations according to Eq. (23). The squares show the calculations of ref. [20].

The number of neutrons of the confined atoms determines the quantum statistics; odd number implies Fermions, and even number implies bosons. Studies of Bose condensates can be done for the bosonic gases, and studies of quantum phenomena including superfluidity and the transition to a Bose-Einstein condensate can be conducted for the Fermi gases.

Since the atom-atom interaction is short ranged and much smaller than the interparticle distance, the atom-atom interaction can be approximated by the δ-interaction,

$$V(r_1 - r_2) = 4\pi \frac{\hbar^2 a}{m} \delta^{(3)}(r_1 - r_2), \tag{24}$$

where a is the s-wave scattering length, that can be externally controlled in size and even in sign through the Feshbach resonance. Also the confinement potential can be externally controlled to create regular as well as chaotic dynamics. Ultracold fermionic gases thus provide excellent conditions for theoretical as well as experimental studies of pairing properties, including pairing fluctuations. Since both particle number and interaction strength are experimentally controlled parameters, the fluctuations may appear in major parts of Fig. 6.

We estimate $\delta = (2E_F)/(3N)$ and $g = \frac{1}{2}(3N)^{2/3}$; in the dilute BCS region $\bar{\Delta} = (2/e)^{7/3} E_F \exp(-\pi/2k_F|a|)$ [22], with k_F the Fermi wavevector, giving

$$D = \frac{2R}{\xi_0} = 2\pi(2/e)^{7/3}(3N)^{1/3} \exp(-\pi/2k_F|a|). \tag{25}$$

Recent experiments using Li^6 reach $k_F|a| = 0.8$ [23], implying negligible fluctuations for typical values of $N \sim 10^6$. Reducing to $k_F|a| = 0.2$ and $N = 10^4$ yields for generic regular systems fluctuations that are on the same magnitude as the mean pairing gap, $\sigma_{\text{reg}} \approx 0.5\bar{\Delta}/\delta$.

SUMMARY

In summary, we have presented a semiclassical theory that provides a generic description of fluctuations and shell structure of the BCS pairing gap in finite Fermi systems. These mesoscopic systems are generically dominated by system specific features not included in purely statistical models like GOE. Different possible regimes, as well as the influence of order/chaos dynamics, were investigated, in particular for the typical size of the fluctuations (Fig. 6). The present theory provides analytic predictions, valid for a wide range of physical situations. It also compares very favorably with available experimental data.

S.Å. thanks the Swedish Science Research Council, and P.L. acknowledges support by grants ANR–05–Nano–008–02, ANR–NT05–2–42103 and by the IFRAF Institute.

REFERENCES

1. A. Bohr, B.R. Mottelson and D. Pines, Phys. Rev. **110**, 936 (1958).
2. H. Olofsson, S. Åberg, P. Leboeuf, Phys. Rev. Lett. 100, 037005 (2008).
3. J. Dobaczewski *et al.*, Phys. Rev. C **63**, 024308 (2001).
4. G. Audi, A.H. Wapstra and C. Thibault, Nucl. Phys. **A729**, 337 (2003).
5. P. Möller, J. R. Nix, W. D. Myers, and W. J. Swiatecki, At. Data and Nucl. Data Tables **59** (1995) 185.
6. M. Samyn, S. Goriely, M. Bender, and J. M. Pearson, Phys. Rev. C **70**, 044309 (2004).
7. J. Duflo and A. P. Zuker, Phys. Rev. C **52**, R23 (1995).
8. D. Brink and R.A. Broglia, *Nuclear Superfluidity: Pairing in Finite Systems* (Cambridge Univ. Press, 2005).
9. M. Brack and R.K. Bhaduri, *Semiclassical Physics* (Addison and Wesley, Reading, 1997).
10. M.V. Berry, Proc. Roy. Soc. Lond. **A 400**, 229 (1985).
11. P. Leboeuf and A.G. Monastra, Ann. Phys. **297**, 127 (2002).
12. P.W. Anderson, J. Phys. Chem. Solids B **11**, 26 (1959).
13. O. Bohigas and P. Leboeuf, Phys. Rev. Lett. **88**, 092502 (2002).
14. R. Hasse, Ann. Phys. **68**, 377 (1971).
15. S. C. Creagh, Ann. Phys. (N.Y.) **248**, 60 (1996); P. Leboeuf, Lect. Notes Phys. **652**, Springer, Berlin Heidelberg 2005, p.245, J. M. Arias and M. Lozano (Eds.).
16. D.C. Ralph, C.T. Black and M. Tinkham, Phys. Rev. Lett. **74**, 3241 (1995); *ibid* **76**, 688 (1996)
17. J. von Delft and D.C. Ralph, Phys. Rep. **345**, 61 (2001).
18. The expression for pairing fluctuations in the universal GOE limit (corresponding to D=0) was first derived in [19].
19. K.A. Matveev and A.I. Larkin, Phys. Rev. Lett. **78**, 3749 (1997).
20. G. Sierra *et al.*, Phys. Rev. B **61**, 11890 (2000).
21. M. Garcia-Garcia et al, arXiv:0710.2286 (2007).
22. L.P. Gor'kov and T.K. Melik-Barkhudarov, Sov. Phys. JETP **13**, 1018 (1961).
23. C.H. Schunck *et al.*, Phys. Rev. Lett. **98**, 050404 (2007).

Thermodynamic analysis of spectra

G. E. Mitchell* and J. F. Shriner, Jr.†

*North Carolina State University, Raleigh, NC 27695 and Triangle Universities Nuclear Laboratory, Durham, NC 27708
†Tennessee Technological University, Cookeville, TN 38505

Abstract.
Although random matrix theory had its initial application to neutron resonances, there is a relative scarcity of suitable nuclear data. The primary reason for this is the sensitivity of the standard measures used to evaluate spectra – the spectra must be essential pure (no state with a different symmetry) and complete (no states missing). Additional measures that are less sensitive to these experimental limitations are of significant value. The standard measure for long range order is the Δ_3 statistic. In the original paper that introduced this statistic, Dyson and Mehta also attempted to evaluate spectra with thermodynamic variables obtained from the circular orthogonal ensemble. We consider the thermodynamic "internal energy" and evaluate its sensitivity to experimental limitations such as missing and spurious levels. Monte Carlo simulations suggest that the internal energy is less sensitive to mistakes than is Δ_3, and thus the internal energy can serve as a addition to the tool kit for evaluating experimental spectra.

Keywords: random matrix theory, thermodynamics, internal energy
PACS: 24.60.Ky,24.60.Lz,05.70.Ce,05.45.Mt

INTRODUCTION

Random matrix theory (RMT) was first introduced into nuclear physics by Wigner [1] in order to describe the statistical properties of spectra – specifically neutron resonances. The early development of the theory is summarized in a collection of reprints edited by Porter [2]. It was some time before data of a sufficient quality were available to confirm or reject the validity of the theory. In fact Dyson and Mehta [3] performed an evaluation of the best available neutron resonance data (from Rainwater's group using the Nevis Synchrocyclotron at Columbia University) and could reach no definite conclusions concerning the applicability of RMT for nuclear spectra. By the early 1970s, Rainwater [4, 5] obtained neutron data which appeared to agree with RMT. However, due to limited statistics this was not considered definitive. By the middle 1970s additional confirmation of RMT was obtained with proton resonances by the TUNL group [6, 7]. However, the statistics for proton resonances were even more limited than in the neutron case.

In the early 1980s Haq, Pandey, and Bohigas [8] provided what was accepted as a definitive result. They combined the best available neutron and proton resonance data; for the nearest neighbor spacing distribution the agreement with RMT was excellent. The spacing data from individual nuclides could be combined because the nearest neighbor spacing (NNS) distribution is "universal". Following this result it was generally accepted that RMT correctly described nuclear resonance data – at least for medium and heavy nuclei.

Another key result was the so-called Bohigas-Giannoni-Schmit conjecture – that quantum analogs of classically chaotic systems obey the statistics of the Gaussian Orthogonal Ensemble (GOE) version of RMT [9]. After this suggestion there has been an explosion of activity of RMT in many different fields: The comprehensive review by Guhr, Müller-Groeling and Weidenmüeller [10] cites over 800 references in areas from mesoscopics to lattice gauge applications. However, activity in nuclear physics has been relatively muted.

The simplest explanation for this relative dearth of activity lies in the practical application of RMT to nuclei. RMT describes spectra which are pure (all quantum numbers correctly assigned) and complete (no missing levels). Implicitly one also requires a sufficient sample size for statistical analysis to apply. In nuclei these requirements are very difficult to satisfy. The sample size issue is almost always present. The purity issue is crucial. Analysis of low-lying spectra may involve many different spins as well as two parities. For example an experiment in ^{30}P to obtain complete spectroscopy for the first 100 or so levels involved 12 different spin-parity combinations. In addition the most robust tests to determine the fraction of missing levels all assume the validity of RMT. This can lead to the employment of circular logic. Although these problems are not insurmountable, they do imply that obtaining suitable data to evaluate the applicability of RMT is very difficult and time-consuming. The net result is a relative scarcity of

CP995, *Nuclei and Mesoscopic Physics - WNMP 2007*, edited by P. Danielewicz, P. Piecuch, and V. Zelevinsky

suitable data.

In this context the relative sensitivity of the various measures used to evaluate experimental data becomes crucial. In practice the measures commonly employed to test for long-range order (which along with level repulsion is one of the two key predictions of RMT) are very sensitive to missing and spurious levels. In the classic 1963 paper [3] by Dyson and Mehta where they attempted to evaluate the early neutron resonance data, they suggested a number of measures. One of them is the now ubiquitous Dyson-Mehta Δ_3 statistic which is used to determine the presence or lack of long range order. Unfortunately a missing (or spurious) level introduces a linear term [11] in addition to the logarithmic term predicted by RMT for a pure and complete spectrum. This statistic is, therefore, extremely sensitive to mistakes, which only accentuates the practical difficulties faced by experimental nuclear physicists.

A more robust statistic that is less sensitive to mistakes is very highly prized. Interestingly enough, in the seminal paper by Dyson and Mehta [3] they also introduced another statistic to evaluate spectra. To our knowledge this statistic has never been applied since to experimental spectra. Here we consider this statistic, the internal energy. In Sect. II we describe the thermodynamic procedure adopted by Dyson and Mehta and apply it to simulated GOE spectra. Then in Sect. III we study the effect of missing and spurious levels on this statistic and compare the sensitivity of the internal energy U with the standard Δ_3 statistic.

THE INTERNAL ENERGY U

General

An aspect of RMT that was introduced over 40 years ago by Dyson [12] but has received significantly less attention than many other areas is that of the thermodynamics of the ensembles. Quantities such as free energy, internal energy, entropy, and specific heat were studied, and expected values were obtained for each of the standard ensembles (orthogonal, unitary, and symplectic). In their evaluation of early neutron resonance data, Dyson and Mehta used several statistics but concluded that the data quality was not sufficient to decide whether the data agreed with GOE or not. One of the statistics they introduced at that time – the Dyson-Mehta Δ_3 statistic [3] – has become the standard measure of long range order in spectra. In that early paper Dyson and Mehta also used one of the thermodynamic variables (the internal energy) as a test. However, this test has apparently not been utilized to evaluate experimental data since. Here we investigate the internal energy U and study its behavior for GOE and Poisson spectra. In particular, we examine how this statistic behaves as a function of spectrum size and how it is affected by the presence of missing or spurious levels in the spectrum. We then compare the behavior of U with that of the Δ_3 statistic.

We follow Dyson and use a 2-D Coulomb gas within the context of a circular orthogonal ensemble (COE) as a starting point for the discussion of internal energy for random matrix ensembles. For N unit charges free to move along a circle of radius 1 whose positions are specified by angular variables $\theta_1, \theta_2 \ldots \theta_N$, the potential energy of the system is given by

$$W = -\sum_{j>i} \ln \left| e^{i\theta_i} - e^{i\theta_j} \right| . \tag{1}$$

The minimum value of W occurs for charges that are equally spaced around the circle; we call this minimum value W_0. In turn the ensemble average of $W - W_0$ is related to the internal energy per particle U:

$$NU = \langle W - W_0 \rangle . \tag{2}$$

In the limit $N \to \infty$, U approaches the value $1 - \frac{1}{2}\gamma - \frac{1}{2}\ln 2 = 0.365$ for the COE (and the GOE).

For the circular ensemble the point charges in the Coulomb gas are confined. Consider a series of energy eigenvalues $E_1, E_2 \ldots E_N$. When one tries to impose the same analogy by placing unit point charges at the appropriate locations along an energy axis, the repulsion among the charges will cause the charges to spread out indefinitely. To apply these ideas for a series of energy eigenvalues, we must include in our definition of W an additional potential V that confines the charges to a finite range of energies. We define

$$W = -\sum_{j>i} \ln \left| E_i - E_j \right| + \sum_i V(E_i). \tag{3}$$

To determine V, the connection between level density ρ and potential V implied by Dyson's mean field equation [13] has been employed [15]:

$$PV \int \frac{\rho(t)}{E-t} dt = \frac{dV}{dE}, \tag{4}$$

where PV indicates the Cauchy principal value. We consider cases where the N levels are confined to an interval $E \in [-N/2, N/2]$ with $\rho(t) = 1$ in an appropriate set of units. We further define $\mathscr{L} \equiv N/2$. Performing the necessary integration in Eq. (4) and choosing the constant of integration such that the average value of V over the energy interval $[-\mathscr{L}, \mathscr{L}]$ is zero give the result

$$\begin{aligned} V(E) &= (\mathscr{L} - E) * \left[\frac{1}{2} + \ln\left(\frac{\mathscr{L} - E}{2\mathscr{L}}\right)\right] \\ &\quad + (\mathscr{L} + E) * \left[\frac{1}{2} + \ln\left(\frac{\mathscr{L} + E}{2\mathscr{L}}\right)\right]. \end{aligned} \tag{5}$$

As with the circular orthogonal ensemble, we define the absolute scale of the energy by subtracting the minimum value W_0. With the formulas of Eqs. (3) and (5), W_0 occurs when the eigenvalues comprise a so-called "picket fence" spectrum, a spectrum which has equally spaced levels located at energies

$$\xi_i = -\mathscr{L} + i - \frac{1}{2}. \tag{6}$$

The combination of Eqs. (2), (3), (5), and (6) then gives an expression for U that is applicable to any series of eigenvalues satisfying the level density and energy range constraints imposed during the derivation of V:

$$NU = -\sum_{j>i} \ln\left|\frac{E_j - E_i}{j - i}\right| + \sum_i [V(E_i) - V(\xi_i)]. \tag{7}$$

The limiting behaviors of the Gaussian ensembles and the circular ensembles are known to be the same. We have confirmed that the average value of the quantity defined in Eq. (7) is indeed ≈0.365 for pure and complete GOE spectra.

It is also of interest to understand the behavior of U for Poisson spectra. A Poisson spectrum has uncorrelated spacings and has been said to show "maximum randomness" [14]; the spacing distribution is an exponential function. The average value of U for a series of N eigenvalues from a Poisson spectrum is [15]:

$$\begin{aligned} < U >_{Poisson} &= \sum_{i=1}^{N} [(N - i)\ln i - V(\xi_i)] \\ &\quad - \frac{1}{4} N(N-1)(2\ln N - 3). \end{aligned} \tag{8}$$

Empirically, this result can be approximated by

$$< U >_{Poisson} \approx 0.499 \ln N + 0.174. \tag{9}$$

Application to simulated spectra

Our first step in determining the usefulness of U is to apply it to pure and complete spectra. Ensembles of 2500 GOE and Poisson spectra ranging in size from $N = 10$ levels to $N = 1000$ levels were generated. Because GOE spectra have an energy-dependent level density and because U (like most eigenvalue statistics) assumes constant level density, it is necessary to "unfold" the level density of the GOE spectra before further analysis. The details of this unfolding procedure are given in [15]. The Poisson spectra already have a constant average level density and do not require an unfolding of this type.

Our confining potential function V assumed an average spacing of 1 (in appropriate units) and that the levels were centered on the interval $[-\mathscr{L}, \mathscr{L}]$. To ensure that these conditions are met exactly, we apply two more transformations to the energy eigenvalues (for both GOE and Poisson spectra), the first to set the average spacing and the second to center the eigenvalues (see ref. [15]). One can now apply Eq. (7) to these eigenvalues.

For each spectrum in each ensemble, a value of U is determined. We characterize each collection of spectra by the median value of U and by the 16th and 84th percentiles of the 2500 values. We utilize the median rather than the mean because that statistic seems more appropriate as a guide of what to expect for a single set of eigenvalues (these distributions are generally not symmetric, especially for Poisson spectra). Similarly, we use the percentiles to provide a better idea of the actual variation expected in a single measurement (instead of a standard deviation).

From an experimental perspective, the key questions often involve a single spectrum of eigenvalues. One needs not only a value of U for a single spectrum, but also an estimate of the uncertainty in that value of U. An approach that has been utilized in the calculation of Δ_3 to deal with this issue is to use overlapping subintervals of the spectrum [9], since one can then use the standard deviation of the values for the different subintervals to estimate an uncertainty. That same method can be applied to estimate uncertainties in U.

To apply this approach, we divide each spectrum into subspectra of size N_s and let each subspectrum overlap the next one by $N_s/2$ levels. For each subspectrum, the energies are renormalized and U is calculated as described above. The value of U for an entire spectrum is taken to be the mean of the values of U for the various subspectra. Again, one can look at averages over an ensemble of GOE matrices by taking the individual values of U for each spectrum and determining their median and uncertainties. Of great importance in the implementation of this method is determining what the size of the subspectra should be. Shorter subspectra means there are more of them to average, but longer subspectra might be expected to reduce the statistical fluctuations.

To establish the best "window" size, we calculated the ensemble medians for a variety of different window sizes N_s. We found that a larger number of windows N_{win} (a smaller value of N_s) biases the calculation toward values of U that are too small. The best choice for all values of N appears to be $N_{win} = 2$. We shall refer to calculations performed with $N_{win} = 2$ as "optimized" calculations, and we use this method from here on. As stated, the motivation for undertaking this approach is to allow an estimate of the uncertainty of U for a single spectrum.

Having specified a method for using subintervals, we characterize the ensembles for different size spectra using the optimized calculations. Results are illustrated in Fig. 1. Several features are clearly evident from the figure. As

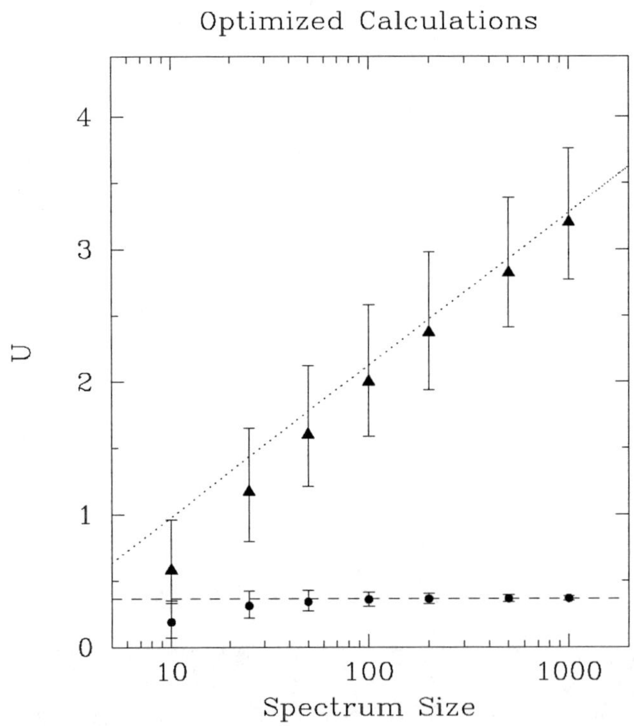

FIGURE 1. The internal energy U as a function of spectrum size. Values for GOE spectra are indicated with circles; those for Poisson spectra are indicated with triangles. The dashed line indicates the expected value of 0.365 for GOE spectra, and the dotted line shows the expected behavior for Poisson spectra.

expected, for very small sample size the ranges of values for Poisson and GOE spectra nearly overlap. As compared with using the entire spectrum, the utilization of overlapping subspectra reduces the spread in values obtained for different Poisson spectra while producing only small increases in the spread for GOE spectra.

For each spectrum size, even $N = 10$, the expected value is well within the range of predicted values for an ideal GOE spectrum. For the GOE spectra, the spread in values decreases as N increases, while the spread for Poisson spectra is (1) relatively insensitive to the value of N and (2) significantly larger than the spread for GOE spectra. We note that large variations over different members of the ensemble seem to be a standard feature for statistics of Poisson spectra, as similar results for Δ_3 were discussed in Ref. [16]. The figure illustrates clearly the asymmetric distributions for the Poisson spectra; for each spectrum size, the average value of U over the ensemble is even closer to the expected value than is the median. Most importantly, there is little chance of mistaking a GOE spectrum for a Poisson spectrum or vice versa once the spectrum size exceeds $N \approx 30-40$.

EFFECTS OF MISSING AND SPURIOUS LEVELS

Effect on U

From an experimental perspective, one of the important issues for any spectral statistic is how it is affected by imperfect data. It is extremely difficult in most experiments to obtain pure and complete data: limited resolution often causes levels to be missed, while difficulties in identifying the appropriate quantum numbers for a level may lead to spurious levels in a spectrum. Discussions of how missing levels affect spectral statistics have been given recently by Bohigas and Pato [11, 17], but quantitative analysis of the effects of spurious levels has not been discussed in as much detail.

To test how missing or spurious levels affect the internal energy, we have simulated both spectra with missing levels and spectra with spurious levels. The details of this procedure are described in ref. [15]. An ensemble of 2500 spectra was generated for each scenario (sample size N and either missing fraction f_m or spurious fraction f_s), and U was determined. The results are illustrated in Fig. 2

Our results show that either missing levels or spurious levels cause both U and the variation in U to increase relative to a pure and complete spectrum. Spurious levels have a slightly larger effect than do missing levels. Over these ranges of f_m and f_s, the deviations in U from the expected values for pure/complete data appear to be linear with either f_m or f_s. Most importantly, very good separation between GOE spectra and Poisson spectra remains; even for an $N = 50$ GOE spectrum missing $\approx 20\%$ of its levels, the difference in U for this spectrum and a Poisson spectrum is still nearly two standard deviations.

Comparison between U and Δ_3

We have shown that the internal energy U does a good job of distinguishing GOE behavior from Poisson behavior as long as the spectrum exceeds a certain minimum size ($\approx$ 50 levels), even when the data are imperfect. We conclude by comparing U to the Dyson-Mehta Δ_3 statistic. For an energy interval $[E_{min}, E_{max}]$, Δ_3 is defined in terms of the "staircase" function $N(E)$ – the number of levels with energy less than or equal to E – as follows:

$$\Delta_3(L) = \min_{A,B} \frac{1}{E_{max} - E_{min}} \int_{E_{min}}^{E_{max}} [N(E) - AE - B]^2 dE. \tag{10}$$

Here L is the number of average spacings between E_{min} and E_{max}. The expected value of Δ_3 for a Poisson spectrum is $L/15$; the result must be evaluated numerically for GOE spectra, but a good approximation for large L is

$$\Delta_{3_{GOE}}(L) \approx \frac{1}{\pi^2} [\ln L - 0.0687]. \tag{11}$$

The large difference between a linear function for Poisson spectra and a logarithmic function for GOE spectra is part of what makes Δ_3 such an appealing choice to characterize spectra. However, the effect on Δ_3 of missing levels from a GOE spectrum is to introduce a linear term [17] that reduces significantly the separation between GOE and Poisson spectra.

For U, we also have a significant difference in expected behavior for GOE and Poisson spectra, although this time it is the difference between a logarithmic function for Poisson spectra and a constant value for GOE spectra. Both U and Δ_3 are measuring long-range order within a spectrum. To compare the behavior of these two statistics, we have calculated Δ_3 for each of the spectra for which we earlier calculated U; although it has become common to study Δ_3 as a function of L, here we just use the value of Δ_3 obtained for the entire spectrum. This approach provides a simple method of comparison between the two statistics. Again, the median and the 16th and 84th percentiles are used to represent the ensemble behavior in each case. The behavior of both U and Δ_3 are shown in Fig. 2.

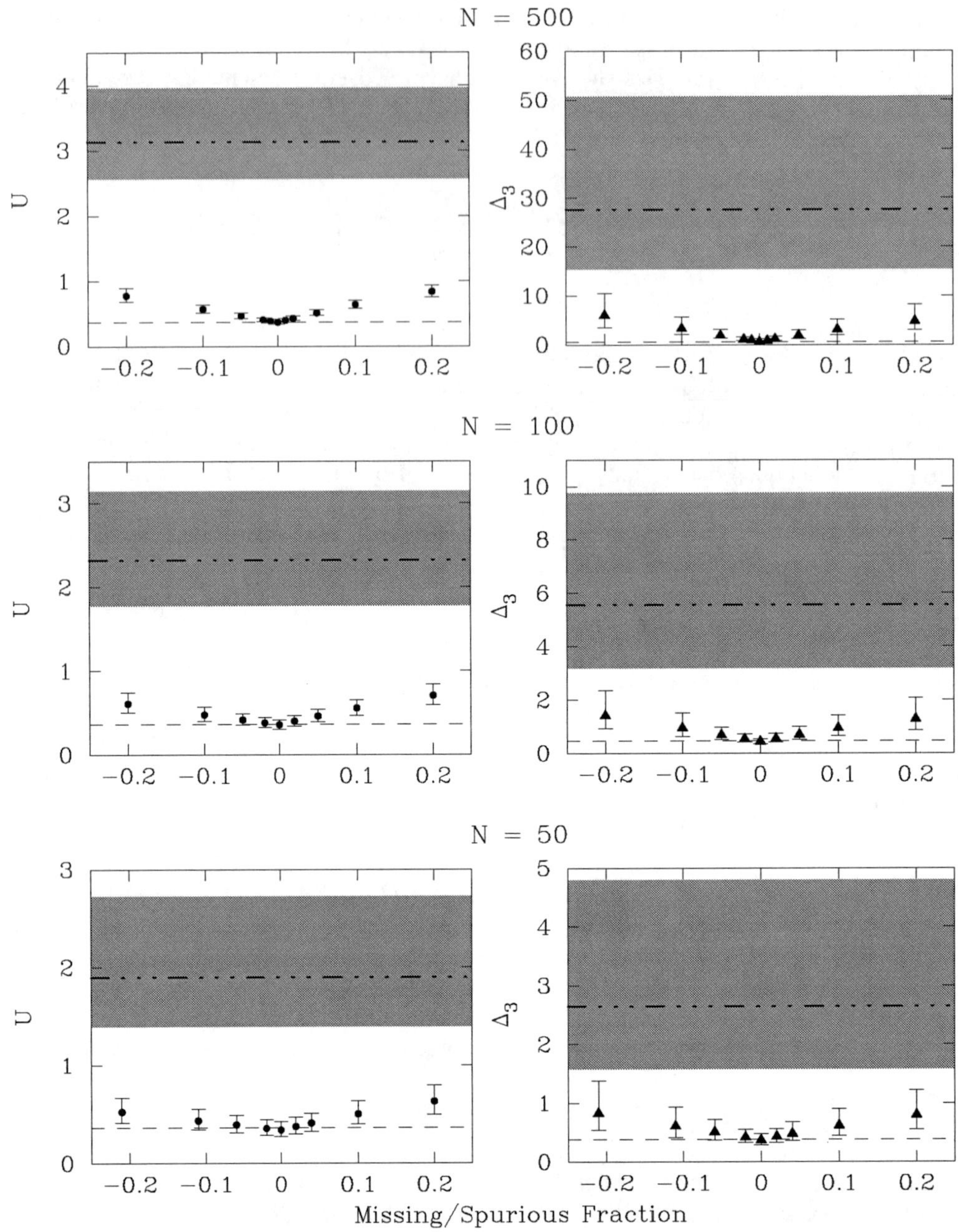

FIGURE 2. U (circles) and Δ_3 (triangles) as a function of missing/spurious fraction. Δ_3 is calculated for the full spectrum, while the U values are obtained as described in the text. The dashed lines indicate the expected value of U and Δ_3 for GOE spectra, while the dash-dot lines show the median behavior for Poisson spectra. The shaded regions show the variations in Poisson behavior observed in the Monte Carlo studies (missing or spurious levels do not affect the statistics of Poisson spectra, so the same values are applicable for any missing or spurious fraction). Note that the scales for U and Δ_3 are very different.

Examination of Fig. 2 shows that the median values of Δ_3 are more sensitive to missing or spurious levels than are the median values of U but that the spread of Δ_3 values is also significantly larger. It can also be seen that the spectrum-to-spectrum variations for Poisson spectra are much larger for Δ_3 than for U. When one considers how likely a GOE spectrum (even if somewhat imperfect) is to be mistaken for a Poisson spectrum or a Poisson spectrum for a GOE spectrum, this figure makes clear that U provides a better separation of the two behaviors than does Δ_3.

In principle, such statistics can be used to estimate the number of missing or spurious levels in a spectrum; this has been done with nearest-neighbor spacing distributions [18, 19]. Mulhall *et al.* [20] have recently formulated such an approach using the Δ_3 statistic and applied it to neutron resonances. Comparison of this new approach with the spacing method [18, 19] yielded excellent agreement for the fraction of missing levels. Establishing and testing such a method for U is beyond the scope of this paper, although such a techniques might be very helpful to address this question. Although the increased sensitivity of Δ_3 to mistakes initially appears to favor it for such tests over U, the smaller uncertainties in the values of U may offset this advantage. In fact, the results shown in Fig. 2 suggest that U would provide an equivalent basis for this estimation.

ACKNOWLEDGMENTS

This work was supported in part by the U. S. Department of Energy Grants No. DE-FG02-97-ER41042 and DE-FG02-96ER40990. The two figures are reprinted from Nucl. Instrum. Methods A **581**, J. F. Shriner, Jr., M. P. Pato, G. E. Mitchell, and A. P. B. Tufaile, "A thermodynamic analysis of energy eigenvalues," pp. 831-838, Copyright (2007), with permission from Elsevier.

REFERENCES

1. E. P. Wigner, “Results and Theory of Resonance Absorption,” in *Statistical Theories of Spectra: Fluctuations*, edited by C. E. Porter, Academic Press, New York, 1965, p. 199.
2. C. E. Porter, *Statistical Theories of Spectra*, Academic Press, New York, 1965.
3. F. J. Dyson, and M. L. Mehta, *J. Math. Phys.* **4**, 701 (1963).
4. H. I. Liou, H. S. Camarda, S. Wynchank, M. Slagowitz, G. Hacken, F. Rahn, and J. Rainwater, *Phys. Rev. C* **5**, 974 (1972).
5. H. I. Liou, H. S. Camarda, and F. Rahn, *Phys. Rev. C* **5**, 1002 (1972).
6. W. M. Wilson, E. G. Bilpuch, and G. E. Mitchell, *Nucl. Phys.* **A245**, 285 (1975).
7. W. A. Watson, III, E. G. Bilpuch, and G. E. Mitchell, *Phys. Rev. C* **24**, 1992 (1981).
8. R. U. Haq, A. Pandey, and O. Bohigas, *Phys. Rev. Lett.* **48**, 1086 (1982).
9. O. Bohigas, M. J. Giannoni, and C. Schmit, *Phys. Rev. Lett.* **52**, 1 (1984).
10. T. Guhr, A. Müller-Groeling, and H. Weidenmüller, *Phys. Rep.* **299**, 189 (1998).
11. O. Bohigas, and M. P. Pato, *Phys. Rev. E* **74**, 036212 (2006).
12. F. J. Dyson, *J. Math. Phys.* **3**, 140 (1962).
13. M. L. Mehta, *Random Matrices*, vol. 142 of *Pure and Applied Mathematics*, Academic, New York, 2004, third edn.
14. O. Bohigas, M.-J. Giannoni, and C. Schmit, “Spectral Fluctuations of Classically Chaotic Quantum Systems,” in *Quantum Chaos and Statistical Nuclear Physics*, edited by T. H. Seligman, and H. Nishioka, Lecture Notes in Physics 263, Springer-Verlag, New York, 1986, p. 18.
15. J. F. Shriner, Jr., M. P. Pato, G. E. Mitchell, and A. P. B. Tufaile, *Nucl. Insrum. Methods* **A581**, 831 (2007).
16. J. F. Shriner, Jr., and G. E. Mitchell, *Z. Phys. A* **342**, 53 (1992).
17. O. Bohigas, and M. P. Pato, *Phys. Lett. B* **595**, 171 (2004).
18. U. Agvaanluvsan, G. E. Mitchell, J. F. Shriner, Jr., and M. Pato, *Nucl. Instrum. Methods* **A498**, 459 (2003).
19. U. Agvaanluvsan, G. E. Mitchell, J. F. Shriner, Jr., and M. Pato, *Phys. Rev. C* **67**, 064608 (2003).
20. D. Mulhall, Z. Huard, and V. Zelevinsky, Ergodicity of the Δ_3 statistic and purity of neutron resonance data, arxiv:physics.nucl-th/0708.0774, to be published in Phys. Rev. C (2007).

Interaction matrix element fluctuations in quantum dots

L. Kaplan* and Y. Alhassid†

*Department of Physics, Tulane University, New Orleans, LA 70118, USA
†Center for Theoretical Physics, Sloane Physics Laboratory, Yale University, New Haven, CT 06520, USA

Abstract. In the Coulomb blockade regime of a ballistic quantum dot, the distribution of conductance peak spacings is well known to be incorrectly predicted by a single-particle picture; instead, matrix element fluctuations of the residual electronic interaction need to be taken into account. In the normalized random-wave model, valid in the semiclassical limit where the number of electrons in the dot becomes large, we obtain analytic expressions for the fluctuations of two-body and one-body matrix elements. However, these fluctuations may be too small to explain low-temperature experimental data. We have examined matrix element fluctuations in realistic chaotic geometries, and shown that at energies of experimental interest these fluctuations generically exceed by a factor of about 3-4 the predictions of the random wave model. Even larger fluctuations occur in geometries with a mixed chaotic-regular phase space. These results may allow for much better agreement between the Hartree-Fock picture and experiment. Among other findings, we show that the distribution of interaction matrix elements is strongly non-Gaussian in the parameter range of experimental interest, even in the random wave model. We also find that the enhanced fluctuations in realistic geometries cannot be computed using a leading-order semiclassical approach, but may be understood in terms of short-time dynamics.

Keywords: Interaction matrix elements, coulomb blockade, quantum chaos, semiclassical methods
PACS: 73.23.Hk, 05.45.Mt, 73.63.Kv, 73.23.-b

INTRODUCTION

There has been much interest in the mesoscopic properties of quantum dots whose single-particle dynamics are chaotic [1]. The generic fluctuation properties of the single-particle spectrum and wave functions in such dots are usually described by random matrix theory (RMT) [2]. However, in almost-isolated dots, electron-electron interactions are important and must also be taken into account. The simplest model is the constant interaction (CI) model, in which the interaction is taken to be the classical charging energy. Charging energy leads to Coulomb blockade peaks in the conductance versus gate voltage. Each peak occurs as the gate voltage is tuned to compensate for the Coulomb repulsion and an additional electron tunnels into the dot. For a fixed number of electrons, the CI model is essentially a single-particle model and RMT can be used to derive the statistical properties of the conductance peak heights [3]. While the CI plus RMT model has explained (at least qualitatively) [3, 4, 5] several observed features of the peak height fluctuations [6, 7, 8], there have been significant discrepancies with experimental data, in particular regarding the peak spacing statistics [9, 10, 11, 12]. Such discrepancies indicate the importance of interactions beyond charging energy.

A more systematic way of treating electron-electron interactions in chaotic ballistic dots is to expand the interaction in a small parameter, the inverse of the Thouless conductance $g_T \sim kL \sim \sqrt{N}$, where k is the Fermi wave number, L is the typical linear size of the dot (i.e. $L = \sqrt{V}$, where V is the area), and N is the number of electrons in the dot. In the limit of large Thouless conductance (equivalently, in the semiclassical or many-electron limit), only a few interaction terms survive, constituting the interacting part of the universal Hamiltonian [13, 14]. These universal interaction terms include, in addition to charging energy, a constant exchange interaction. The inclusion of an exchange interaction has explained the statistics of peak heights at low and moderate temperatures as well as the suppression of the peak spacing fluctuations [15, 16]. However, at low temperatures, the peak spacing distribution remains bimodal even when the exchange interaction is included, while none of the experimental distributions are bimodal [9, 11, 10, 12].

For finite Thouless conductance, residual interactions beyond the universal Hamiltonian must be taken into account. In a Hartree-Fock-Koopmans [17] approach (assuming the Hartree-Fock single-particle wave functions do not change as electrons are added to the dot), the peak spacings can be expressed directly in terms of certain (diagonal) two-body interaction matrix elements [18]. Sufficiently large fluctuations of these interaction matrix elements can explain the absence of bimodality in the peak spacing distribution [18, 19]. In a diffusive dot, the variance of the matrix elements of the screened Coulomb interaction was shown to behave as Δ^2/g_T^2 to leading order in $1/g_T$ [20], where the single-

CP995, *Nuclei and Mesoscopic Physics - WNMP 2007*, edited by P. Danielewicz, P. Piecuch, and V. Zelevinsky

electron mean level spacing Δ sets the energy scale. However, dots studied in the experiments are usually ballistic.

An additional contribution to the peak spacing fluctuations originates in surface charge effects [20]. In a finite size system, screening leads to the accumulation of charge on the surface of the dot. The confining one-body potential is then modified upon the addition of an electron to the dot.

In this paper we will summarize some of our recent results on fluctuations of the two-body interaction matrix elements and of the surface charge one-body matrix elements in ballistic dots [21]. We begin by defining the matrix elements of interest, and note that their fluctuations can be expressed to leading order in $1/g_T$ in terms of spatial correlations within single-electron wave functions. Berry's random wave conjecture [22] provides the first approximation for these correlations, which (in contrast with the situation for diffusive dots) is geometry independent. However, the spatial correlator of wave function intensity obtained from the Berry conjecture is not consistent with the normalization requirement of the wave functions [23]. We discuss the importance of normalization corrections to the random wave correlator, and show how the variances of the two-body and one-body interaction matrix elements may be computed in a normalized random wave model. We also find that the distribution of interaction matrix elements may be very far from Gaussian, even in a normalized random wave model where the wave functions are very close to Gaussian.

An interesting quantity that we refrain from discussing here is the covariance of interaction matrix elements, relevant for understanding spectral scrambling when several electrons are added to the dot [24].

We then proceed to study matrix element fluctuations in actual chaotic systems, using a family of modified quarter-stadium billiards as an example. We find strongly enhanced fluctuations in comparison with the normalized random wave model. Semiclassical corrections due to bounces from the dot's boundaries lead to only a modest increase in the fluctuations, and do not correctly predict the scaling with kL in the experimentally relevant range. Insight into the underlying mechanism of fluctuation enhancement is obtained by examining a family of quantum maps. An important conclusion is that the expansion in $1/g_T$, while asymptotically correct, can be problematic in quantifying matrix element fluctuations in the regime relevant to experiments. Finally, in the last section we study systems beyond the chaotic regime, i.e., billiards dominated by marginally-stable bouncing-ball modes as well as billiards with mixed dynamics (partly regular and partly chaotic).

INTERACTION MATRIX ELEMENTS

The general two-body interaction matrix element for potential $v(\vec{r}\,',\vec{r})$ is given by

$$v_{\alpha\beta;\gamma\delta} \sim \int_V \int_V d\vec{r}\, d\vec{r}'\, \psi^*_\alpha(\vec{r})\, \psi^*_\beta(\vec{r}')\, v(\vec{r}',\vec{r})\, \psi_\gamma(\vec{r})\, \psi_\delta(\vec{r}')\,, \tag{1}$$

where ψ_α, ψ_β, ... are single-electron orbital wave functions. In practice, screening of the residual electron-electron interaction causes the range of the interaction to be much smaller than the size of the dot. Modeling the residual interaction as a contact interaction, we obtain $v_{\alpha\beta;\gamma\delta} = \Delta V \int_V d\vec{r}\, \psi^*_\alpha(\vec{r})\, \psi^*_\beta(\vec{r})\, \psi_\gamma(\vec{r})\, \psi_\delta(\vec{r})$, where the mean single-particle level spacing Δ sets the energy scale. We note that for a contact interaction, exchange terms have precisely the same form and need not be considered separately.

Three distinct situations must be treated. First, the diagonal matrix elements

$$v_{\alpha\beta} = v_{\alpha\beta;\alpha\beta} = \Delta V \int_V d\vec{r}\, |\psi_\alpha(\vec{r})|^2\, |\psi_\beta(\vec{r})|^2 \tag{2}$$

($\alpha \neq \beta$) appear when two electrons are found on the same two orbitals α and β before and after the interaction. Secondly, we have the double-diagonal matrix elements $v_{\alpha\alpha} = v_{\alpha\alpha;\alpha\alpha}$, where two electrons are annihilated from one orbital and created on the same orbital (for a contact interaction, $v_{\alpha\alpha}$ is simply an inverse participation ratio in position space). Finally, we may consider the off-diagonal matrix elements $v_{\alpha\beta\gamma\delta}$ where the four orbitals are all distinct. We will focus mostly on the diagonal matrix elements $v_{\alpha\beta}$, but the other two situations are treated similarly.

The mean $\overline{v_{\alpha\beta}}$, averaged over all pairs of orbitals or over an ensemble, may always be absorbed into the mean field part of the Hamiltonian. Thus, we are primarily interested not in the average, but in the fluctuations. To leading order in $g_T \sim kL$, the dominant contribution to the variance arises from correlations between the intensities of a *single* wave function at different points [24]:

$$\overline{\delta v^2_{\alpha\beta}} = \Delta^2 V^2 \int_V \int_V d\vec{r}\, d\vec{r}'\, C^2(\vec{r},\vec{r}') + \cdots \tag{3}$$

where

$$C(\vec{r},\vec{r}') = \overline{|\psi(\vec{r})|^2\,|\psi(\vec{r}')|^2} - \overline{|\psi(\vec{r})|^2}\;\overline{|\psi(\vec{r}')|^2}\,. \tag{4}$$

Thus, the problem of two-body matrix elements statistics has been reduced to an apparently simpler problem of understanding the statistics of individual single-electron wave functions in a chaotic potential. Terms we have omitted in Eq. (3) involve spatial correlators between $|\psi_\alpha(\vec{r})|^2$ and $|\psi_\beta(\vec{r}')|^2$; such correlators are essential for computing the covariance $\overline{\delta v_{\alpha\beta}\delta v_{\alpha\gamma}}$, but are subleading in the calculation of the variance.

RANDOM WAVE APPROXIMATION

For a two-dimensional billiard system, the random wave model implies that a typical chaotic wave function may be written locally as a random superposition of plane waves at fixed energy $\hbar^2k^2/2m$. Adopting the usual normalization $\overline{|\psi(\vec{r})|^2} = 1/V$, we obtain an amplitude correlator $\overline{\psi^*(\vec{r})\psi(\vec{r}')} = \frac{1}{V}J_0(k|\vec{r}-\vec{r}'|)$ and intensity correlator

$$C^{\text{rw}}(\vec{r},\vec{r}') = \frac{1}{V^2}\frac{2}{\beta}J_0^2(k|\vec{r}-\vec{r}'|)\,, \tag{5}$$

where $\beta = 1$ or 2 represents the presence or absence of time reversal symmetry, i.e. the absence or presence of an external magnetic field, respectively.

The intensity correlator $C^{\text{rw}}(\vec{r},\vec{r}')$ is valid to leading order in $|\vec{r}-\vec{r}'|/L$, but becomes problematic when applied to to all $\vec{r}$, $\vec{r}'$ in the finite area V. Indeed wave function normalization requires the correlator to vanish on average,

$$\int_V d\vec{r}\, C(\vec{r},\vec{r}') = 0\,. \tag{6}$$

However, $C^{\text{rw}}(\vec{r},\vec{r}')$ in (5) manifestly does not satisfy the condition (6). The reason for this failure is that in the random wave model, normalization is satisfied only for the ensemble average, i.e., $\overline{\int_V d\vec{r}\,|\psi(\vec{r})|^2} = 1$, but not for each individual random superposition of plane waves $\psi(\vec{r})$.

This deficiency can be corrected by introducing the *normalized* random wave model, in which each "random" wave function is normalized in area V, i.e.,

$$\psi^{\text{norm}}(\vec{r}) = \psi(\vec{r})/\int_V d\vec{r}\,|\psi(\vec{r})|^2\,. \tag{7}$$

Assuming the deviation from exact normalization in area V is small for the original random waves $\psi(\vec{r})$, we apply a perturbative scheme [21] and obtain

$$C^{\text{norm}}(\vec{r},\vec{r}') = \tilde{C}(\vec{r},\vec{r}') + O\left(\frac{1}{(kL)^{3/2}}\right)\,, \tag{8}$$

where

$$\tilde{C}(\vec{r},\vec{r}') = C^{\text{rw}}(\vec{r},\vec{r}') - \frac{1}{V}\int_V d\vec{r}_a C^{\text{rw}}(\vec{r},\vec{r}_a) - \frac{1}{V}\int_V d\vec{r}_a C^{\text{rw}}(\vec{r}_a,\vec{r}') + \frac{1}{V^2}\int_V\int_V d\vec{r}_a d\vec{r}_b C(^{\text{rw}}\vec{r}_a,\vec{r}_b)\,. \tag{9}$$

The leading-order normalized correlator $\tilde{C}(\vec{r},\vec{r}')$ was previously derived in Ref. [23] by adding a weak smooth disorder and using the non-linear supersymmetric sigma model. Higher-order contributions may be computed systematically; the full expression for the normalized two-point correlator $C^{\text{norm}}(\vec{r},\vec{r}')$ involves all unnormalized n-point correlators $\overline{(|\psi(\vec{r}_1)|^2 - \frac{1}{V})\cdots(|\psi(\vec{r}_n)|^2 - \frac{1}{V})}$, starting with the unnormalized three-point correlator that gives rise to the $O((kL)^{-3/2})$ correction in Eq. (8). In practice the higher-order terms are small; the approximation $\tilde{C}$ already satisfies (6) and can be proven to imply normalization of individual wave functions [21]. However, n-point correlators will be important below, when we discuss the shape of matrix element distributions.

Two-body matrix elements

Substituting the normalized random wave correlator $\tilde{C}$ into Eq. (3), we obtain

$$\overline{\delta v_{\alpha\beta}^2} = \Delta^2\frac{3}{\pi}\left(\frac{2}{\beta}\right)^2\frac{\ln kL + b_g}{(kL)^2} + O\left(\frac{\Delta^2}{(kL)^3}\right)\,, \tag{10}$$

where the leading $\ln kL/(kL)^2$ term may be obtained already from the unnormalized correlator C^{rw} and depends only on the symmetry class, while the shape-dependent coefficient b_g may be easily evaluated by an integral involving Bessel functions over the area V of interest.

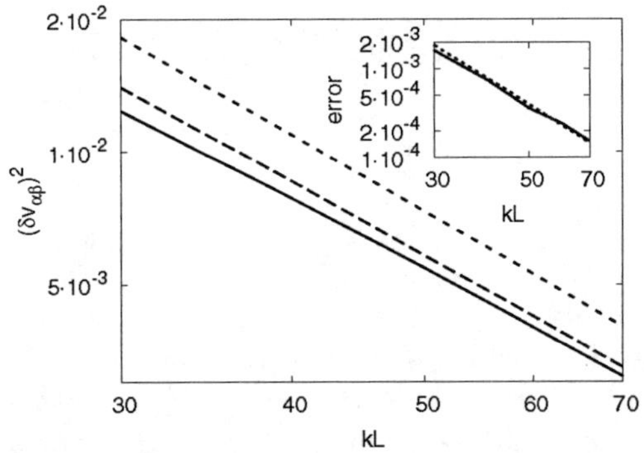

FIGURE 1. The variance of the two-body matrix element $v_{\alpha\beta}$ versus kL in the real ($\beta = 1$) random wave model: (a) the solid curve is the result of exact numerical simulations; (b) the long-dashed line is the result of substituting the normalized random wave correlator $\tilde{C}(\vec{r},\vec{r}')$ into Eq. (3); (c) the short-dashed line is the result of using the unnormalized correlator $C^{\mathrm{rw}}(\vec{r},\vec{r}')$. In the inset, the solid line is the difference between the full result (a) and the approximation (b); the dashed line indicates that omitted terms scale as $\frac{1}{(kL)^3}$. Here and in all following figures, the single-particle level spacing Δ, which sets the overall energy scale, is set to unity.

In Fig. 1 we show the results for a disk geometry ($b_g = -0.1$) in the range $30 \le kL \le 70$, corresponding roughly to the parameter range relevant for experiments (i.e., a few hundred electrons in the dot). The discrepancy between the analytic result (10) and exact numerics is indeed $O(1/(kL)^3)$, resulting in an accuracy of $\sim 10\%$ for the analytic formula. In practice, the dependence of b_g on the geometry is weak, e.g. deforming the disk to an ellipse with an aspect ratio of 16 while keeping the area fixed produces a change of only $\sim 5-6\%$ in the $v_{\alpha\beta}$ variance for the experimental range of kL. Thus, for all practical purposes, shape effects on the $v_{\alpha\beta}$ variance can be ignored (at least within the normalized random wave model).

Also, at this subleading order, we must in principle take into account the wave number difference $\delta k = k_\alpha - k_\beta$. For $(k_\alpha + k_\beta)/2 \gg \delta k \gg 1/L$ (i.e. for a wave number difference that is quantum mechanically large but classically small), this results merely in a modification of the geometry-dependent coefficient in Eq. (10): $b_g \to b_g - (1/3)\ln \delta kL$. In practice, for reasonable separations δkL, the consequent reduction in the variance is at most 10%, and may be safely ignored compared to the much larger dynamical effects to be discussed later.

The variance of double-diagonal or off-diagonal matrix elements may be computed similarly. To leading order, $O\left(\ln kL/(kL)^2\right)$, all the results arise from integrating the unnormalized correlator (5) and differ only by geometry-independent combinatorial prefactors. However, normalization-related subtraction gives different results in the three situations, and thus the subleading coefficient b_g in (10) must be replaced with b'_g or b''_g in the case of $v_{\alpha\alpha}$ or $v_{\alpha\beta\gamma\delta}$, respectively. Thus the variance ratios converge to universal shape-independent constants in the $kL \to \infty$ limit, but the convergence is logarithmically slow. For example, in the presence of time-reversal symmetry,

$$\overline{\delta v_{\alpha\alpha}^2}/\overline{\delta v_{\alpha\beta}^2} = 6 + \frac{b'_g - b_g}{\ln kL} + \cdots = 6 - \frac{2.15}{\ln kL} + \cdots, \tag{11}$$

to leading order in $1/\ln kL$. For a disk, this ratio barely reaches 3 in the kL range of experimental interest. This is our first indication that beautiful analytic results valid as $kL \to \infty$ may not always have relevance to experiments, even when the dot contains hundreds or thousands of electrons.

One-body matrix elements

When an electron is added to the finite dot, charge accumulates on the surface and its effect can be described by a one-body potential energy $\mathcal{V}(\vec{r})$. The diagonal matrix elements of $\mathcal{V}(\vec{r})$ are given by

$$v_\alpha \equiv \mathcal{V}_{\alpha\alpha} = \int_V d\vec{r}\, |\psi_\alpha(\vec{r})|^2\, \mathcal{V}(\vec{r}). \tag{12}$$

Again, we wish to express the variance in terms of the wave function intensity correlator:

$$\overline{\delta v_\alpha^2} = \int_V \int_V d\vec{r}\, d\vec{r}'\, \mathscr{V}(\vec{r}) C(\vec{r},\vec{r}') \mathscr{V}(\vec{r}'), \tag{13}$$

Because only one power of C appears in the variance, the integral is dominated by distant pairs of points $|\vec{r}-\vec{r}\,'| \sim L$, and scales as $1/kL$:

$$\overline{\delta v_\alpha^2} = \frac{c_g}{\beta} \frac{\Delta^2}{kL} + O\left(\frac{\Delta^2}{(kL)^2}\right), \tag{14}$$

where c_g is a shape-dependent dimensionless coefficient. We note that normalization-related subtraction of the correlator, which had only a moderate effect on the two-body matrix element fluctuations, here reduces the variance by a full order of magnitude, resulting in a very small dimensionless coefficient: $c_g = 0.035$ for a disk, and even smaller for less symmetric shapes. This is due the fact that the integrand (13) ceases to be everywhere positive after subtraction (9), in contrast with the integrand in (3). The analytic expression (14) is in excellent agreement with exact Monte Carlo simulations (not shown). Due to the small size of the coefficient c_g, the numerical value of $\overline{\delta v_\alpha^2}$ is smaller in the physically interesting kL regime than the corresponding result for the two-body matrix element variance $\overline{\delta v_{\alpha\beta}^2}$, despite the fact that the former is parametrically larger in a $1/kL$ expansion.

Matrix element distributions

The central limit theorem implies that all interaction matrix elements in the random wave model must be distributed as Gaussian random variables as $kL \to \infty$. This justifies our focus so far on the variance (and covariance) of these matrix elements, to the neglect of higher moments. However, we have seen above that non-universal finite kL effects are sometimes significant in the experimentally relevant regime $kL \le 70$, e.g., for the variance ratio (11). Thus, we should look explicitly at the shape of matrix element distributions for finite kL.

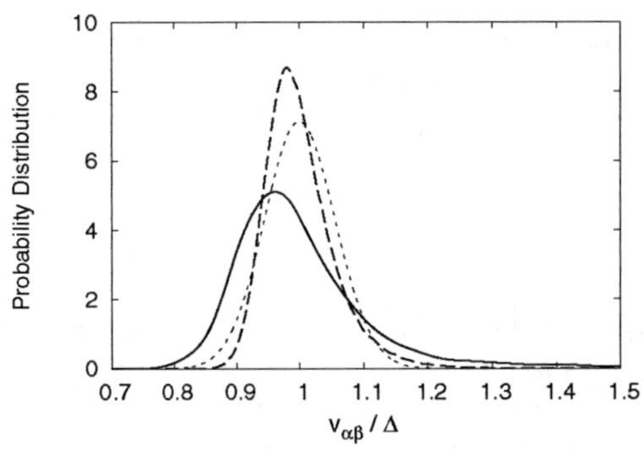

FIGURE 2. The distribution of diagonal interaction matrix elements $v_{\alpha\beta}$ is shown for real random waves in a disk (dashed curve) and for actual eigenstates in a modified quarter-stadium billiard geometry (solid curve, see below) at $kL = 70$. A Gaussian distribution with the same mean and variance as the random wave distribution is shown as a dotted curve for comparison.

Data for the diagonal matrix elements $v_{\alpha\beta}$ in shown in Fig. 2. Results for other shapes and for other matrix elements (e.g., $v_{\alpha\alpha}$, $v_{\alpha\beta\gamma\delta}$, or v_α) are qualitatively similar. We note that the numerically obtained interaction matrix element distribution for random waves (dashed curve) has a long tail on the right side as compared with a Gaussian distribution of the same mean and variance (dotted curve). In other words, there is an excess of anomalously large matrix elements, compensated for by a reduction in the median to a value slightly below Δ.

Deviations from a Gaussian shape can be quantified by considering higher moments. For large kL, we may estimate these higher moments in a manner analogous to our estimate for the variance in Eq. (3), but using $n-$point correlators. After some calculation [21], we obtain the skewness

$$\gamma_1 = \frac{\overline{\delta v_{\alpha\beta}^3}}{\left[\overline{\delta v_{\alpha\beta}^2}\right]^{3/2}} = b_{3g}\, c_{3\beta}^2 \left(\frac{\beta}{2}\right)^3 \left(\frac{\pi}{3}\right)^{3/2} (\ln kL)^{-3/2} \tag{15}$$

and excess kurtosis

$$\gamma_2 = \frac{\overline{\delta v^4_{\alpha\beta}} - 3\left[\overline{\delta v^2_{\alpha\beta}}\right]^2}{\left[\overline{\delta v^2_{\alpha\beta}}\right]^2} = b_{4g}\left(c^2_{4\beta} + \left(\frac{2}{\beta}\right)^4\right)\left(\frac{\pi^2}{3}\right)(\ln kL)^{-2}. \tag{16}$$

Because the decay is only logarithmic, γ_1 and γ_2 always remain ≥ 1 for values of kL relevant in the experiments. The same holds true for other matrix elements and for the higher moments. Therefore, matrix element distributions are predicted to be strongly non-Gaussian, even within the random wave model.

CHAOTIC BILLIARDS

We now investigate how dynamical effects may modify the fluctuations of interaction matrix elements beyond the (normalized) random wave model. As an example, we use a modified quarter-stadium billiard geometry [25], where the quarter-circle has radius R and the straight edge of length a has been replaced by a parabolic bump to eliminate bouncing-ball modes (Fig. 3). The system has been verified numerically to be fully chaotic for the range of parameters used. Variation of the bump size s allows us to check the sensitivity of the results to details of the billiard geometry while maintaining the chaotic character of the classical dynamics. Furthermore, variation of the parameter a allows us to control the degree of classical chaos.

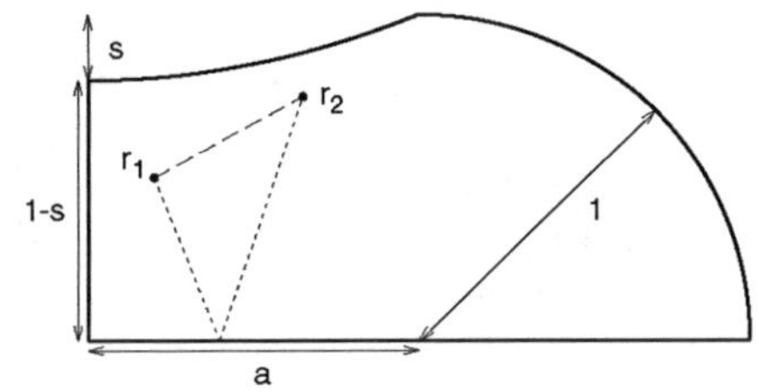

FIGURE 3. A modified quarter-stadium geometry with parameters a and s is used to illustrate dynamical effects on matrix element fluctuations. In the Figure, we set the quarter-circle radius $R = 1$. The random wave contribution to the correlator, $C^{\mathrm{rw}}(\vec{r}_1, \vec{r}_2)$ is schematically indicated by a dashed line, and a typical dynamical contribution by a dotted line.

We first study the variance of the diagonal interaction matrix elements $v_{\alpha\beta}$. Typical numerical results are shown in Fig. 4. We note the large enhancement of the billiard results over the random wave model. To understand this discrepancy, we first compare the exact numerical result for $\overline{\delta v^2_{\alpha\beta}}$ with the first term on the right hand side of Eq. (3), where the intensity correlator for normalized random waves $\tilde{C}(\vec{r}, \vec{r}')$ is replaced by the billiard correlator $C^{\mathrm{bill}}(\vec{r}, \vec{r}')$ (calculated numerically for the appropriate billiard system). The discrepancy is immediately reduced to a $\sim 5 - 10\%$ level, which is comparable to the $O(1/(kL)^3)$ discrepancy observed in the random wave model (see inset of Fig. 1). Thus, the large enhancement of $v_{\alpha\beta}$ fluctuations over the random wave prediction can be traced directly to an enhancement in the single-particle intensity correlator $C^{\mathrm{bill}}(\vec{r}, \vec{r}')$ over $\tilde{C}(\vec{r}, \vec{r}')$.

Can this dynamical enhancement of the intensity correlator (as compared with a random wave model) be obtained using a semiclassical approach? The random wave correlator $C^{\mathrm{rw}}(\vec{r}, \vec{r}')$ may be interpreted semiclassically as arising from straight-line free propagation, as indicated by the dashed line in Fig. 3. As discussed by several authors [26, 27], additional contributions to the correlator in a specific dynamical system can be associated with classical trajectories that bounce off the boundary n times on their way from $\vec{r}$ to $\vec{r}'$, such as the one indicated by a dotted line in the same figure. We obtain [21]

$$\overline{\delta v^2_{\alpha\beta}} = \Delta^2 V^2 \int_V \int_V d\vec{r}\, d\vec{r}'\, (C^{\mathrm{bill}}(\vec{r}, \vec{r}'))^2 + O\left(\frac{\Delta^2}{(kL)^3}\right) \tag{17}$$

$$= \Delta^2 \frac{3}{\pi}\left(\frac{2}{\beta}\right)^2 \frac{(\ln kL + b_g) + b_{\mathrm{sc}}}{(kL)^2} + O\left(\frac{\Delta^2}{(kL)^3}\right) \tag{18}$$

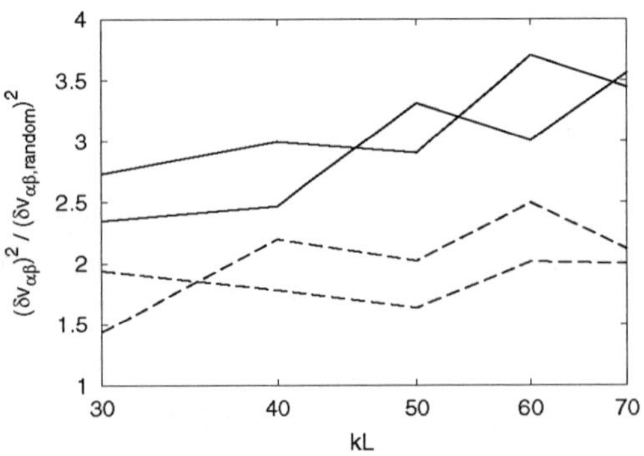

FIGURE 4. The enhancement of the $v_{\alpha\beta}$ variance over the random wave prediction is shown for sevarl billiards with Neumann boundary conditions. Solid lines: $a = 0.25$, $s = 0.1, 0.2$. Dashed lines: $a = 1.0$, $s = 0.1, 0.2$.

where $b_{\rm sc}$ is a classical constant that in practice must be determined numerically by performing the integral in Eq. (17). Random wave and semiclassical contributions to $C^{\rm bill}(\vec{r},\vec{r}')$ are of the same order except for $|\vec{r}-\vec{r}'| \ll L$; it is these short-distance pairs that result in a logarithmic enhancement of the random-wave term.

The coefficient $b_{\rm sc}$ may in practice be quite large, even for generic chaotic systems, such as the modified stadium billiard. In a diagonal approximation, $b_{\rm sc} \sim T^2_{\rm clas}$ where $T_{\rm clas}$ is the decay time of classical correlations. Qualitatively, this is consistent with Fig. 4, as fluctuations are observed to be consistently larger for the less chaotic $a = 0.25$ billiard, as compared with the $a = 1.00$ billiard. We note that both billiards are "generic", in the sense that they are not fine-tuned to obtain an anomalously long time scale $T_{\rm clas}$. We also note that varying the bump size s has a very weak effect on the matrix element statistics (as long as s is large enough to destroy the bouncing-ball modes) and serves instead to provide an estimate of the statistical fluctuations.

However, a closer look at the data shows that the numerical results cannot be explained fully by semiclassical arguments, since (18) converges to the random wave limit (10) as $kL \to \infty$, while the enhancement factor in Fig. 4 *grows* with kL in the parameter range of experimental interest.

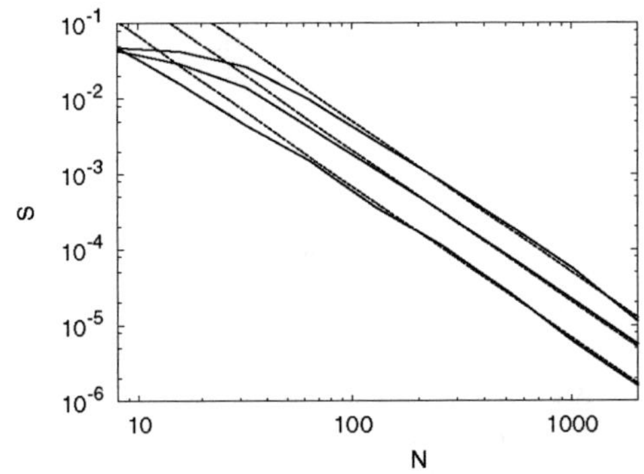

FIGURE 5. The two-body matrix element variance S for a quantum map, as a function of Hilbert space dimension N. From top to bottom, the three solid lines represent data for dominant orbit stability exponent $\lambda_0 = 0.25, 0.50, 1.00$. The three dashed lines indicate the asymptotic $1/N^2$ behavior for each case in the semiclassical regime of large N.

This anomalous behavior results from a combined effect of two factors: the large numerical value of $b_{\rm sc}$ for generic dynamical systems, and saturation of the $1/(kL)^2$ behavior at moderate (< 100) values of kL. As the classical system becomes less unstable and the correlation time $T_{\rm corr}$ increases, $b_{\rm sc}$ also increases, leading to greatly enhanced matrix element variance at very large values of kL (18). Because the variance is bounded above independent of kL, the $(kL)^{-2}$ growth in the variance necessarily breaks down at moderate values of kL. This saturation sets in at ever larger values of kL as the system becomes less unstable. Alternatively, one may note that the natural expansion parameter for interaction matrix element fluctuations in a dynamical system is not $1/kL$ but rather the inverse Thouless conductance $g_T^{-1} \sim T_{\rm corr}/kL$, and the semiclassical contribution with prefactor $b_{\rm sc}$ in Eq. (18) is the leading $O(g_T^{-2})$ effect in such an expansion. Terms of third and higher order in g_T^{-1}, although formally subleading and not included in a semiclassical calculation, become quantitatively as large as the leading $O(g_T^{-2})$ term when g_T falls below some characteristic value.

Furthermore, if one considers chaotic billiards with a long correlation decay time $T_{\rm corr}$, the importance of formally subleading terms in the g_T^{-1} expansion will extend to quite large values of kL.

The above assertions are explicitly confirmed for a quantum map model, which has scaling behavior analogous to that of a two-dimensional billiard, with the number of states $N = 2\pi/\hbar$ playing the role of semiclassical parameter $kL = pL/\hbar$ in the billiard [30, 31]. As in the billiard, a free parameter λ_0 in the definition of the map allows for control of the classical correlation decay time $T_{\rm corr}$. We see in Fig. 5 that the expected N^{-2} behavior of the variance is observed at sufficiently large N, for all three families of maps considered. Furthermore, the prefactor multiplying N^{-2} in each case agrees with that obtained from a semiclassical calculation, and as expected this prefactor grows with increasing classical correlation time $T_{\rm corr}$ (corresponding to a decrease in the chaoticity of the system). We also see in Fig. 5 that even for a "typical" chaotic system (i.e., $T_{\rm corr} \sim 1$), strong deviations from the $1/N^2$ law appear already below $N \approx 80$. Such deviations extend to even larger N for chaotic systems with slower classical correlation decay. This suggests that the large-N or large-kL expansion, though theoretically appealing and asymptotically correct, is problematic in describing the quantitative behavior of interaction matrix element fluctuations for real chaotic systems in the physically interesting energy range.

The behavior of $v_{\alpha\alpha}$, $v_{\alpha\beta\gamma\delta}$, and v_α in real dynamical systems may be studied similarly, again using the normalized random wave predictions as the baseline for comparison. The enhancement of the variance at large kL is particularly dramatic in the case of double-diagonal matrix element fluctuations. This is consistent with the reasonable expectation that dynamical effects lead to particularly strong deviations from random wave behavior in a modest fraction of the total set of single-particle states, such as those associated with scarring on unstable periodic orbits [28]. Such deviations lead to a significant tail in the $v_{\alpha\alpha}$ distribution, but have a minimal effect on the distribution of off-diagonal matrix elements, since it is unlikely for all four wave functions ψ_α, ψ_β, ψ_γ, and ψ_δ to be strongly scarred or antiscarred on the same orbit.

We can also go beyond the variance to investigate higher moments of the matrix element distribution for actual chaotic systems. A typical distribution for diagonal two-body matrix elements $v_{\alpha\beta}$ in a modified quarter-stadium billiard with $a = 0.25$ and $s = 0.1$ is shown in Fig. 2 (solid line). Since the approach to Gaussian behavior is already very slow in the case of random waves, it is not surprising to find even stronger deviations from a Gaussian shape for matrix elements in real chaotic systems at the same energies. For example, the excess kurtosis γ_2 increases from 8.3 at $kL = 70$ to 20.9 at $kL = 140$, while dropping from 3.7 to 3.3 in the random wave model. Similar behavior is obtained for the skewness and for other matrix elements. Clearly, the distribution tails are very long, and the assumption of Gaussian matrix element distributions is even less justified for real chaotic systems than it was in the random wave model.

BEYOND THE CHAOTIC REGIME

Finally, we consider fluctuations of matrix elements in systems that are not fully chaotic. Here no universal behavior is expected but we shall see that in such systems the variance can be enhanced much more than in fully chaotic systems [29]. We use the modified quarter-stadium billiard with $s = 0$ or $a < 0$. The choice $s = 0$ corresponds to the original Bunimovich stadium, whose quantum fluctuation properties are dominated by the marginally-stable bouncing-ball modes, while $a < 0$ corresponds to a lemon billiard, which has a classically mixed, or soft chaotic, phase space.

In contrast with the $\ln kL/(kL)^2$ falloff in the $v_{\alpha\beta}$ variance predicted for fully chaotic dynamics by Eq. (18), in the case of regular or mixed dynamics we expect kL-independent matrix element fluctuations of order unity [21]. Equivalently, we have very large enhancement, scaling as $(kL)^2/\ln kL$, of the matrix element variance in mixed dynamical systems, over the random wave prediction.

The diagonal matrix element variance $\overline{\delta v^2_{\alpha\beta}}$ is computed as a function of kL for two typical mixed phase-space quarter-lemon billiards and compared with the normalized random wave prediction (dashed lines in Fig. 6). As expected, the enhancement becomes more pronounced at larger kL. Enhancement of an order of magnitude or more over random wave behavior can easily be obtained for physically interesting values of kL. The most dramatic enhancement is observed for the $a = -0.25$ quarter-lemon billiard, which is closer to integrability.

Behavior intermediate between hard chaos and mixed chaotic/regular phase space is obtained in the presence of families of marginally stable classical trajectories, such as the "bouncing ball" orbits of the stadium billiard. In the quarter stadium billiard ($s = 0$ in Fig. 3), exceptional states associated with such orbits are concentrated in the rectangular region of the billiard and constitute a fraction $\sim 1/(kL)^{1/2}$ of the total set of states [32]. When α and β are both bouncing ball states, $\delta v_{\alpha\beta} \sim \overline{v_{\alpha\beta}} \sim \Delta$, just as would be the case for regular states concentrated in a finite fraction

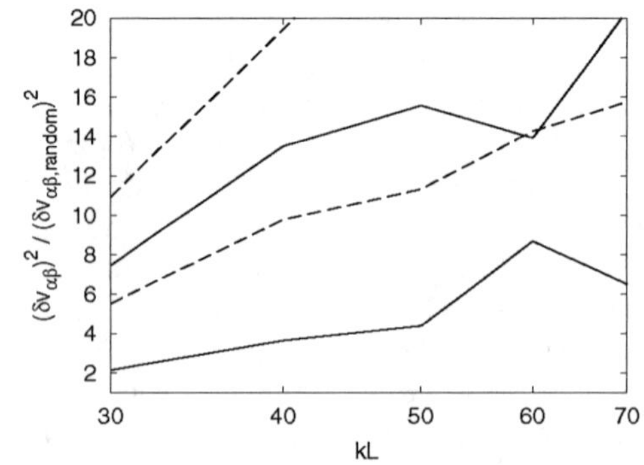

FIGURE 6. Enhancement of $v_{\alpha\beta}$ variance as compared with the random wave prediction for $a = 0.25$, 1.00 quarter-stadium billiards (solid lines); $a = -0.25$, -0.50 quarter-lemon billiards (dashed lines).

of the available coordinate space. These special matrix elements dominate the variance, leading to

$$\overline{\delta v_{\alpha\beta}^2} \sim \frac{\Delta^2}{kL}, \tag{19}$$

Numerical data for quarter-stadium billiards is shown by solid lines in Fig. 6. The stronger fluctuations are observed in the less chaotic $a = 0.25$ stadium.

Again, analogous results are obtained for the other matrix elements (not shown).

SUMMARY

We have studied fluctuations of two-body and one-body matrix elements in ballistic quantum dots as a function of semiclassical parameter kL. Understanding the quantitative behavior of these fluctuations is important for the proper analysis of peak spacing statistics in the Coulomb blockade regime.

We find that the variance and higher cumulants of two-body and one-body matrix elements may be expressed in terms of spatial correlations within single-particle Hartree-Fock wave functions. For the purpose of computing the variance of two-body matrix elements $v_{\alpha\alpha}$, $v_{\alpha\beta}$, and $v_{\alpha\beta\gamma\delta}$, these correlations may be approximated, to leading order in kL for a chaotic system, by correlations given by Berry's random wave model. The calculation results in a variance scaling as $\ln kL/(kL)^2$, with universal prefactors depending only on the symmetry class of the system. Shape-dependent effects on the variance enter at $O(1/(kL)^2)$, where the random wave intensity correlator must be corrected to satisfy individual wave function normalization in finite volume. In the normalized random wave model, ratios such as $\overline{\delta v_{\alpha\alpha}^2}/\overline{\delta v_{\alpha\beta}^2}$ converge only with a logarithmic rate in the $kL \to \infty$ limit; as a result, the asymptotic values of such ratios are of little relevance in the regime of experimental interest.

The variance of one-body matrix elements v_α is affected by normalization even at leading order, resulting in $O(1/kL)$ scaling in a random wave model, with a shape-dependent prefactor. Both two-body and one-body matrix elements follow, already within the random wave model, a strongly non-Gaussian distribution, for all physically reasonable values of kL. Thus, higher cumulants of these matrix elements will be important in peak spacing calculations, especially in the case of two-body matrix elements where we show that the approach to a Gaussian distribution is logarithmically slow.

Dynamical effects, associated with nonrandom short-time behavior in actual chaotic systems, are formally subleading for two-body matrix elements, and of the same order as the random wave prediction for one-body matrix elements. In practice, however, these effects can easily lead to enhancement by a factor of 3 or 4 of the variance in both one-body and two-body matrix elements, for experimentally relevant values of kL in reasonable hard chaotic geometries. The size of these effects scales in each case as a power of T_{clas}, a time scale associated with approach to ergodicity in the corresponding classical dynamics. Random wave behavior is recovered as $T_{\mathrm{clas}} \to 0$. In typical geometries, dynamical effects on matrix element fluctuations cannot be properly computed in a semiclassical approximation, as higher-order terms are quantitatively of the same size as the leading-order semiclassical expression in the kL range of experimental interest. The approach to semiclassical scaling at very large values of kL as well as saturation of matrix element fluctuations at moderate to small kL are investigated in the context of a quantum map model.

Systems with a mixed chaotic-regular phase space or with families of marginally stable classical orbits show even stronger enhancement of matrix element fluctuations as compared with the random wave model. The expected asymptotic scaling with kL of the fluctuations in these cases is discussed, and is very different from the scaling found in chaotic systems.

Our calculations strongly indicate that statistics of actual chaotic single-particle systems, including dynamical effects, are needed to make a proper comparison between Hartree-Fock-Koopmans theory and experiment. A better understanding of correlations in single-particle systems is then essential to compute observable properties of the interacting many-electron system. Furthermore, these correlations need to be understood beyond the naive leading order semiclassical approximation, to allow comparison with experiments, which are generally performed at moderate values of the semiclassical parameter.

ACKNOWLEDGMENTS

We acknowledge useful discussions with Y. Gefen, Ph. Jacquod, and C. H. Lewenkopf. This work was supported in part by the U.S. Department of Energy Grants No. DE-FG03-00ER41132 and DE-FG-0291-ER-40608.

REFERENCES

1. Y. Alhassid, Rev. Mod. Phys. **72**, 895 (2000).
2. For a review of random matrix theory, see T. Guhr, A. Müller-Groeling, and H. A. Weidenmüller, Phys. Rep. **299**, 190 (1998).
3. R. A. Jalabert, A. D. Stone, and Y. Alhassid, Phys. Rev. Lett. **68**, 3468 (1992).
4. Y. Alhassid and H. Attias, Phys. Rev. Lett. **76**, 1711 (1996).
5. Y. Alhassid, Phys. Rev. B **58**, R 13383 (1998).
6. J. A. Folk, S. R. Patel, S. F. Godijn, A. G. Huibers, S. M. Cronenwett, C. M. Marcus, K. Campman, and A. C. Gossard, Phys. Rev. Lett. **76**, 1699 (1996).
7. A. M. Chang, H. U. Baranger, L. N. Pfeiffer, K. W. West, and T. Y. Chang, Phys. Rev. Lett. **76**, 1695 (1996).
8. J. A. Folk, C. M. Marcus, and J. S. Harris, Jr., Phys. Rev. Lett. **87**, 206802 (2001).
9. U. Sivan, R. Berkovits, Y. Aloni, O. Prus, A. Auerbach, and G. Ben-Yoseph, Phys. Rev. Lett. **77**, 1123 (1996).
10. F. Simmel, T. Heinzel, and D. A. Wharam, Europhys. Lett. **38**, 123 (1997).
11. S. R. Patel, S. M. Cronenwett, D. R. Stewart, A. G. Huibers, C. M. Marcus, C. I. Duruöz, J. S. Harris, Jr., K. Campman, and A. C. Gossard, Phys. Rev. Lett. **80**, 4522 (1998).
12. S. Lüscher, T. Heinzel, K. Ensslin, W. Wegscheider, and M. Bichler, Phys. Rev. Lett. **86**, 2118 (2001).
13. I. L. Kurland, I. L. Aleiner, and B. L. Altshuler, Phys. Rev. B **62**, 14886 (2000).
14. I. L. Aleiner, P. W. Brouwer, and L. I. Glazman, Phys. Rep. **358**, 309 (2002).
15. Y. Alhassid and T. Rupp, Phys. Rev. Lett. **91**, 056801 (2003).
16. G. Usaj and H. U. Baranger, Phys. Rev. B **67**, 121308(R) (2003).
17. T. Koopmans, Physica (Amsterdam) **1**, 104 (1934).
18. Y. Alhassid and S. Malhotra, Phys. Rev. B **66**, 245313 (2002).
19. G. Usaj and H. U. Baranger, Phys. Rev. B **66**, 155333 (2002).
20. Ya. M. Blanter, A. D. Mirlin, and B. A. Muzykantskii, Phys. Rev. Lett. **78**, 2449 (1997).
21. L. Kaplan and Y. Alhassid, in preparation.
22. M. V. Berry, J. Phys. A **10**, 2083 (1977).
23. I. V. Gornyi and A. D. Mirlin, Phys. Rev. E **65**, 025202 (2002); J. Low Temp. Phys. **126**, 1339 (2002).
24. Y. Alhassid and Y. Gefen, preprint.
25. L. A. Bunimovich, Commun. Math. Phys. **65**, 295 (1979).
26. S. Hortikar and M. Srednicki, Phys. Rev. Lett. **80**, 1646 (1998).
27. J. D. Urbina and K. Richter, Phys. Rev. E **70**, 015201(R) (2004).
28. E. J. Heller, Phys. Rev. Lett. **53**, 1515 (1984).
29. D. Ullmo, T. Nagano, and S. Tomsovic, Phys. Rev. Lett **90**, 176801 (2003); **91**, 179901(E) (2003).
30. S. Fishman, D. R. Grempel, and R. E. Prange, Phys. Rev. Lett. **49**, 509 (1984).
31. A. Altland and M. R. Zirnbauer, Phys. Rev. Lett. **77**, 4536 (1996).
32. A. Bäcker, R. Schubert and P. Stifter, J. Phys. A **30**, 6783 (1997).

Superscars, Doorway States and Nodal Domains in a Barrier Billiard

A. Richter

Institut für Kernphysik, Technische Universität Darmstadt, D-64289 Darmstadt, Germany

Abstract. The recently measured superscars in a pseudointegrable barrier billiard [Phys. Rev. Lett. **97**, 254102 (2006)] serve as a paradigm for strength function phenomena known in a variety of open mesoscopic quantum systems ranging from atoms to nuclei. The superscarring state is thereby of a simple structure whose strength is spread over the many other nonscarring chaotic states in the billiard, i.e. it acts as a so called doorway state. In this talk the observed spreading of four different superscars is briefly discussed in terms of a model based on Random Matrix Theory and also by investigating the spatial correlations of the measured wave functions. It is shown that the results derived are fully consistent with each other. Furthermore, some properties of the nodal domains of the experimental wave functions are analyzed. It is found that the mean value of the rescaled nodal domains shows a deviation from what is expected from a percolation model. This deviation can, however, be qualitatively understood in terms of the doorway state picture.

Keywords: quantum billiards; superscarring wave functions; strength function phenomena; random matrix theory; spatial correlations of wave functions; nodal domains; percolation model
PACS: 05.45.Mt, 03.65.Sq, 42.25.Fx, 24.30.Cz

INTRODUCTION

Strength function phenomena [1] in open mesoscopic quantum systems are an important object of study in atomic and molecular physics, in atomic clusters and solid state devices (quantum dots), and in nuclear physics. Such systems are often characterized by a "simple" single-particle or collective excitation whose amplitude may be spread over many "complicated" or chaotic states [1, 2]. In nuclei the simple state is usually called "doorway state" – prime examples for it are isobaric analog states (IAS) and multipole giant resonances (GR) – and its strength function is expected to be of Breit-Wigner shape with a particular spreading width [3, 4]. Recent studies of the doorway state mechanism in connection with isospin symmetry breaking of IAS and the fine structure of GR can be found in [5, 6].

In the present talk we discuss in the first part the recently predicted [7] and subsequently experimentally identified [8] superscars in a pseudointegrable quantum barrier billiard simulated through a quasi two-dimensional microwave billiard of a rectangular shape with a barrier in it (Fig. 1) as a very clean model for doorway states. Superscars are

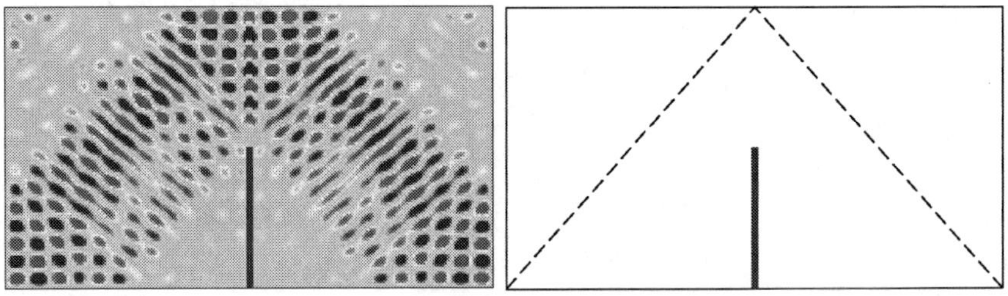

FIGURE 1. Example of a measured superscar in the barrier billiard with side lengths of $l_x = 1000$ mm and $l_y = 585$ mm, and a height of 25 mm. The barrier has a length of 297 mm. A clear wave function structure connected with the family of classical periodic orbits is observed as well as a distinct localization of excitation strength (from [8]).

associated with certain scarring wave functions related to families of classical periodic orbits. Four of those measured are shown in Fig. 2, they do not disappear at large quantum numbers (unlike ordinary scars [9]) and are embedded into but clearly distinguishable from a large number of ordinary nonscarred wave functions. For their analysis we introduce two observables. The first one arises from a model based on Random Matrix Theory (RMT) [11], and the second one from spatial correlations functions with Berry's random wave ansatz [12]. It should already be emphasized here, however, that unlike in atoms, molecules and nuclei, where strength function phenomena can only be studied through measured intensities or cross sections, we show here for the first time how measured wave functions and their

CP995, *Nuclei and Mesoscopic Physics - WNMP 2007*, edited by P. Danielewicz, P. Piecuch, and V. Zelevinsky

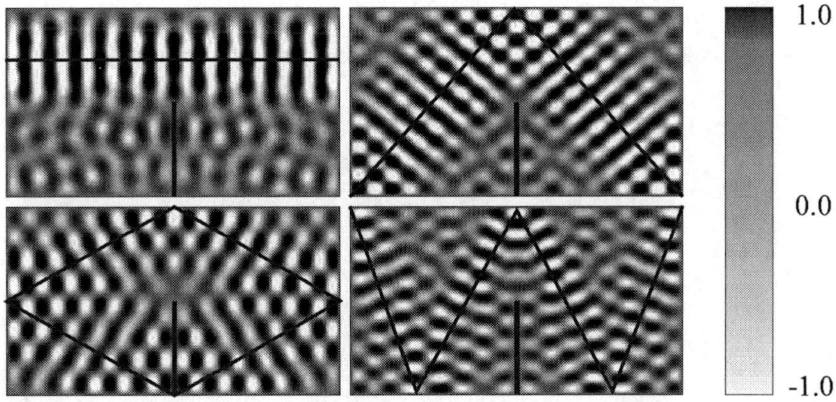

FIGURE 2. Examples for measured superscars. Top row: horizontal Bouncing Ball BB and Inverted V superscars; bottom row: Diamond D and the W superscars. The superscars are concentrated along the indicated classical periodic orbits. The gray level corresponds to the measured amplitudes (from [10]).

spatial correlation corroborate the doorway state phenomenon. We thereby arrive at a consistent description of this phenomenon in terms of RMT and spatial correlation functions. A more detailed account of this is given in [10].

In the second part of the talk we discuss some experimental results from our investigations of nodal domains of barrier billiard wave functions. Nodal domains are connected regions of the same sign of the wave function, separated by nodal lines where the wave function vanishes. The interest in the subject of nodal domains of the eigenfunction of a self-adjoint operator dates back to the seminal work of Courant [13]. The statistical properties of nodal domains are system dependent and are likely connected to the statistical properties of the wave function. For example, it has recently been shown in [14] that the distribution of the number of nodal domains is different for regular and chaotic quantum billiards, and it was furthermore conjectured in [15] on the basis of a percolation model that the distribution is universal for chaotic quantum billiards. There is no theoretical prediction available yet for the family of polygonal pseudointegrable billiards [16] to which the barrier billiard belongs [7, 17]. We thus set out to study this problem and found a clear deviation from expectations of the percolation model which can, however, be understood qualitatively with result from the doorway state model of the first part of the talk. A more extensive discussion as is possible here will be given in [18].

SUPERSCARS

Let us first repeat briefly some salient features about wave functions and their particular structure in the barrier billiard from [7, 8]. In the measurements [8], among a large numbers of nonscarring wave functions several superscarring ones were observed (Fig. 1 and 2) with a clear structure obviously related to the corresponding periodic orbit (PO). Each family of these superscars lives within a so called (infinitely long) periodic orbit channel (POC) in which superscar states can be constructed. Due to diffraction on the tip of the barrier, the amplitude of the scarred wave function tends to zero along the boundary of the POC. We recall from [7, 8] that they are of the form

$$\Psi^{scar}_{m,j}(\xi,\eta) = e^{-ik_m\xi}\sin\left(\frac{\pi j}{w}\eta\right)\chi(\eta). \tag{1}$$

Here ξ is the coordinate along the POC and η ($0 < \eta < w$) the one perpendicular to it, and w is the channel width. The quantity $\chi(\eta)$ is the characteristic function of the POC ($\chi(\eta) = 1$ if $0 < \eta < w$, and $\chi(\eta) = 0$ otherwise). The frequency of the eigenfunction (Eq. (1)) of the POC is

$$f_{m,j} = \frac{c}{2\pi}\sqrt{k_m^2 + \left(\frac{\pi j}{w}\right)^2}, \tag{2}$$

where $k_m = 2\pi m/l$ or $k_m = \pi(2m+1)/l$, depending upon the fact if the PO of length l experiences an even or odd numbers of crossings with the Dirichlet boundary while traveling through the POC, respectively. The quantum numbers

(m, j) can be counted from the number of wave maxima along the POC and perpendicular to it. Thus, once again, the superscarred wave function is approximated by a constructed superscar state, defined as an eigenfunction $\Psi^{(\mathsf{F})}_{m,j}(\vec{r})$ in the infinitely long POC [7, 8]. Here $\mathsf{F}\in\{\mathsf{BB},\mathsf{V},\mathsf{D},\mathsf{W}\}$ stands for the superscar families, as defined in Fig. 2. A typical

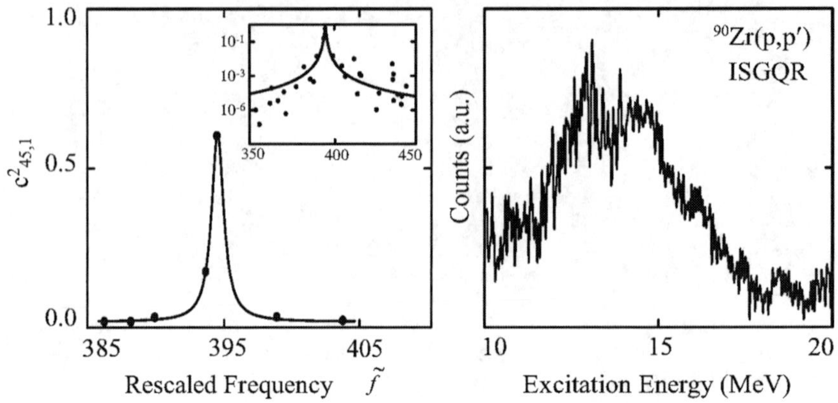

FIGURE 3. Left: the overlap between the constructed superscar state of the V family with quantum numbers $m = 45$, $n = 1$ and the experimentally obtained wave functions of the barrier billiard as function of the rescaled frequency. The solid curve is a Breit-Wigner function. The inset shows the overlap on a logarithmic scale over a large frequency interval. Right: spectrum of the (p, p') reaction at 200 MeV on ^{90}Zr in the region of the Isoscalar Giant Quadrupole Resonance (ISGQR) as an example for a nuclear doorway state [6] (from [6])

nonscarred background state $\Psi_{\tilde{f}}(\vec{r})$ at (rescaled) frequency $\tilde{f}$ in the barrier billiard has an overlap,

$$c_{m,j} = \langle \Psi^{(\mathsf{F})}_{m,j} | \Psi_{\tilde{f}} \rangle \tag{3}$$

with many of the constructed superscars [7, 8]. As an example, the distribution of the overlaps with the constructed V superscar family with quantum numbers $(m, j) = (45, 1)$ is depicted in Fig. 3. The superscar state strength spreads into a few neighboring nonscarring background states following a Breit–Wigner shape. This nicely confirms our doorway interpretation [10] which will be discussed next. The mechanism is indeed similar to what is typically observed in nuclear giant resonances where the excitation strength of the simple collective state is spread over many complicated underlying states. The response function therefore shows a spreading width and fine structure [6] as seen in an example of a GR plotted on the r.h.s. of Fig. 3.

SUPERSCARS AS A MODEL FOR DOORWAY STATES

In order to model the observed features of the superscars we follow very closely Ref. [10] and set up a simple model based on RMT fully in the spirit of models in nuclear physics [1, 4, 11]. As is illustrated schematically in Fig. 4 the doorway picture implies that the strength of one particular state, that we shall assume as the constructed superscar, is distributed over a background of states. This process can be represented by the Hamiltonian,

$$\hat{H} = \hat{H}_s + \hat{H}_b + \hat{V}, \tag{4}$$

where $\hat{H}_s$ and $\hat{H}_b$ describe the constructed superscar and background states, respectively, and $\hat{V}$ couples the two classes of states. Eigenstates of the uncoupled Hamiltonians are,

$$\hat{H}_s|s\rangle = e_s|s\rangle \text{ and } \hat{H}_b|b\rangle = e_b|b\rangle, \tag{5}$$

and matrix elements of the interaction are assumed as,

$$\langle s|\hat{V}|s'\rangle = \langle b|\hat{V}|b'\rangle = 0 \text{ and } \langle b|\hat{V}|s\rangle = v_{bs}, \tag{6}$$

for any s and s', and b and b'. We interpret the constructed superscars, $\Psi^{(\mathsf{F})}_{m,j}(\vec{r})$ for a given family F but with different (m, j) as the doorway states s. Due to the interaction $\hat{V}$ the doorway state is not an eigenstate of the Hamiltonian

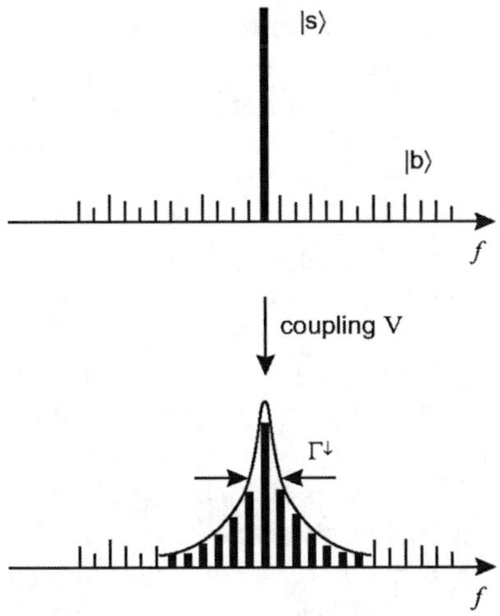

FIGURE 4. Doorway state picture for the spreading of the superscar state $|s\rangle$ into the background states $|b\rangle$ through the coupling V between them.

$\hat{H}$. In accordance with ideas from RMT the interaction matrix elements, $v_{bs} = v_{sb}$, are taken as Gaussian distributed random numbers with zero mean and variance v^2. If the average distance between the energy levels is d, the parameter of the model is defined by v/d. Since only a few states are seen to carry superscar strength with given values of (m, j), and superscar states with different (m, j) are assumed not to mix, it is sufficient to consider only *one* superscar state, s, coupled to a surrounding of many (N) background states, b. To resemble the experimental situation, we shall generally include $N = 294$ background states. The pseudointegrability of the studied barrier billiard implies that the eigenstates are semi-Poisson distributed [19]. We thus consider an ensemble of $N+1$ semi-Poissonian distributed energy states. The constructed superscar is chosen as the middle state and interacts with the surrounding N states. For each realization, energies and wave functions are numerically obtained, and the mixture of the superscar on surrounding states is calculated. The aim is to see how well this theoretical picture can model the observed features, and then try to determine the coupling constant, v/d, for each family of superscars BB, V, D and W.

The eigenvalue problem

$$\hat{H}|n> = E_n|n>, \tag{7}$$

can be solved by diagonalization of the $(N+1) \times (N+1)$ matrix or from the equation

$$E_n = e_s - \sum_{b=1}^{N} \frac{v_{bs}^2}{e_b - E_n}, \tag{8}$$

that gives the eigenenergies, E_n. The wave functions are then given by

$$|n> = c_s \left(|s> - \sum_{b=1}^{N} \frac{v_{bs}}{e_b - E_n} |b> \right). \tag{9}$$

The superscar component of each eigenstate is therefore

$$c_s(n) = \left(1 + \sum_{b=1}^{N} \frac{v_{bs}^2}{(e_b - E_n)^2} \right)^{-1/2}. \tag{10}$$

The strength of the superscar over the different eigenstates $|n>$, given by c_s^2, follows a Breit-Wigner distribution [1] with the width

$$\Gamma^{\downarrow} = 2\pi v^2 d^{-1}, \tag{11}$$

or in dimensionless variables, $\Gamma^{\downarrow}/d = 2\pi(v/d)^2$. In the left panel of Fig. 3 we show an example of an observed distribution of the superscar strength over different states. It is seen that the superscar strength is distributed over a small number of states (v/d is smaller than or of the order of one), and the fit of a Breit-Wigner distribution to it is not always overwhelming, there is a quite large variation of the fitted shape as well as the width of the distribution over the ensemble of observed superscars in a given superscar family. We shall therefore use other measures than the width, $\Gamma^{\downarrow}$, when determining the coupling strength, v/d. The state with the largest superscar component

$$c^2_{\text{max}} = \max(c^2_{m,j}) \tag{12}$$

for a given superscar is directly obtained from experiment. Notice that since a rather small number of states carry strength from the doorway state (i.e. the constructed superscar), the peak of the fitted Breit-Wigner shaped strength function usually deviates from the measured largest superscar strength; the discretely measured state does not appear exactly at the peak. We may, however, directly compare the maximal measured value to the corresponding calculated value, $\max(c_s^2(n))$, where c_s is obtained from Eq. (10). In addition, not only the average value of c^2_{max} may be studied but also its higher moments. We shall here study the *full distribution* of the maximal component of each constructed superscar for a given family of superscars. This quantity has, as far as we know, never been considered before, neither from experimental data nor from theory. The shape of the c^2_{max}-distribution strongly depends on the interaction strength, v/d, and it is a particularly sensitive measure for small values of v/d, i.e. of the order one or smaller. The c^2_{max}

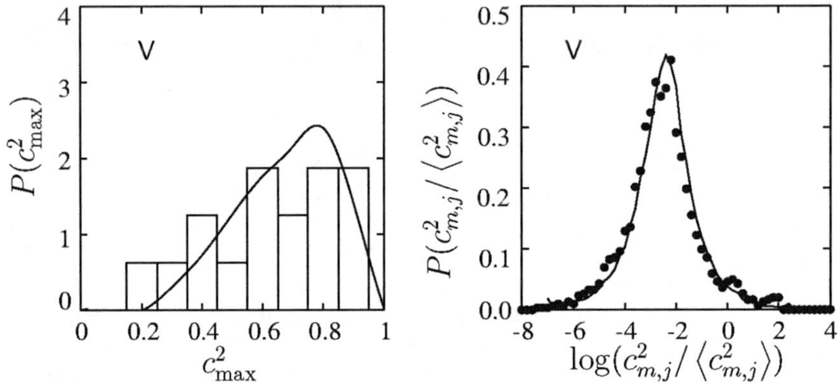

FIGURE 5. l.h.s.: c^2_{max} distributions of measured V superscars (histogram), and the fit of the RMT model predictions (solid line). r.h.s.: normalized distributions of superscar strength spread over all states on a logarithmic scale. Experimental distributions (dots) are compared with the RMT model prediction (from [10]).

values are limited to the interval between zero and one, and limiting cases of distributions are obtained for $v/d = 0$ and $v/d \to \infty$, when strongly peaked distributions appear at $c^2_{max} = 1$ and $1/N$, respectively. As an example, we show in Fig. 5 (l.h.s.) the measured distribution of c^2_{max} with the best fit curve of the RMT model for the V superscar. The fit gives the value $v/d = 0.35$ for the interaction strength, and the first moment (mean value) of measured and calculated c^2_{max} values of 0.63 ± 0.05 and 0.68, respectively. The corresponding widths $\Gamma^{\downarrow}$ are 0.81 ± 0.1 and 0.8 in dimensionless units.

Another measure to compare experimental data and the model prediction is to study the distribution of all superscar components of eigenstates. The strength of each constructed superscar is measured (and calculated) over 295 states, where the major part of the strength is concentrated in a few states only. The r.h.s. of Fig. 5 shows the measured distribution compared to the calculation for the interaction strengths obtained from the fit to the c^2_{max} distributions for the considered V of superscars. Since the strength (x-axis) spans several orders of magnitude, it is shown on a logarithmic scale. Also here we clearly see that the RMT model reproduces the experimentally obtained distribution. Similar very good agreement between experiment and the RMT model is also obtained for the other members BB, D and W of the superscar families [10].

SPATIAL CORRELATIONS OF THE WAVE FUNCTIONS

We now turn to the spatial correlations of the wave functions. They provide new insights to the doorway mechanism which are experimentally not accessible in nuclei, atoms or molecules, as the wave functions cannot be measured in these systems. Berry [12] introduced the correlator

$$C(kr) = \frac{\langle \psi_k(\vec{R}-\vec{r}/2)\psi_k^*(\vec{R}+\vec{r}/2)\rangle}{\langle |\psi_k(\vec{R})|^2\rangle} \tag{13}$$

of the wave functions $\psi_k(\vec{r})$ where the average is performed *isotropically* over all vectors $\vec{R}$ and, for fixed moduli of wave vector $\vec{k}$ and r, over all directions of the vector $\vec{r}$. In our context, all wave functions are real and no complex conjugation is needed in the definition (13). Berry argued that the spatial correlations of a wave function in an ergodic system should be indistinguishable from those of superimposed plane waves. In two dimensions this yields the universal prediction $C(kr) = J_0(kr)$, if possible boundary effects are ignored. Here J_0 is the Bessel function of order zero. Indeed, this behavior was confirmed in numerous systems, mostly in quantum billiards [11, 20, 21], but also for classical waves in optics [22] and in elasticity [23]. The superscars, however, clearly have non–ergodic features.

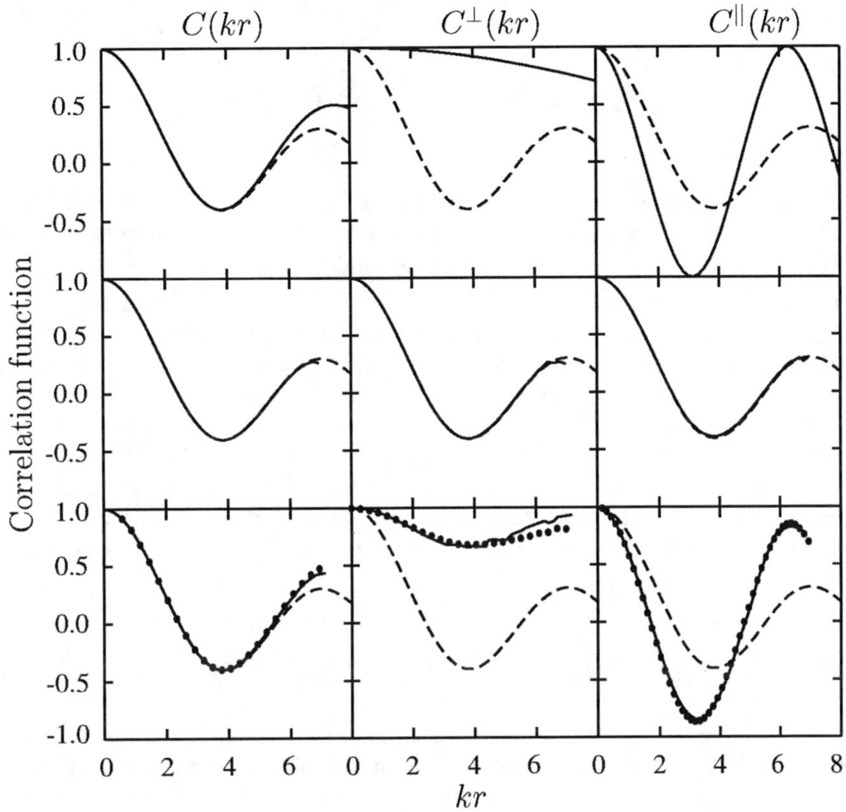

FIGURE 6. The wave function correlators. The $J_0(kr)$ prediction is always given as dashed line. The top row shows the correlation function of the constructed superscar state as solid lines: $C(kr)$ averaged over all directions, $C^{\perp}(kr)$ perpendicular to the channel and $C^{||}(kr)$ parallel to the channel, respectively. In the middle row, the same observables are depicted as solid lines for the averages over all experimental wave functions in the barrier billiard. In the bottom row, the correlations functions averaged over all observed V superscars are displayed as solid lines and the correlators resulting from Eq. (14) with $c^2_{\max} = 0.69$ are shown as dots (from [10]).

To analyze their correlations we define new, especially tailored observables to which we refer as *directed* correlators. Instead of averaging isotropically as for $C(kr)$, we now carry out the averages either only across or only along the channel in which the superscar exists. We thereby obtain the correlators $C^{\perp}(kr)$ and $C^{||}(kr)$, respectively. In the top row of Fig. 6 the three correlators of a constructed V superscar $\Psi^{(\mathrm{V})}_{m,j}(\vec{r})$ are depicted. While the isotropic correlator $C(kr)$ follows the $J_0(kr)$ prediction up to a certain scale, the directed correlators strongly deviate from it. The results for $C^{\perp}(kr)$ and $C^{||}(kr)$ show that the constructed superscar fills the channel and moves through it as a sine wave whose wave front is perpendicular to the channel, see also Figs. 1 and 2 above. This information about the form of the waves, however, is washed out when averaging over all wave functions in the billiard. As the middle row of Fig. 6 shows, each of the three correlators worked out for all measured wave functions agrees with the $J_0(kr)$ prediction.

We now use these observations to extract information about the superscar couplings from the correlators of the measured superscar families. Correlators averaged over all experimentally observed V superscars are displayed in the bottom row of Fig. 6. They are similar to, but slightly different from those for the constructed V superscars in the top row. The difference is due to the leaking of the superscar out of the channel or, in the language of the doorway description, due to the coupling of the background states to the superscar. We thus can model the measured superscars

$\Psi_{\tilde{f}}^{(\mathsf{F})}(\vec{r})$ for family F as a linear combination of a constructed superscar $\Psi_{m,j}^{(\mathsf{F})}(\vec{r})$, which only contributes in the channel, and a state $\widetilde{\chi}_k(\vec{r})$ which is ergodically distributed everywhere in the billiard,

$$\Psi_{\tilde{f}}^{(\mathsf{F})}(\vec{r}) = c_{\max}\Psi_{m,j}^{(\mathsf{F})}(\vec{r}) + \sqrt{1-c_{\max}^2}\,\widetilde{\chi}_k(\vec{r})\,. \tag{14}$$

This ansatz is fully consistent with the RMT model set up above and extends it by also modeling the spatial dependence. The states describing the background should, first, have $J_0(kr)$ correlations and, second, be orthogonal to the constructed superscar. Both requirements are achieved by choosing the "scarless" plane waves

$$\widetilde{\chi}_k(\vec{r}) = \frac{\chi_k(\vec{r}) - \langle\Psi_{m,j}^{(\mathsf{F})}|\chi_k\rangle\Psi_{m,j}^{(\mathsf{F})}(\vec{r})}{\sqrt{1-\langle\Psi_{m,j}^{(\mathsf{F})}|\chi_k\rangle^2}}\,, \tag{15}$$

obtained by orthonormalizing plane waves $\chi_k(\vec{r})$ in the spirit of Berry to the constructed superscar $\Psi_{m,j}^{(\mathsf{F})}(\vec{r})$. The superscar contribution in the plane waves is small (but not negligible); the distribution of the overlap integrals $\langle\Psi_{m,j}^{(\mathsf{F})}|\chi_k\rangle$ has a standard deviation of 0.13. We convinced ourselves that the correlator of the "scarless" plane waves (15) follows the $J_0(kr)$ prediction very closely. We work out the three correlators for the model (14). They depend on the parameter $c_{\max}$ which is, just as in the RMT model above, the coupling to the doorway, i.e. to the superscar. By fitting these correlators to the measured superscar families we determine the couplings $c_{\max}$. The fits for the V superscar are shown as well in Fig. 6. The resulting value of $c_{\max}^2$ is 0.69, very close to the value obtained from the RMT model prediction of 0.68 discussed in the previous section. The same statement can be made with respect to the other investigated superscar families, *i.e.* BB, D and W, as is discussed in [10].

We thus conclude that our doorway interpretation yields a thorough understanding of the experimental findings. We also achieve a deeper insight into the statistical features of the doorway mechanism as such by our new observables, the $c_{\max}^2$ distributions and the spatial correlators. It is encouraging to see how well the two data analyses based on the two new observables agree.

NODAL DOMAINS AND THEIR STATISTICS IN THE BARRIER BILLIARD

Let us finally discuss briefly the properties of the (real) wave functions $\psi_k(x,y)$ in the two-dimensional barrier billiard in the context of nodal domains, which are – as already mentioned above – connected regions of the same sign of the wave functions $\psi_k(x,y)$, separated by nodal lines on which ψ_k vanishes. Typical nodal domain structures for a regular rectangular, a chaotic stadium and a pseudointegrable barrier billiard are shown in Fig. 7 It is obvious from an

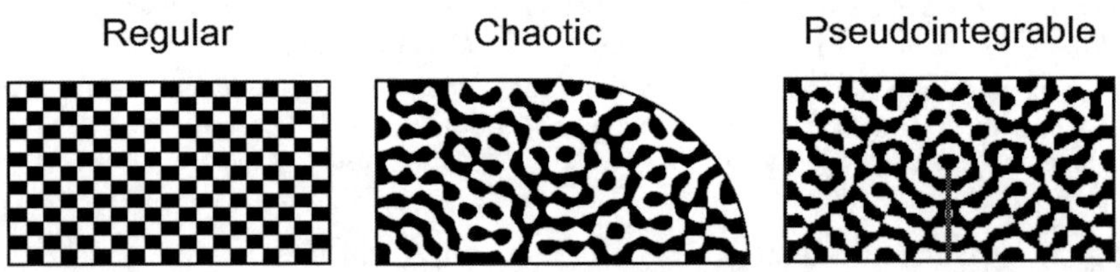

FIGURE 7. Typical nodal domain patterns for a regular rectangular, a chaotic stadium and a pseudointegrable barrier billiard. Black (white) regions indicate regions of minima (maxima) of wave functions.

inspection of the pattern in Fig. 7 that there is a clear difference between the number of nodal domains for regular and chaotic quantum billiards [14]. This statement can be quantified somewhat more through the relation [14]

$$n(N) = aN + b\sqrt{N}. \tag{16}$$

Here n is the number of nodal domains which grows with the state number N proportional to a term representing the area of the billiard (first term) and a correction due to the boundary of the billiard (second term). While the coefficient a is non-universal for regular systems it was shown theoretically in [15] that $a \equiv 0.0624$ for chaotic systems with wave functions whose generic properties are generally well reproduced by Berry's random plane wave model (RPWM) wave functions [12]. This has been confirmed numerically for the chaotic stadium and the Sinai billiard [14] and experimentally for the rough circular [24] and open chaotic billiard [25].

Since nothing is known about a in Eq. (16) for pseudointegrable billiards whose eigenvalue statistics is intermediate between those of regular and chaotic systems [17] we have set out to determine a for the pseudointegrable barrier billiard and have compared it to the value for chaotic billiards. In fact inspection by eye of the nodal domain patterns in Fig. 7 for a chaotic stadium billiard and the barrier billiard already suggests similarities between them if there was not the presence of a strong superscarring behaviour of particular wave functions in case of the latter. For this comparison we generated within the RPWM a set of wave functions in a rectangle of the same size as the barrier billiard in the experiment and at the experimental resonance frequencies. We further consider the rescaled number of nodal domains, *i.e.*

$$\frac{n(N)}{N} = a + \frac{b}{\sqrt{N}}, \tag{17}$$

and fit this expression to the experimental data for the barrier billiard (l.h.s. of Fig. 8) as well as the equivalent "data" from the RPWM (r.h.s. of Fig. 8).

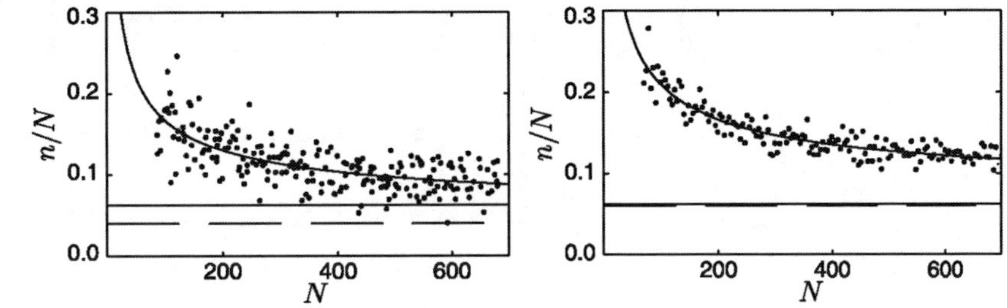

FIGURE 8. l.h.s.:The rescaled number of nodal domains as function of the state number for the measured barrier billiard wave functions. r.h.s.: The rescaled number of the nodal domains as function of the state number for RPWM wave functions. The solid curves correspond to the fit of Eq. (17) to the data, the solid straight line indicates the percolation model prediction for a and the dashed line results from the value of a obtained from the fit (from [18]).

The resulting values for a are 0.040 ± 0.005 and 0.062 ± 0.003, respectively. The difference between the two values must clearly be attributed to the presence of the superscarring wave functions in the barrier billiard, which influence the scaling properties of the nodal domains. To support this conjecture let us again look back at the back folded superscarring wave function (cf. Eq. (1) above) in the periodic orbit channel (POC),$\Psi^{\mathrm{F}}_{m,j}(x,y)$, where $\mathrm{F} \in \{\mathrm{BB}, \mathrm{V}, \mathrm{D}, \mathrm{W}\}$ stands for the four investigated superscar families, as defined in Fig. 2. (Recall, that the quantum numbers (m, j) are the wave maxima along and perpendicular to the POC.) In the case of the horizontal bouncing ball BB superscar, the back folding procedure is straightforward, and in Cartesian coordinates the superscarring wave function is given by the expression

$$\Psi^{\mathrm{BB}}_{m,j}(x,y) = \sin\left(\frac{2\pi m}{l}x\right)\sin\left(\frac{\pi i}{w}y\right)\chi(y), \tag{18}$$

where the length of the POC $l = 2l_x$, its width $w = l_y/2$, and $\chi(y)$ is a characteristic function of the POC ($\chi(y) = 1$ if $0 < y < w$, and $\chi(y) = 0$ otherwise). The approximate state number is equal to

$$N \approx \frac{A}{4\pi}\left(\left(\frac{2\pi m}{l}\right)^2 + \left(\frac{\pi i}{w}\right)^2\right) \tag{19}$$

with a A representing the area of the barrier billiard.

We now divide the billiard into two areas, one inside the POC and one outside of it, and estimate roughly the contribution of the nodal domains inside the POC to the quantity a in Eq. (17). The ansatz of Eq. (1) works well only if the wave momentum along the POC is much bigger than the one perpendicular to it [7] *i.e.*

$$\frac{2\pi m}{l} \gg \frac{\pi j}{w}. \tag{20}$$

For a given m and j the number of nodal domains inside the POC is equal to $n = m \cdot j$ and the contribution of the billiard part covered by the POC is

$$a_{\mathrm{POC}} = \lim_{N\to\infty}\frac{n(N)}{N} = \lim_{N\to\infty}\frac{m \cdot j}{N} \propto \frac{m}{m^2} \to 0. \tag{21}$$

This means that the nodal domains inside the POC indeed yield only an asymptotically small contribution to the scaling quantity in Eqs.(16, 17).

This statement can be rephrased by saying that the main contribution to a must come from the nonscarring wave functions outside the periodic orbit channel. Those wave functions we have, however, already considered in the context of the doorway state model discussed above, cf.Eqs. (14,15). We recall that in this model it has been assumed that

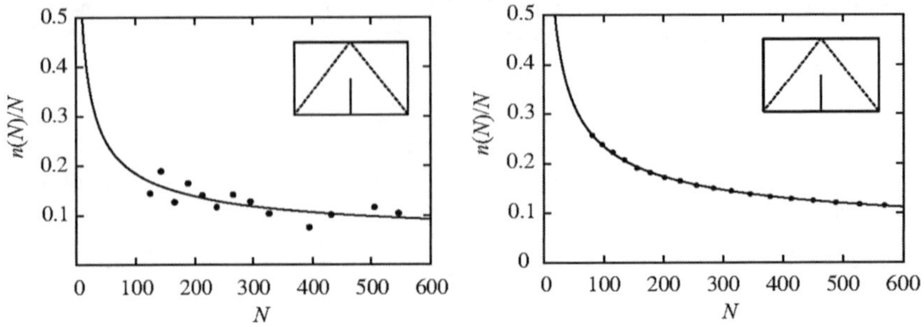

FIGURE 9. Rescaled number of nodal domains for the V superscars, from the experiment (l.h.s.) and from the model ansatz (r.h.s.) for the superscarring wave function of Eq. (22).

the measured superscar $\Psi'^{\mathsf{F}}_{\tilde{f}}(x,y)$ of the family F is represented as a linear combination of the constructed superscar $\Psi^{\mathsf{F}}_{m,j}(x,y)$ which is non-zero in the POC only, and a state $\tilde{\chi}_{\tilde{f}}(x,y)$ which is uniformly distributed everywhere in the billiard

$$\Psi'^{(\mathsf{F})}_{\tilde{f}}(x,y) = c_{\max}\Psi^{(\mathsf{F})}_{m,j}(x,y) + \sqrt{1-c^2_{\max}}\,\tilde{\chi}_{\tilde{f}}(x,y)\,, \tag{22}$$

where $\tilde{\chi}_{\tilde{f}}(x,y)$ is obtained by orthonormalization of the RPWM wave function to the constructed superscar state $\Psi^{(\mathsf{F})}_{m,j}(x,y)$. In this way we are able to determine the quantity a separately for the four observed superscar families by fitting the expression (17) to their nodal domain structure. As an example, the rescaled number of nodal domains determined experimentally for the V superscars is shown in Fig. 9 in comparison with the prediction based on the model ansatz of Eq. (22).

The parameter varied is $c_{\max}$, and the best fit yields $a = 0.026 \pm 0.003$ for the experiment, in excellent agreement with the model prediction of $a = 0.024$, for $c_{\max} = 0.78$. The square of this value *i.e.* $c^2_{\max} = 0.61$ is close to the one from the RMT model prediction (0.68) and from the correlation function analysis (0.69) discussed above. This remark applies also to the other members of the investigated superscar families, *i.e.* BB, D and W. The threefold validation of the assumption that the superscars represents the typical strength function phenomenon, known in mesoscopic quantum systems as doorway state, in the particular quantum barrier billiard is thus very satisfactory indeed.

We finally compare also the particular distribution of the areas s of the nodal domains of the pseudointegrable barrier billiard with predictions of a model [15] based on percolation theory. In the case of chaotic wave functions one expects a power law behaviour, *i.e.*

$$n_s \propto (s/s_{\min})^{-\tau}, \tag{23}$$

where $\tau = 187/91$ is the so called Fisher exponent [26] and $s_{\min} = \pi(j_1/k)^2$ is the smallest possible domain area for a given state with $k = 2\pi f/c$ and j_1 being the first zero of the Bessel function J_0. We have constructed the nodal domain area distribution for each measured wave function in the barrier billiard, performed the average over all wave functions and compared it to the one expected from Berry's RPWM for chaotic systems. As the log – log plot in Fig. 10 shows, both distributions surprisingly show the same power behaviour. On the basis of what has been said in the talk about the subtle differences in the behaviour of the wave functions in the pseudointegrable barrier quantum billiard as compared to fully chaotic quantum billiards this is still another riddle which needs to be solved.

ACKNOWLEDGMENTS

This work was supported by grants from the SFB 634 and SFB/TR 12 of the Deutsche Forschungsgemeinschaft and from the Det Svenska Vetenskapsrådet. I am furthermore very grateful for all my collaborators listed in Refs. [8, 10, 18] for sharing their insight into the physics of this presentation with me. My particular thanks go to Maksym Miski-Oglu who not only within his doctoral thesis ingeniously has designed and built the apparatus to measure the billiard wave

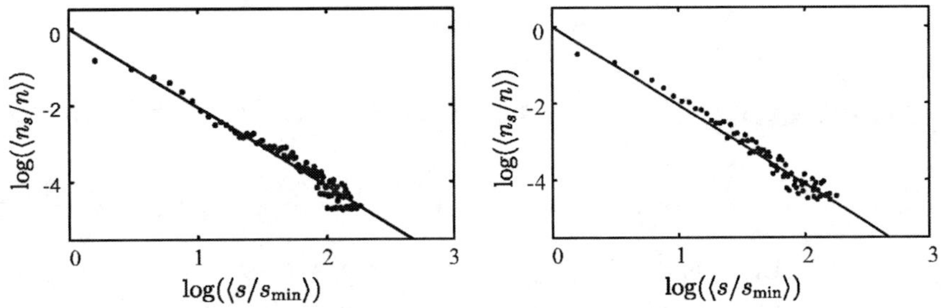

FIGURE 10. The nodal domain area distribution averaged over all RPWM functions (l.h.s.) and averaged over all measured wave functions in the barrier billiard (r.h.s.). In both cases the slope (cf. Eq. (23)) is the same, *i.e.* $\tau = 187/91$ as expected from the percolation model for chaotic systems (from [18]).

functions but also performed all the analysis and comparisons with model predictions which have let to the results I have presented in this work.

REFERENCES

1. A. Bohr and B. Mottelson, *Nuclear Structure Vol. 1*, W.A. Benjamin, INC, 1969.
2. V. V. Sokolov and V. Zelevinsky, *Phys. Rev. C* **56**, 311–323 (1997).
3. H. L. Harney, A. Richter, and H. A. Weidenmüller, *Rev. Mod. Phys.* **58**, 607–645 (1986).
4. V. Zelevinsky, *Ann. Rev. Nucl. Part. Sci.* **46**, 237 (1996).
5. S. Åberg, A. Heine, G. Mitchell, and A. Richter, *Phys. Lett. B* **598**, 42 (2004).
6. A. Shevchenko, J. Carter, R. W. Fearick, S. V. Förtsch, H. Fujita, Y. Fujita, Y. Kalmykov, D. Lacroix, J. J. Lawrie, P. von Neumann-Cosel, R. Neveling, V. Y. Ponomarev, A. Richter, E. Sideras-Haddad, F. D. Smit, and J. Wambach, *Phys. Rev. Lett.* **93**, 122501 (2004).
7. E. Bogomolny and C. Schmit, *Phys. Rev. Lett.* **92**, 244102 (2004).
8. E. Bogomolny, B. Dietz, T. Friedrich, M. Miski-Oglu, A. Richter, F. Schäfer, and C. Schmit, *Phys. Rev. Lett.* **97**, 254102 (2006).
9. E. J. Heller, *Phys. Rev. Lett.* **53**, 1515 (1984).
10. S. Åberg, T. Guhr, M. Miski-Oglu, and A. Richter, *Phys. Rev. Lett., submitted.*
11. T. Guhr, A. Müller-Groeling, and H. A. Weidenmüller, *Phys. Rep.* **299**, 189 (1998).
12. M. V. Berry, *J. Phys. A* **10**, 2083 (1977).
13. R. Courant, *Nachr. Ges. Wiss. Götingen, Math. Phys. Kl.,* 81 (1923).
14. G. Blum, S. Gnutzmann, and U. Smilansky, *Phys. Rev. Lett.* **88**, 114101 (2002).
15. E. Bogomolny and C. Schmit, *Phys. Rev. Lett.* **88**, 114102 (2002).
16. P. J. Richens and M. V. Berry, *Physica D* **2**, 495 (1981).
17. J. Wiersig, *Phys. Rev. E* **65**, 046217 (2002).
18. E. Bogomolny, B. Dietz, T. Friedrich, M. Miski-Oglu, A. Richter, F. Schäfer, and C. Schmit, *in preparation* (2008).
19. E. B. Bogomolny, U. Gerland, and C. Schmit, *Phys. Rev. E* **59**, R1315 (1999).
20. H.-J. Stöckmann, *Quantum Chaos: An Introduction*, Cambridge University Press, Cambridge, 2000.
21. F. Haake, *Quantum Signatures of Chaos*, Springer-Verlag, Heidelberg, 2001.
22. V. Doya, O. Legrand, F. Mortessagne, and C. Miniatura, *Phys. Rev. E* **65**, 056223 (2002).
23. K. Schaadt, T. Guhr, C. Ellegaard, and M. Oxborrow, *Phys. Rev. E* **68**, 036205 (2003).
24. N. Savytskyy, O. Hul, and L. Sirko, *Phys. Rev. E* **70**, 056209 (2004).
25. U. Kuhl, R. Höhmann, H.-J. Stöckmann, and S. Gnutzmann, *Phys. Rev. E* **75**, 036204 (2007).
26. D. Stauffer and A. Aharony, *Introduction to Percolation Theory*, Taylor and Francis, London, 1994.

Symmetry Breaking Study with Random Matrix Ensembles

M. S. Hussein[*,†], J. X. de Carvalho[*,†], M. P. Pato[†] and A. J. Sargeant[†]

[*]*Max-Planck-Institut für Physik komplexer Systeme*
Nöthnitzer Straβe 38, D-01187 Dresden, Germany
[†]*Instituto de Física, Universidade de São Paulo*
C.P. 66318, 05315-970 São Paulo, S.P., Brazil

Abstract. A random matrix model to describe the coupling of m-fold symmetry is constructed. The particular threefold case is used to analyze data on eigenfrequencies of elastomechanical vibration of an anisotropic quartz block. It is suggested that such experimental/theoretical study may supply a powerful means to discern intrinsic symmetry of physical systems.

Keywords: Quantum chaos, acoustic resonators
PACS: 05.30.Ch, 05.45.Mt, 05.40.-a, 24.60.-k

The standard ensembles of Random Matrix Theory (RMT) [1] have had wide application in the description of the statistical properties of eigenvalues and eigenfunctions of complex many-body systems. Other ensembles have also been introduced [2], in order to cover situations that depart from universality classes of RMT. One such class of ensembles is the so-called Deformed Gaussian Orthogonal Ensemble (DGOE) [3, 4, 5, 6] that proved to be particularly useful when one wants to study the breaking of a discrete symmetry in a many-body system such as the atomic nucleus.

In fact, the use of spectral statistics as a probe of symmetries in physical systems has been a subject of intensive experimental and theoretical investigation following the pioneering work of Bohigas, Giannoni and Schmit [7] which showed that the quantal behaviour of classically chaotic systems exhibits the predictions supplied by the RMT. Examples of symmetry breaking in physical systems that have been studied include nuclei [8, 9], atoms [10, 11] and mesoscopic devices such as quantum dots [12].

In the case of nuclei, the Mitchell group at the Triangle Universities Nuclear Laboratory [8, 9], studied the effect of isospin symmetry breaking, in odd-odd nuclei such as ^{26}Al. They detected the breakdown of this important symmetry by the applications of two statistics: the short-range, nearest neighbor level spacing distribution (NND) and the long range Dyson's Δ-statistics [8, 9]. These results were well described by a DGOE in which a pair of diagonal blocks is coupled. The strength of the coupling needed to account for the symmetry breaking can be traced to the average matrix element of the Coulomb interaction responsible for this discrete symmetry breaking [4, 13]. The justification for the use of block matrices to describe the statistics of a superposition of R spectra with different values of the conserved quatum number can be traced to Refs. [1, 14]. In the case of non-interacting spectra, *i.e.* if the quantum number is exactly conserved, the answer is a superposition of the R spectra. Since the level repulsion is present in each one of the R spectra, their superposition does not show this feature. Thus, we can say that for each spectra of states of a given value of the quantum number, one attaches a random matrix (GOE). For R spectra each of which has a given value of the conserved quantum number, one would have an $R \times R$ block diagonal matrix. Each block matrix will have a dimension dictated by the number state of that spectra. If the quantum number is not conserved then the $R \times R$ block matrix acquires non-diagonal matrices that measure the degree of the breaking of the associated symmetry. This idea was employed by Guhr and Weidenmüller [13] and Hussein and Pato [3] to discuss isopin violation in the nucleus ^{26}Al. In reference [3], the random block matrix model was called the Deformed Gaussian Orthogonal Ensemble (DGOE).

In order to study transitions amongst universal classes of ensembles such as order-chaos (Poisson$\rightarrow$GOE), symmetry violation transitions (2GOE$\rightarrow$1GOE), experiments on physical systems are more complicated due to the difficulty of tuning the interaction (except, e.g. in highly excited atoms where the application of a magnetic field allows the study of GOE-GUE transitions). To simulate the microscopic physical systems, one relies on analog computers such as microwave cavities, pioneered by A. Richter and collaborators [15] and acoustic resonators of Ellegaard and collaborators[16, 17, 18]. It is worth mentioning at this point that the first to draw attention to the applicability of RMT to accoustic waves in physical system was Weaver [19].

In the experiment of Ellegaard *et al.* [17] what was measured were eigenfrequencies of the elastomechanical vibrations of an anisotropic crystal block with a D3 point-group symmetry. The rectangular crystal block employed by Ellegard was so prepared as to have only a two-fold flip symmetry retained. Then, to all effects, the quartz specimen

CP995, *Nuclei and Mesoscopic Physics - WNMP 2007*, edited by P. Danielewicz, P. Piecuch, and V. Zelevinsky

resembles a system of two three-dimensional Sinai billiards. The statistical treatment of the eigenfrequencies of such a block would follow that of the superposition of two uncoupled GOE's.

Then, by removing octants of progressively larger radius from a corner of the crystal block this remnant two-fold symmetry was gradually broken. The spectral statistics show a transition towards fully a chaotic system as the octant radius increases. What was then seen was that the measured NND is compatible with a two block DGOE description but the Δ-statistics was discrepant. This discrepancy was attributed to pseudo integrable behavior and this explanation was later implemented with the result that the long-range behavior was fitted at the cost, however, of loosing the previous agreement shown by the NND[20].

Here we reanalyse this experiment following the simpler idea of extending the DGOE matrix model [5] to consider the coupling of three instead of two GOE's [6]. We show that within this extension both, the short- and the long-range statistics, are reasonably fitted suggesting that the assumption of the reduction of the complex symmetries of anisotropic quartz block may not be correct. Our findings have the potential of supplying very precise means of testing details of symmetry breaking in pysical systems.

To define the ensembles of random matrices we are going to work with, we recall the construction based on the Maximum Entropy Principle [3], that leads to a random Hamiltonian which can be cast into the form

$$H = H_0 + \lambda H_1, \tag{1}$$

where the block diagonal H_0 is a matrix made of m uncoupled GOE blocks and λ $(0 \leq \lambda \leq 1)$ is the parameter that controls the coupling among the blocks represented by the H_1 off-diagonal blocks. For $\lambda = 1$, the H_1 part completes the H_0 part and $H = H^{GOE}$.

These two matrices H_0 and H_1 are better expressed introducing the following m projection operators

$$P_i = \sum_{j \in I_i} |\, j >< j\,|, \tag{2}$$

where I_i defines the domain of variation of the row and column indexes associated with ith diagonal block of size M_i. Since we are specifically interested in the transition from a set of m uncoupled GOE's to a single GOE, we use the above projectors to generalize our previous model [3, 4] by writing

$$H_0 = \sum_{i=1}^{m} P_i H^{GOE} P_i \tag{3}$$

and

$$H_1 = \sum_{i=1}^{m} P_i H^{GOE} Q_i \tag{4}$$

where $Q_i = 1 - P_i$. It is easily verified that $H = H^{GOE}$ for $\lambda = 1$.

The joint probability distribution of matrix elements can be put in the form [3, 21]

$$P(H, \alpha, \beta) = Z_N^{-1} \exp\left(-\alpha tr H^2 - \beta tr H_1^2\right) \tag{5}$$

with the parameter λ being given in terms of α and β by

$$\lambda = (1 + \beta/\alpha)^{-1/2}. \tag{6}$$

Statistical measures of the completely uncoupled m blocks have been derived. They show that level repulsion disappears which can be understood since eigenvalues from different blocks behave independently. In fact, as m increases the Poisson statistics are gradually approached. In the interpolating situation of partial coupling, some approximate analytical results have been derived. In Ref. [21], for instance, it has been found that the density $\rho(E)$ for arbitrary λ and m is given by

$$\rho(E) = \sum_{i=1}^{m} \frac{M_i}{N} \rho_i(E) \tag{7}$$

where

$$\rho_i(E) = \begin{cases} \frac{2}{\pi a_i^2} \sqrt{a_i^2 - E^2}, & |\, E \,| \leq a \\ 0, & |\, E \,| > a \end{cases} \tag{8}$$

is Wigner's semi-circle law with $a = \sqrt{N/\alpha}$ and

$$a_i = a^2 \left[\frac{M_i}{N} + \lambda^2 \left(1 - \frac{M_i}{N}\right)\right]. \tag{9}$$

The transition parameter utilized in the following is defined as [22]

$$\Lambda = \lambda^2 \rho(0)^2 = \lambda^2 \frac{N^2}{\pi a/2}. \tag{10}$$

Eq. (5) can be used to calculate exactly analytically the NND for 2×2 and 3×3 matrices [6]. For the 2×2 case the DGOE, Eq. (5), gives

$$P_{2\times2}(s,\Lambda) = s\sqrt{\frac{\pi}{8\Lambda}} I_0(\frac{s^2}{16\Lambda}) \exp\left(-\frac{s^2}{16\Lambda}\right), \tag{11}$$

where I_0 is the modified Bessel function, whose asymptotic form is

$$I_0(x) \to \frac{e^x}{\sqrt{2\pi x}}. \tag{12}$$

Thus, $P_{2\times2}(s,0) = 1$, and there is no level repulsion for $\Lambda \to 0$. In the opposite limit, $\Lambda \to \infty$, $I_0(x) \approx 1 - x^2/4$ and one obtains

$$P_{2\times2}(s,\Lambda \gg 1) \approx s\sqrt{\frac{\pi}{8\Lambda}} e^{-s^2/16\Lambda}. \tag{13}$$

For higher dimensions Eq. (5) can only be used for numerical simulations, however, using appropriate perturbative methods Leitner [23] was able to find a formula for the NND. He started basically with the formula for the nearest neigbhour spacing distribution for the superposition of m GOE's block matrices [1]

$$P_m(s) = \frac{d^2}{ds^2} E_m(s) \tag{14}$$

where, for the case of all block marices having the same dimension one has

$$E_m(s) = \left(E_1(\frac{s}{m})\right)^m, \tag{15}$$

$$E_1(x) = \int_x^\infty (1 - F(t))\, dt, \tag{16}$$

$$F(t) = \int_0^t P_1(z)\, dz. \tag{17}$$

In the above $P_1(z)$ is the normalized nearest neighbour spacing distribution of one block matrix. It is easy to find for $P_m(s)$, the following

$$\begin{aligned} P_m(s) &= \frac{1}{m}\left[(E_1(s/m))^{m-1} P_1(s/m) + (m-1)(E_1(s/m))^{m-2}(1 - F(s/m))^2\right] && (18)\\ &\equiv P_m^{(1)}(s) + P_m^{(2)}(s) && (19) \end{aligned}$$

If all the block matrices belong to the GOE, then one can use the Wigner form for $P_1(z)$

$$P_1(z) = \frac{\pi}{2} z e^{-\frac{\pi}{4}z^2} \approx \frac{\pi}{2} z, \tag{20}$$

thus

$$F_1(z) = 1 - e^{-\frac{\pi}{4}z^2} \approx \frac{\pi}{4} z^2, \tag{21}$$

$$E_1(z) = erfc\left(\frac{\sqrt{\pi}}{2} z\right) \approx 1 - z. \tag{22}$$

where the large-z limits of Eqs. (20)-(22) are also indicated above. It is now clear that the above expression for $P_m(s)$, (18) and (19), contains a term $P_m^{(1)}(s)$ with level repulsion, indicating short-range correlation among levels pertaining to the same block matrix and a second term $P_m^{(2)}(s)$ with no level repulsion, implying short-range correlation among NND levels pertaining to different blocks. Notice that for very small spacing, $P_m(s)$ behaves as

$$P_m(s) \approx \frac{\pi}{2m^2}s + \frac{m-1}{m} \tag{23}$$

for $m = 1$, we get the usual $P_1(0) = 0$, while for $m > 1$, we get $P_m(0) = (m-1)/m$. To account for symmetry breaking, Leitner [23] considered the mixing between levels pertaining to nearest neigbhour block matrices. This amount to constrain the mixing to be of the form given by Eq. (11) for the 2×2 DGOE, thus, he found

$$P_m(s,\Lambda) = P_m^{(1)}(s) + P_{2\times 2}(s,\Lambda)P_m^{(2)}(s). \tag{24}$$

Though $P_m(s)$ is normalized, $P_m(s,\Lambda)$ is not. Accordingly one supplies coefficients c_N and c_D, such that

$$P_m(s,\Lambda,c_N,c_D) \equiv c_N P_m(c_D s,\Lambda) \tag{25}$$

is normalized to unity. Similarly, $<s>$ should be unity too. Eq. (24) can certainly be generalized to consider the effect of mixing of levels pertaining to next to nearest neighbour blocks, and accordingly, $P_{3\times 3}(s,\Lambda)$, given in Ref. [6] would be used in Eq. (24) instead of $P_{2\times 2}(s,\Lambda)$. In the following, however, we use Eqs. (24), (25) as Leitner did [23].

In Ref. [23], Leitner also obtained approximate expression for the spectral rigidity $\Delta_3(L)$ using results derived by French *et al.* [24]. Leitner's approximation to Δ_3 is equal to the GOE spectral rigidity plus perturbative terms, that is

$$\begin{aligned}\Delta_3^{(m)}(L;\Lambda) \quad &\approx \quad \Delta_3(L;\infty) + \frac{m-1}{\pi^2}\left[\left(\frac{1}{2} - \frac{2}{\varepsilon^2 L^2} - \frac{1}{2\varepsilon^4 L^4}\right)\right. \\ &\left. \times \ln(1+\varepsilon^2 L^2) + \frac{4}{\varepsilon L}\tan^{-1}(\varepsilon L) + \frac{1}{2\varepsilon^2 L^2} - \frac{9}{4}\right],\end{aligned} \tag{26}$$

where

$$\varepsilon \quad = \quad \frac{\pi}{2(\tau + \pi^2\Lambda)} \tag{27}$$

For the cut off parameter we use the value [25] $\tau = c_m e^{\pi/8-\gamma-1}$, where $c_m = m^{m/(m-1)}$ and $\gamma \approx 0.5772$ is Euler's constant. This choice guarantees that when the symmetry is not broken, $\Lambda = 0$, $\Delta_3^{(m)}(L,0) = m\Delta_3(L/m,\infty)$. In Ref. [26], Leitner fitted Eq. (25) for m=2 to the NND from Ref. [17], however, he did not fit the spectral rigidity. It is often the case that there are some missing levels in the statistical sample analysed. Such a situation was addressed recently by Bohigas and Pato [27] who have shown that if g fraction of the levels or eigenfrequencies is missing, the $\Delta_3(L)$ becomes

$$\Delta_3^g(L) = g\frac{L}{15} + (1-g)^2\Delta_3\left(\frac{L}{1-g}\right). \tag{28}$$

The presence of the linear term, even if small, could explain the large L behavior of the *measured* $\Delta_3(L)$. We call this effect the Missing Level (ML) effect. Another possible deviation of Δ_3 from Eq. (26) could arise from the presence of pseudo-integrable effect (PI) [20, 28]. This also modifies Δ_3 by adding a Poisson term just like Eq. (28).

We now apply our model to analyse the eigenfrequency data of the elastomechanical vibrations of an anisotropic quartz block used in [17]. In this reference in order to break the flip symmetry of the crystal block gradually they removed an octant of a sphere of varying size at one of the corners. The rectangular quartz block has the dimensions $14 \times 25 \times 4\,mm^3$. The radii of the spheres containing the octants are $r = 0.0, 0.5, 0.8, 1.1, 1.4$ and $1.7\,mm$ representing figures $(a)-(f)$. Figs. $1x$ and $2x$ of Ref. [17] correspond to the octant $r \gg 1.7$. They found 1424, 1414, 1424, 1414, 1424 and 1419 frequency eigenmodes, respectively. The histograms and circles in the two figures of Ref. [17] represent the short-range nearest-neighbor distributions $P(s)$ (Fig. 1) and the long range $\Delta_3(L)$ statistics (Fig. 2).

The results of our analysis are shown in the two figures. In Fig. 1, the sequence of six measured NNDs were fitted for $m = 2$ and $m = 3$. It can be seen that the DGOE model with three coupled GOE's give a comparable and in some cases even better fit than the $m = 2$ one. Figure 1a in fact shows a rather sharp peak in our calculated $P(s)$ for

TABLE 1. Values of Λ obtained by fitting Eqs. (25) and (26) respectively to the experimental NNDs and spectral rigidities from Ref. [17].

	$P(s)$			$\Delta_3(L)$
Data Set	Ref. [26]	Eq. (25) m=2	Eq. (25) m=3	Eq. (26) m=3
(a)	0.0013	0.0030	0.0067	0.0056
(b)	0.0054	0.0063	0.0098	0.0016
(c)	0.0096	0.010	0.017	0.0017
(d)	0.0313	0.032	0.064	0.027
(e)	0.0720	0.070	0.13	0.050
(f)	0.113	0.12	0.30	0.16
(x)	0.138	0.13	0.34	2.4

$m = 3$, $P_3(s, 0.0056)$. We consider this a failure of our formula (25) for the uncut crystal. In fact, a more appropriate description of the uncut crystal is to take $\Lambda = 0$, namely a superposition of 3 uncoupled GOE's, which works almost as good as the 2 uncoupled GOE's description. The other parts of figure 1, $(b)-(x)$ seem to show the same insensitivity of $P_m(s,\Lambda)$ to m; the number of matrix blocks used in DGOE description. It is this insensitivity of the short-range nearest neighbour level correlation, measured by the spacing distribution, to the assumed symmetry inherent in the uncut crystal (and thus the number uncoupled GOE's employed to describe it) that forces us to examine the long-range level correlation, namely spectral rigidity, "measured" by Dyson's Δ_3 statistics.

In Fig. 2. the Δ-statistic was fitted with equation (26). It is clear from the figure that a good fit to the data of Ref. [17] is obtained with $m = 3$ for the values of Λ given in table 1. This is to be contrasted with the case of $m = 2$ which, according to Eq. (26) results in $\Delta_3^{(2)}(L,\Lambda)$ that is *always* below the one with $\Lambda = 0$, $\Delta_3^{(2)}(L,0)$, which itself is always below the data points of Ref. [17]. For this reason, only the $\Delta_3^{(2)}(L,0)$ is shown in the figure. It should be noted that the Δ-statistics of the uncut crystal, Fig. 2a is very well described by that of 3 uncoupled GOE's, namely $\Delta_3^{(3)}(L) = 3\Delta_3^{(1)}(L/3)$ which is always larger than the above mentioned $\Delta_3^{(2)}(L) = 2\Delta_3^{(1)}(L/2)$. The most conspicuous exception is Fig. 2b which corresponds to $r = 0.5\,mm$ and where 1414 frequency eigenvalues were found. We consider this a potential ML case and take for Δ_3, the expression given in Eq. (28) and use it in Eq. (26). We find perfect fit to the *data*, if g is taken to be 6%, namely only 94% of the eigenfrequencies were in fact taken into account in the statistical analysis. In contrast, if 2GOE is used we still do not get very good agreement even if 18% of the levels are taken to be missing, as shown in Fig 3. There is, threfore, room to account much better for all cases (Fig. 2*a*, 2*c*, ...) by appropiately choosing the correponding value of g.

In conclusion, a random matrix model to describe the coupling of m-fold symmetry is constructed. The particular threefold case is used to analyse data on eigenfrequencies of elastomechanical vibration of a anisotropic quartz block. By properly taking into account the ML effect we have shown that the quartz block could very well be described by 3 uncoupled GOE's , which are gradually coupled by the breaking of the three-fold symmetry (through the gradual removal of octants of increasing sizes), till a 1GOE situation is attained. This, therefore, indicates that the unperturbed quartz block may posses another symmetry, besides the flip one. We have also verified that if a 2GOE description is used, namely, $m = 2$, then an account of the large-L behaviour of Δ_3 can also be obtained if a much larger number of levels were missing in the sample. In our particular case of Fig. 2b, we obtained $g = 0.18$. This is 3 times larger than the ML needed in the 3GOE description. We consider the large value of g needed in the 2GOE description, much too large to conform to the reported data in Ref [17]. A preliminary version of the formal aspect of this work has appeared in [29].

ACKNOWLEDGMENTS

This work was supported in part by the CNPq and FAPESP (Brazil). M.S.H. is the Martin Gutzwiller Fellow, 2007/2008.

REFERENCES

1. M.L. Mehta, *Random Matrices* 2nd Edition (Academic Press, Boston, 1991); T.A. Brody et al., Rev. Mod. Phys. **53**, 385 (1981); T. Guhr, A. Müller-Groeling and A. Weidenmüller, Phys. Rep. **299,** 189 (1998).
2. F.J. Dyson, J. Math. Phys. **3**, 1191 (1962).
3. M. S. Hussein, and M.P. Pato, Phys. Rev. Lett. **70,** 1089 (1993).
4. M. S. Hussein, and M.P. Pato, Phys. Rev. **C 47,** 2401 (1993).
5. M. S. Hussein, and M.P. Pato, Phys. Rev. Lett. **80,** 1003 (1998).
6. C. E. Carneiro, M. S. Hussein e M. P. Pato, in H. A. Cerdeira, R. Ramaswamy, M. C. Gutzwiller and G. Casati (eds.) *Quantum Chaos*, p. 190 (World Scientific, Singapore) (1991).
7. O. Bohigas, M. J. Giannoni and C. Schmit, Phys. Rev. Lett. **52**, 1 (1984). See also O. Bohigas and M. J. Giannoni, in Mathematical and Computational Methods in Nuclear Physics, edited by J. S. DeHesa, J. M. Gomez and A. Polls, Lecture Notes in Physics Vol. 209 (Springer-Verlag, New York, 1984).
8. G. E. Mitchell, E. G. Bilpuch, P. M. Endt, and F. J. Shriner, Phys. Rev. Lett. **61**, 1473 (1988).
9. A.A. Adams, G.E. Mitchell, and J.F. Shriner, Jr., Phys. Lett. **B 422,** 13 (1998).
10. B. D. Simons, A. Hashimoto, M. Courtney, D. Kleppner and B. L. Altshuler, Phys. Rev. Lett. 71, 2899 (1993).
11. G. R. Welch, M. M. Kash, C-h Iu, L. Hsu and D. Kleppner, Phys. Rev. Lett. 62, 893 (1989).
12. See, e.g. Y. Alhassid, Rev. Mod. Phys. 72, 895 (2000) and references therein.
13. T. Guhr and, H.A. Weidenmüller. Ann. Phys. (NY), **199**, 412 (1990).
14. N. Resenzweig and C. E. Porter, Phys. Rev. **120**, 1698 (1960).
15. H.-D. Graf, H. L. Harney, H. Lengeler, C. H. Lewnkopf, C. Rangacharyulu, A. Richter, P. Schardt and H. A. Weidenmuller, Phys. Rev. Lett. 69, 1296(1992); H. Alt, H. -D. Graf, H. L. Harney, R. Hofferbert, H. Lengeler, A. Richter, P. Schardt and H. A. Weidenmuller, Phys. Rev. Lett. 74, 62 (1995); H. Alt, C. Dembowski, H. -D. Graf, R. Hofferbert, H. Rehfeld, A. Richter, R. Schuhmann and Weiland, Phys. Rev. Lett. 79, 1029 (1997); H. Alt, C. I. Brabosa, H. -D. Graf, T. Guhr, H. L. Harney, R. Hofferbert, H. Rehfeld and A. Richter, Phys. Rev. Lett. 81, 4847 (1998); C. Dembowski, H. -D. Graf, A. Heine, R. Hofferbert, H. Rehfeld and A. Richter, Phys. Rev. Lett. 84, 867 (2000); C. Dembowski, H. -D. Graf. H. L. Harney, A. Heine, W. D. Heiss, H. Rehfeld and A. Richter, Phys. Rev. Lett. 86, 787 (2001); C. Dembowski, H. -D. Graf, A. Heine, T. Hesse, H. Rehfeld and A. Richter, Phys. Rev. Lett. 86, 3284 (2001); C. Dembowski, B. Dietz, H. -D. Graf, A. Heine, T. Papenbrock, A. Richter and C. Richter, Phys. Rev. Lett. 89, 064101-1 (2002); C. Dembowski, B. Dietz, A. Heine, F. Leyvraz, M. Miski-Oglu, A. Richter and T. H. Seligman, Phys. Rev. Lett. 90, 014102-1 (2003); C. Dembowski, B. Dietz, H. -D. Graf, H. L. Harney, A. Heine, W. D. Heiss and A. Richter, Phys. Rev. Lett. 90, 034101-1 (2003); C. Dembowski, B. Dietz, T. Friedrich, H. -D. Graf, A. Heine, C. Mejia-Monasterio, M. Miski-Oglu, A. Richter and T. H. Seligman, Phys. Rev. Lett. 93, 134102-1 (2004); B. Dietz, T. Guhr, H. L. Harney and A. Richter, Phys. Rev. Lett. 96, 254101 (2006); E. Bogomolny, B. Dietz, T. Friedrich, M. Miski-Oglu, A. Richter, F. Schafer and C. Schmit, Phys. Rev. Lett. 97, 254102 (2006); B. Dietz, T. Friedrich, H. L. Harney, M. Miski-Oglu, A. Richter, F. Schafer and H. A. Weidenmuller, Phys. Rev. Lett. 98, 074103 (2007).
16. C. Ellegaard, T. Guhr, K. Lindemann, H. Q. Lorensen, J. Nygard and M. Oxborrow, Phys. Rev. Lett. 75, 1546 (1995).
17. C. Ellegaard, T. Guhr, K. Lindemann, J. Nygard and M. Oxborrow, Phys. Rev. Lett. 77, 4918 (1996).
18. P. Bertelsen, C. Ellegaard, T. Guhr, M. Oxborrow and K. Schaadt, Phys. Rev.Lett. 83, 2171 (1999).
19. R. L. Weaver, J. Acoustic. Soc. Am. **85**, 1005 (1989).
20. A. Abd El-Hady, A. Y. Abul-Magd, and M. H. Simbel, J. Phys. A **35**, 2361 (2002).
21. A. C. Bertuola, J. X. de Carvalho, M. S. Hussein, M. P. Pato, and A. J. Sargeant, Phys. Rev. E **71**, 036117 (2005).
22. A. Pandey, Chaos, Solitons and Fractals 5 (1995) 1275.
23. D. M. Leitner, Phys. Rev. E **48**, 2536 (1993).
24. J. B. French, V. K. B. Kota, A. Pandey and S. Tomsovic, Ann. Phys. (NY) **181**, 198 (1988).
25. A.Y. Abul-Magd and M.H. Simbel, Phys. Rev. E 70 (2004) 046218.
26. D.M. Leitner, Phys. Rev. E 56 (1997) 4890.
27. O. Bohigas and M. P. Pato, Phys. Lett. B, **595**, 171 (2004).
28. D. Biswas and S. R. Jain, Phys. Rev. A **42**, 3170 (1990).
29. M. S. Hussein, J. X. de Carvalho, M. P. Pato and A. J. Sargeant, Few-Body Systems, **38**, 209 (2006).

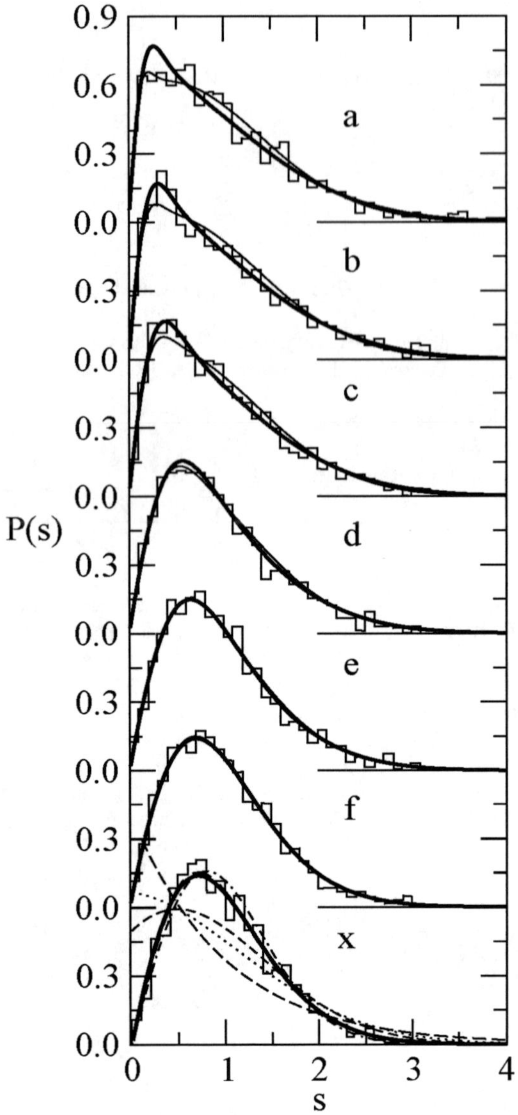

FIGURE 1. Nearest Neighbour Distributions. Histograms show data (a)-(x) from Ref. [17]. Thin and thick solid lines show fits to the data carried out using Eq. (25) with m=2 and m=3 respectively. In graph (x) the long-dashed line is the Poisson distribution, the dot-dashed line is the Wigner distribution and the dashed and dotted lines are the respective distributions for superpositions of 2 and 3 uncoupled GOEs. See Table I for the values of Λ obtained from the fits and the text for details.

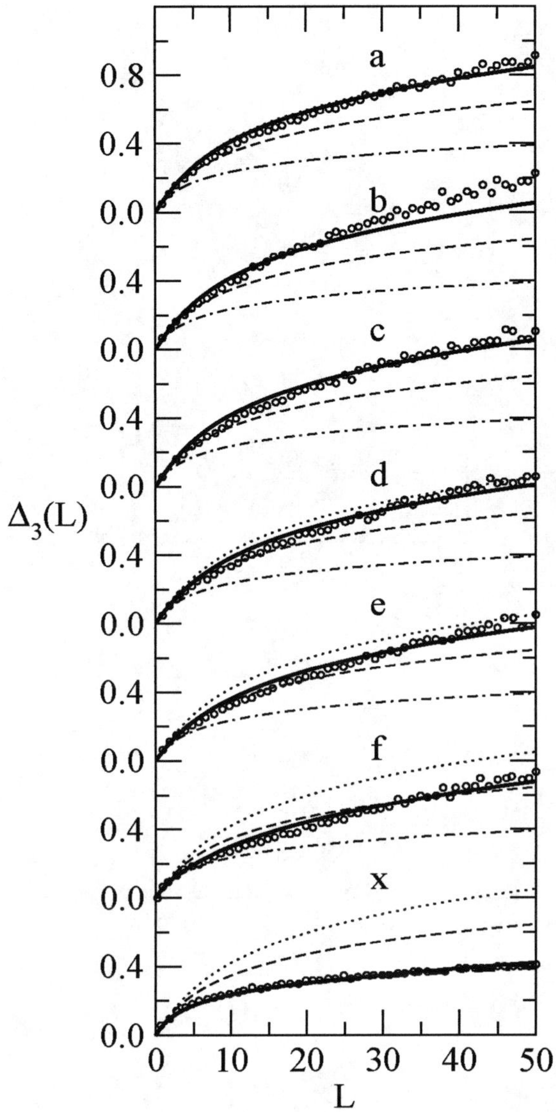

FIGURE 2. Spectral Rigidities. Circles show data (a)-(x) from Ref. [17]. Thick solid lines show fits to the data carried out using Eq. (25) with m=3. The dot-dashed line is the GOE spectral rigidity and the dashed and dotted lines are the respective rigidities for superpositions of 2 and 3 uncoupled GOEs. See Table I for the values of Λ obtained from the fits.

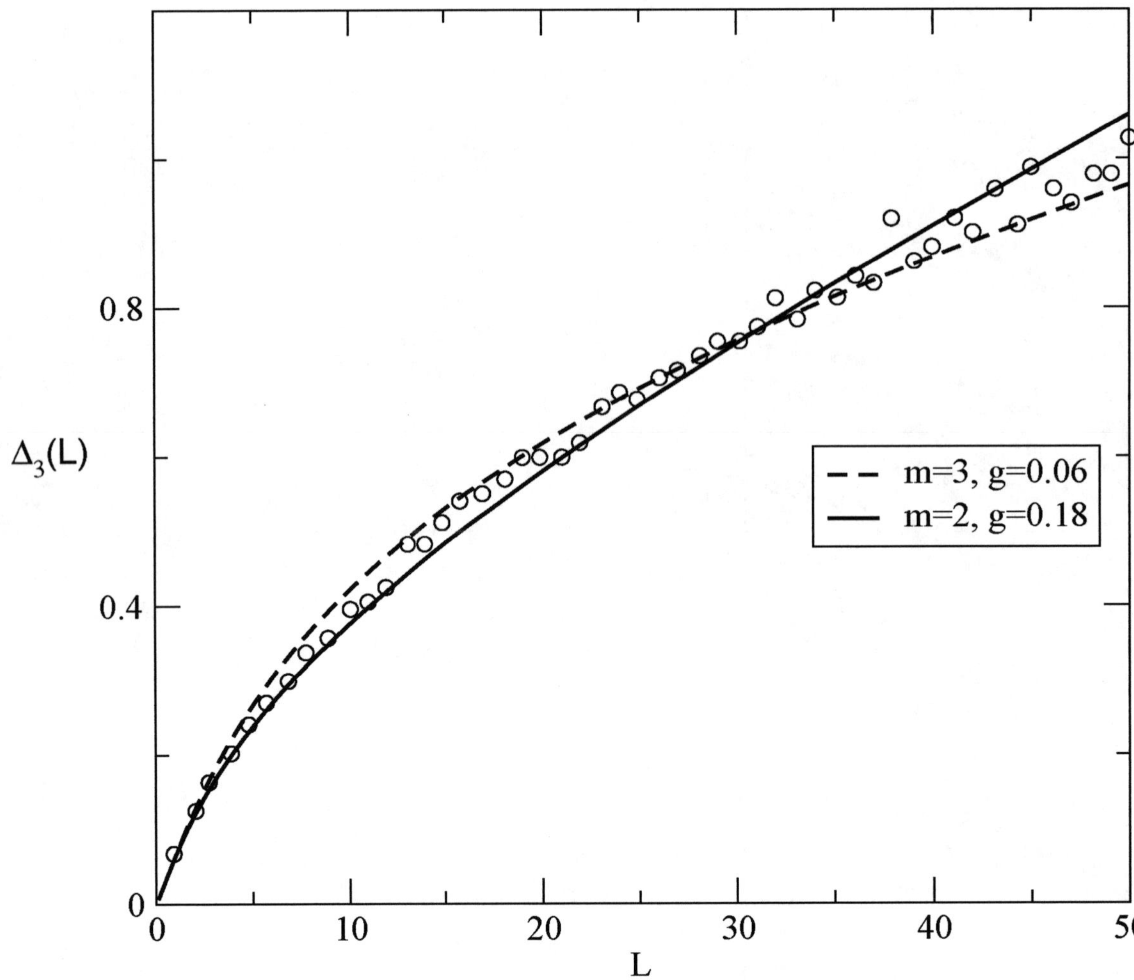

FIGURE 3. The Missing Level Effect: The full curve is the 2GOE Δ_3 for case 2b with 18% of the levels missing, while the dashed curve corresponds to 3GOE with only 6% of the levels missing. The data points (open circles) represent fig. 2b. See text for details.

WNMP07 - PROGRAM

Saturday, October 20. Morning session

9:00. **Konrad Gelbke** (NSCL, Director). Opening Comments.

9:10 - 9:45. **Edward Shuryak** (Stony Brook University). Understanding the Strongly Coupled Quark-Gluon Plasma.

9:50 - 10:25. **Achim Richter** (Technical Univerisity, Darmstadt, Germany). Superscars, Doorway States and Nodal Domains in a Barrier Billiard.

10:50 - 11:25. **Daniel Kosov** (University of Maryland). Electron Transport Through Molecules.

11:30 - 12:05. **John R. Barker** (University of Glasgow, UK). Quantum Transport in Mesoscopic Semiconductor Devices: Vortices, Flows and Atomistic Effects.

12:10 - 12:45. **Oleg Sushkov** (University of New South Wales, Sydney, Australia). Correlated Electron Current and Temperature Dependence of the Conductance of a Quantum Point Contact.

Afternoon session

2:30 - 3:05. **Yoram Alhassid** (Yale University). The mesoscopic competition between superconductivity and ferromagnetism in small metallic grains.

3:10 - 3:45. **Michael Harrison** (Michigan State University). Interplay of Orbital Paramagnetism and Diamagnetism in Nanometer Scale 2DEG Strips.

4:10 - 4:45. **Gary Mitchell** (North Carolina State University). Thermodynamic Analysis of Spectra.

4:50 - 5:25. **Lev Kaplan** (Tulane University). Interaction Matrix Element Fluctuations in Quantum Dots.

5:30 - 6:05. **Mahir Hussein** (Max Planck Institute for Physics of Complex Systems, Dresden, Germany). Symmetry Breaking Study with Deformed Ensembles.

7:00 *Dinner in the Cyclotron Atrium and Poster Session*

Sunday, October 21. Morning session

9:00 - 9:35. **Victor V. Flambaum** (University of New South Wales, Sydney, Australia). Variation of Fundamental Constants in Space and Time: Theory and Observations.

9:40 - 10:15. **Josef Paldus**, together with X. Li (University of Waterloo, Canada). Non-Dynamic Correlation and Coupled-Cluster Methods.

10:40 - 11:15. **Marcos Dantus** (Michigan State University). Coherent and Incoherent Control of Laser-Matter Interactions.

11:20 - 11:55. **Linda Reichl** (University of Texas at Austin). Quantum Control of Cold Atomic Systems Using Nonlinear Dynamics.

12:00 - 12:35. **Garnet Kin-Lic Chan** (Cornell University). Ab-initio Density Matrix Renormalization Group and Canonical Transformation Theories of Strongly Interacting Electrons.

Afternoon session

2:10 - 2:45. **Naftali Auerbach** (Tel Aviv University, Israel, and MSU). Doorway States and the Super-Radiant Mechanism.

2:50 - 3:25. **Felix Izrailev** (University Autonoma de Puebla, Mexico, and MSU). Ericson versus Conductance Fluctuations in a Model of Interacting Fermions.

3:50 - 4:25. **Raymond F. Bishop** (University of Manchester, UK). A Microscopic Treatment of Spin-Lattice Models of Magnetism and Their Quantum Phase Transitions via the Coupled Cluster Method.

4:30 - 5:05. **William A. Lester Jr.** (University of California, Berkeley). Some Applications of Quantum Monte Carlo to Molecular Systems.

5:10 - 5:45. **Hiroshi Nakatsuji** (Quantum Chemistry Research Institute, Kyoto, Japan). Towards Accurately Predictive Quantum Chemistry with the Solution of the Schrödinger equation.

7:00 *Dinner at the Kellogg Center*

Monday, October 22. Morning session

9:00 - 9:35. **Stephanie Reimann** (University of Lund, Sweden). Finite Quantum Systems Set Rotating.

9:40 - 10:15. **Joseph A. Carlson** (Los Alamos National Laboratory). Simulations of Dilute Fermions and Low-Density Neutron Matter.

10:40 - 11:15. **Michael Guidry** (University of Tennessee). Instabilities of Doped Mott Insulators and the Properties of High-Temperature Superconductors.

11:20 - 11:55. **Sven Åberg** (University of Lund, Sweden). Mesoscopic Fluctuations of the Pairing Gap.

12:00 - 12:35. **Maxim Mostovoy** (University of Groningen, The Netherlands). Magnetoelectric Properties of Ferromagnetic Domains.

Afternoon session

2:10 - 2:45. **Debashis Mukherjee** (Indian Association for Cultivation of Science, Mumbai). Applications of a Novel Spin-free Combinatoric Open-shell Coupled Cluster Theory to Single-reference Doublets.

2:50 - 3:25. **Stuart Tessmer** (Michigan State University). Scanning Probe Microscopy of Semiconductor Donor Molecules.

3:50 - 4:25. **Michael Moore** (Michigan State University). Optimized Double-Well Quantum Interferometry with Gaussian Squeeze-States.

4:30 - 5:05. **Masahiro Ehara** (Kyoto University, Japan). Photochemistry of Biological Chemosensors, Organic Light-Emitting Diodes, and Inner-shell Electronic Processes.

5:10 - 5:45. **Valentin V. Sokolov** (Budker Institute of Nuclear Physics, Novosibirsk, Russia). Statistics of Resonance Widths and Decay Rates Distribution.

5:50 - 6:25. **Vaclav Spicka** (Institute of Physics, Prague, Czech Republic). Mesoscopic Systems and Nonequilibrium Green's functions.

PARTICIPANTS

Yoram Alhassid, Yale University
e-mail: yoram.alhassid@yale.edu

Jeremy Armstrong, Michigan State University
e-mail: armstron@nscl.msu.edu

Naftali Auerbach, Tel Aviv University/MSU
e-mail: auerbach@post.tau.ac.il

John R. Barker, University of Glasgow
e-mail: jbarker@elec.gla.ac.uk

Wolfgang Bauer, Michigan State University
e-mail: bauer@pa.msu.edu

Norman Birge, Michigan State University
e-mail: birge@pa.msu.edu

Raymond F. Bishop, The University of Manchester
e-mail: raymond.bishop@manchester.ac.uk

Scott Bogner, Michigan State University
e-mail: bogner@nscl.msu.edu

Ivan Brida, Michigan State University
e-mail: brida@nscl.msu.edu

B. Alex Brown, Michigan State University
e-mail:brown@nscl.msu.edu

David M. Cardamone, Simon Fraser University
email: David_Cardamone@sfu.ca

Joseph A. Carlson, Los Alamos National Laboratory
e-mail: carlson@tmail.lanl.gov

Luca Celardo, Tulane University
e-mail: nicedirac@gmail.com

Garnet Chan, Cornell University
e-mail: gc238@cornell.edu

Shari Conroy, Michigan State University
e-mail: conroy@nscl.msu.edu (*Secretary of the Workshop*)

Pawel Danielewicz, Michigan State University
e-mail: danielewicz@nscl.msu.edu (*Organizing Committee*)

Marcos Dantus, Michigan State University
e-mail: dantus@msu.edu

Helio Dias, Universidade de Sao Paulo
e-mail: heliodia@usp.br

Thomas Duguet, Michigan State University
e-mail: duguet@nscl.msu.edu

Masahiro Ehara, Kyoto University
e-mail: ehara@sbchem.kyoto-u.ac.jp

Leonid Fedichkin, Michigan State University
e-mail: leonid@msu.edu

Victor V. Flambaum, University of New South Wales
e-mail: flambaum@phys.unsw.edu.au

Chasity Fudella, Michigan State University
e-mail: fudella@nscl.msu.edu (*Secretary of the Workshop*)

C. Konrad Gelbke, Michigan State University
e-mail: gelbke@nscl.msu.edu

Jeffrey R. Gour, Michigan State University
e-mail: gourjeff@msu.edu

Michael Guidry, University of Tennessee
e-mail: guidry@utk.edu

Michael Harrison, Michigan State University
e-mail: harrison@pa.msu.edu

Katharine Hunt, Michigan State University
e-mail: klch@chemistry.msu.edu

Mahir Hussein, Max-Planck-Institut für Physik Komplexer Systeme/MSU
e-mail: mhussein@mpipks-dresden.mpg.de

Felix Izrailev, University Autonoma de Puebla/MSU
e-mail: felix.izrailev@gmail.com

Lev Kaplan, Tulane University
e-mail: lkaplan@tulane.edu

Trupti Khaire, Michigan State University
e-mail: khaire@pa.msu.edu

Daniel Kosov, University of Maryland
e-mail: dkosov@umd.edu

William A. Lester Jr., University of California, Berkeley
e-mail: WALester@lbl.gov

Wei Li, Michigan State University
e-mail: weili@msu.edu

Xiaoping Li, Michigan State University
e-mail: lix@msu.edu

Jesse Lutz, Michigan State University
e-mail: lutzjess@chemistry.msu.edu

Bhanu Mahanti, Michigan State University
e-mail: mahanti@pa.msu.edu

Anirban Mandal Sr., Michigan State University
e-mail: mandalan@msu.edu

Gary Mitchell, North Carolina State University
e-mail: mitchell@tunl.duke.edu

Michael Moore, Michigan State University
e-mail: moore@pa.msu.edu

Maxim Mostovoy, University of Groningen
e-mail: M.Mostovoy@rug.nl

Debashis Mukherjee, Indian Association for the Cultivation of Science
email: pcdm@iacs.res.in

Hiroshi Nakatsuji, Quantum Chemistry Research Institute, Kyoto,
e-mail: h.nakatsuji@qcri.or.jp and hiroshi@sbchem.kyoto-u.ac.jp

Filomena Nunes, Michigan State University
e-mail: nunes@nscl.msu.edu

Eivind Osnes, University of Oslo,

e-mail: eivind.osnes@fys.uio.no

Josef Paldus, University of Waterloo
e-mail: paldus@scienide.uwaterloo.ca

Afra Panahi, Michigan State University
e-mail: panahiaf@msu.edu

Piotr Piecuch, Michigan State University
e-mail: piecuch@chemistry.msu.edu (*Organizing Committee*)

Carlo Piermarocchi, Michigan State University
e-mail: piermaro@gmail.com

Linda Reichl, The University of Texas at Austin
e-mail: reichl@mail.utexas.edu

Stephanie Reimann, Lund University
e-mail: stephanie.reimann@matfys.lth.se

Achim Richter, Technische Universitäat Darmstadt
e-mail: richter@ikp.tu-darmstadt.de

Arnau Rios, Michigan State University
e-mail: rios@nscl.msu.edu

Vincent Rotival, CEA, Bruyeres le Chatel
e-mail: vincent.rotival@polytechnique.org

Roman Sen'kov, Michigan State University
e-mail: senkov@nscl.msu.edu

Edward Shuryak, University of Stony Brook
e-mail: shuryak@tonic.physics.sunysb.edu

Valentin V. Sokolov, Budker Institute of Nuclear Physics, Novosibirsk
e-mail: V.V.Sokolov@academ.org

Suren Sorathia, University of Puebla
e-mail: surensorathia@gmail.com

Vaclav Spicka, Institute of Physics, Academy of Sciences of the Czech Republic
e-mail: spicka@fzu.cz

Krzysztof Starosta, Michigan State University

e-mail: starosta@nscl.msu.edu]

Oleg Sushkov, University of New South Wales
e-mail: sushkov@phys.unsw.edu.au

Stuart Tessmer, Michigan State University
e-mail: tessmer@msu.edu

Alexander Volya, Florida State University
e-mail: volya@nucmar.physics.fsu.edu (*Organizing Committee*)

Vladimir Zelevinsky, Michigan State University
e-mail: zelevinsky@nscl.msu.edu (*Organizing Committee*)

Sven Åberg, Lund University
e-mail: Sven.Aberg@matfys.LTH.SE

AUTHOR INDEX

A

Åberg, S., 173
Alhassid, Y., 192
Armstrong, J. R., 12
Auerbach, N., 92

B

Bargi, S., 25
Barker, J. R., 104
Berman, G. P., 75

C

Cardamone, D. M., 135
Carlson, J., 17
Celardo, G. L., 75
Christensson, J., 25

D

Danielewicz, P., 98
Datta, D., 53
de Carvalho, J. X., 212
Domin, D., 72

E

Ehara, M., 145

F

Flambaum, V. V., 1

G

Gelin, M. F., 115
Gezerlis, A., 17
Guidry, M., 160

H

Harrison, J. F., 130
Harrison, M. J., 152
Holder, B. P., 34
Hussein, M. S., 212

I

Izrailev, F. M., 75

K

Kaplan, L., 192
Kaplan, T. A., 130
Kärkkäinen, K., 25
Kavoulakis, G. M., 25
Kirczenow, G., 135
Kosov, D. S., 115
Kuljanishvili, I., 130

L

Leboeuf, P., 173
Lester, Jr., W. A., 72
Li, X., 43
Lutz, J. J., 62

M

Manninen, M., 25
Milstein, A. I., 122
Mitchell, G. E., 185
Mukherjee, D., 53

N

Nakatsuji, H., 145

O

Olofsson, H., 173

P

Paldus, J., 43
Pato, M. P., 212
Piecuch, P., 62
Piermarocchi, C., 130

R

Reddy, S., 17
Reichl, L. E., 34
Reimann, S. M., 25
Richter, A., 202
Rios, A., 98

S

Sakharuk, A. A., 12
Sargeant, A. J., 212
Shriner, Jr., J. F., 185
Shuryak, E. V., 1
Sloggett, C., 122
Sokolov, V. V., 85
Sorathia, S., 75
Sun, Y., 160
Sushkov, O. P., 122

T

Tessmer, S. H., 130

W

Wu, C.-L., 160

Z

Zelevinsky, V. G., 12, 75